Chemical Mutagens

ENVIRONMENTAL EFFECTS ON BIOLOGICAL SYSTEMS

ENVIRONMENTAL SCIENCES

An Interdisciplinary Monograph Series

ARTHUR C. STERN, editor, AIR POLLUTION, Second Edition. In three volumes, 1968

L. FISHBEIN, W. G. FLAMM, and H. L. FALK, CHEMICAL MUTAGENS: Environmental Effects on Biological Systems, 1970

DOUGLAS H. K. LEE and DAVID MINARD, editors, PHYSIOLOGY, ENVIRONMENT, AND MAN, 1970

KARL D. KRYTER, THE EFFECTS OF NOISE ON MAN, 1970

R. E. MUNN, BIOMETEOROLOGICAL METHODS, 1970

M. M. KEY, L. E. KERR, and M. BUNDY, PULMONARY REACTIONS TO COAL DUST: "A Review of U. S. Experience," 1971

DOUGLAS H. K. LEE, editor, METALLIC CONTAMINANTS AND HUMAN HEALTH, 1972

DOUGLAS H. K. LEE, editor, ENVIRONMENTAL FACTORS IN RESPIRATORY DISEASE, 1972

H. ELDON SUTTON and MAUREEN I. HARRIS, editors, MUTAGENIC EFFECTS OF ENVIRONMENTAL CONTAMINANTS, 1972

DOUGLAS H. K. LEE and PAUL KOTIN, editors, MULTIPLE FACTORS IN THE CAUSATION OF ENVIRONMENTALLY INDUCED DISEASE, 1972

In preparation

MOHAMED K. YOUSEF, STEVEN M. HORVATH, and ROBERT W. BULLARD, PHYSIOLOGICAL ADAPTATIONS: Desert and Mountain

Chemical Mutagens

ENVIRONMENTAL EFFECTS ON BIOLOGICAL SYSTEMS

L. FISHBEIN, W. G. FLAMM, and H. L. FALK

NATIONAL INSTITUTE OF ENVIRONMENTAL HEALTH SCIENCES
NATIONAL INSTITUTES OF HEALTH
PUBLIC HEALTH SERVICE AND DEPARTMENT OF HEALTH, EDUCATION AND WELFARE
RESEARCH TRIANGLE PARK, NORTH CAROLINA

1970

ACADEMIC PRESS ☐ New York and London

ACADEMIC PRESS, INC.
111 Fifth Avenue, New York, New York 10003

United Kingdom Edition published by
ACADEMIC PRESS, INC. (LONDON) LTD.
Berkeley Square House, London W1X 6BA

LIBRARY OF CONGRESS CATALOG CARD NUMBER: 71-117078

Second Printing, 1972

PRINTED IN THE UNITED STATES OF AMERICA

Contents

5. *Test System for the Detection and Scoring of Mutants*

6. *Tabular Summaries of Chemical Mutagens*

7. *Alkylating Agents I (Aziridines, Mustards, Nitrosamines, and Related Derivatives)*

8. *Alkylating Agents II (Epoxides, Aldehydes, Lactones, Alkyl Sulfates, Alkane Sulfonic Esters, and Related Derivatives)*

9. *Drugs, Food Additives, Pesticides, and Miscellaneous Mutagens*

Foreword

There is growing concern over the possibility that future generations may suffer from genetic damage by mutation-inducing chemical substances to which the population at large is unwittingly exposed. The hazards of radiation mutagenesis have, of course, been long recognized, and both scientific awareness and public health concern have been extensive. A belief persists that potentially irreversible damage could occur without warning due to the some 10,000 or more natural and synthetic chemical agents in the environment as well as to the chemicals continually being synthesized by man, which assuredly will find widespread use in the future. The induction of mutations in man by environmental agents is of sufficient concern to warrant major efforts directed to improving existing techniques and to develop new ones for detecting mutagenic agents.

Concurrently, a battery of procedures using a spectrum of microbial and more complex submammalian species is being used to identify compounds with mutagenic properties. Compounds chosen for study have been primarily selected by virtue of chemical similarities to established mutagens, carcinogens, or other toxicants. While these have served well as broad guidelines, each has important limitations. Contributing to the acuity of the problem has been the recent observation that the traditional concepts that mutagenic agents are also carcinogenic or toxic is not necessarily always correct. The most vexing frustration is the difficulty inherent in attempting to relate experimental findings of mutagenicity to man.

The demonstration of a mutagenic effect in a microbial system fails to provide important information on whether the responsible environmental agents can reach the genetic material in the nucleus of responsible organs and whether the metabolic capabilities of man would permit the expression of a mutational event. The provocative studies on cell culture and submammalian test systems which have demonstrated mutagenicity, however, make it especially urgent that comparable studies be undertaken on intact mammals and that simultaneously the feasibility of epidemiologic studies be assessed to seek out human data which might indicate the existence of an environmental chemical mutagen.

Numerous advisory groups are addressing themselves to the problems of environmental chemical mutagenesis, and despite differences in sponsorship, organizational titles, and discipline emphasis the problems they face and the questions they ask are similar. First, what guidelines should be established in developing a systematic approach to selecting compounds for testing? Second, how can one best test for mutagenicity; what are adequate tests or test systems? This is a particularly difficult problem since test systems which are relatively simple are those with least relevance to man. The closer one approaches man in the assay of chemical mutagens, the more complex the problems. Third, recognizing the limitations of simple test systems, can chromosome breakage be correlated with mutagenesis and is it feasible to monitor the human population using this gross observation so that mutations due to environmental causes can be detected sufficiently early to prevent catastrophic consequences—a genetic disaster? Recognizing that this genetic monitoring would demand an unachievable effort of monumental proportions in scope and time, there appears to be at present universal agreement that at the very least a registry of environmental mutagens can be established and maintained with ready availability of its information.

The use of different species of progressively increasing phylogenetic complexity for bioassay systems, though sound from a theoretical and feasible from a protocol viewpoint, has contributed little to our understanding of the extent of the mutagenic hazard to man associated with environmental chemicals. This is so for deficiencies already noted plus the additional factor to which this book addresses itself, and that is the need to systematically assay for chemical mutagens even with our less than ideal systems. A crucial problem facing those with public health responsibility is the difficulty in assessing the magnitude of the influence of agents whose potential effects occur rarely and with extreme variability. As an example there is the argument that inasmuch as man has been drinking coffee for 200 years and we have been unable to identify any differences between coffee drinkers and noncoffee drinkers in terms of chromosomal damage, let alone phenotypic variations, there is no hazard associated with drinking coffee. While this may be so, we can only speculate that it is so, particularly since recent evidence has destroyed the shibboleth that "What doesn't show, doesn't hurt." Prior to discussing such problems as the mutagenic effect of coffee drinking, it is imperative that we ask: "How hard has the look been both experimentally and epidemiologically?" Delayed effects due to minimal insults ultimately appear responsible for the diverse epidemiological characteristics of arteriosclerotic cardiovascular disease, degenerative pulmonary disease, and neoplasia. For the future it is crucial to determine how our techniques using the many models available can be united into a constellation of approaches yielding maximum capability for judgment as to whether an environmental mutagen is indeed a hazard to man. To further confound the problem, one must always bear in mind the mutagenic potential resulting from natural phenomena, including differences in the metabolic characteristics of man himself.

In organizing the text, the authors have wisely divided the book into two parts. The first is concerned with a lucid discussion of the modern concepts of the gene at the molecular and biochemical levels. This simple, though emphatically not superficial section, neatly demonstrates important acquisitions of knowledge during the past decade while at the same time noting the many important deficiencies in our knowledge. This part relates importantly to the second part in that it assures that judgments and conclusions have taken into account the existing balance between knowledge and ignorance. Further, the part dealing with individual chemical mutagens provides within one text an opportunity for comparative evaluation of environmental mutagens in addition to assessing their intrinsic hazard. The authors have wisely chosen to review environmental chemical mutagens in a multifactorial manner integrating chemical structure, biological availability and biochemical behavior, environmental presence, and economic use thereby enhancing its single source value. This assures its utility to the student, the investigator, and those charged with public health responsibility. The extensive bibliography precludes the necessity for an encyclopedic text while at the same time assuring comprehensiveness. The intent of the authors and the very nature of the text assure its becoming less all inclusive with time. Serving as an information source for past research and a critical review of current efforts, it should provide inspiration and guidance for the future. It will most assuredly increase intelligent awareness of existing problems and expand interests and efforts in monitoring this growing hazardous area and contribute to its own obsolescence.

The authors have successfully completed a major effort. It provides in a comprehensive manner an authoritative and responsible assessment of chemical mutagens in man's environment, a worthy benchmark for future studies.

Paul Kotin
Director
NIEHS

Preface

This work has of necessity touched upon a number of different and often unrelated subjects: e.g., industrial application, genetics, cytogenetics, molecular biology, biochemistry, and organic and analytical chemistry. The depth to which any particular area could be explored was understandably limited, but we have tried to provide the reader with a key to the literature so that additional information could be obtained as required.

Though the central theme and overriding purpose of this book was to bring together relevant facts about mutagenic chemicals (both synthetic and naturally occurring) which are a part of man's environment today, we also believed it important to explain the theories, ideas, and facts concerning the mechanism by which all types of mutations arise, and we have described, in some detail, how such mutations have been detected and studied.

We have stressed the importance of knowing as many of the salient facts as possible, such as the stability of the substance under various conditions, their degradation products, if any, the frequency and physical form in which they are encountered by man, whether they are systemically absorbed and by what route, how they are distributed among the cells, tissues, and organs of the body, and how quickly they are metabolized, which alternate pathways are used, what their effects on the genetic material (DNA) may be, and whether these effects can be repaired.

Although we have considered chemical mutagens as discrete entities, it is recognized that man is exposed to a broad galaxy of environmental agents, and hence considerations relative to possible synergistic, potentiating, comutagenic, and/or antagonistic interactions of mutagenic and nonmutagenic chemicals are of vital importance.

All of the above factors must be envoked before valid assessments regarding the genetic hazards presented by these environmental chemicals can be made. Since the organism we are most concerned with is man, the difficulty is enhanced. Though results of studies on other mammals often are considered extrapolatable to man, much more evidence must be brought forth to determine if this extrapolation is done correctly.

To solve some of the basic problems, it is imperative that we improve our understanding of the underlying molecular mechanisms by which these effects are mediated. It is hoped that this work will serve such a purpose.

L. Fishbein
W. G. Flamm
H. L. Falk

Chemical Mutagens

Environmental Effects on Biological Systems

CHAPTER 1

Introduction

Some appreciation of heredity is older than written history, although the science concerned with this subject is in its infancy—dating back no further than to Mendel's discoveries in the latter part of the nineteenth century (1). The discoveries, however, went unnoticed for 34 years until independently rediscovered by de Vries, Correns, and Tschermark.

The failure of Mendel's observations to have an impact upon the biologists of his time is attributed to his having been years ahead of his contemporaries; however, by 1900 more biology was known, and the role of the cell nucleus and the chromosomes was beginning to be understood (2). Their part in the process of inheritance was beginning to be realized, and thus the climate was right when Mendel's work was brought to the attention of biologists.

Mendel succeeded in unlocking one of nature's most important secrets because he chose to consider only a few of the physical characters of the offsprings he bred, and this allowed him the opportunity to examine all the individuals of each generation and to look for quantitative differences. Hence he was able to establish that hereditary traits segregate among the progeny in a mathematically predictable way, and thus he opened the door to the study of genetics.

Initially, the concept of a gene envisaged a stable structure since hereditary characters are passed from one generation to the next, i.e., they breed true. If, however, genes were absolutely immutable, how would organic evolution be mediated? Obviously, the stability of a gene is less than absolute; occasionally, though infrequently, a change will arise, but, like the predecessor, the new product is also highly stable. The new gene is a mutant, and the process is known as mutation, the latter mediating evolution (3). Because the development of a mutant is ordinarily a rare and unpredictable event, attempts to study the process are greatly handicapped. One could look at the result, but an understanding of mutation depends upon the development of methods which can produce mutants on demand. H. J. Muller, whose contributions to genetics are well known, discovered that mutants could be induced artificially by x-rays, and thus he opened a new approach which has revealed many important genetic facts (4).

The nature of the gene and of mutation can be understood better when chemicals are used for the induction of mutants, since it is through knowledge of the properties of the chemicals that inferences can be drawn regarding the structure of the other reactant (the gene) and (since we now know the genes to be composed of nucleic acid) of the characteristics of the intragenic site that has undergone mutation.

The search for chemical mutagens covered a period of 20 years (1920–1940) before one such substance was found. The discovery was not entirely accidental. Robson and Carr, who had been studying the effects of mustard gas, noted that the inflammation and interference with cell division caused by a chemical were similar to that caused by x-rays. Believing that the substance might be mutagenic, they enlisted the help of Charlotte Auerbach (5) who confirmed their suspicions, but the work was classified and it'was not until after the war that Dr. Auerbach's paper appeared.

Soon many chemical mutagens were known, but the interest in them was attributable to their usefulness as tools for unveiling genetic facts. Only recently has much attention been paid to the possibility that some of these agents might pose a threat to man. In fact, it has been argued that mutagenic agents are more helpful than harmful and without them evolution of species would have been arrested at a very primitive stage.

As long as the natural environment was entirely responsible for the existence and production of physical and chemical mutagens, this argument has justification. However, within the last two or three decades man has introduced into his environment a variety of chemical substances which are known to be mutagenic in certain test organisms.

It is the purpose of this book to focus on such agents. For convenience the book is divided into two parts: the first is instructional and begins with the modern concepts of the gene in molecular terms, then deals with the different types of mutations and how they form as well as the biological systems used for their detection. The metabolic events which are concerned with the repair of gene damage are outlined in Chapter 4 and the relevance of these metabolic events to mutation frequency are discussed. The second part of this book deals with the individual chemical mutagens of environmental significance, with their manufacture and occurrence, their method of detection, their degradation and metabolism (when known), and with the types of mutation they induce in the various test systems which have been utilized.

We cannot assess as yet the magnitude of their genetic impact on man, but we will show that the techniques for doing so are still in the developmental stage. There is no doubt that important new chemical mutagens have been added to man's environment, although, perhaps, the mutagenic agents of greatest importance are still contributed by nature or even produced by our own normal metabolism.

REFERENCES

1. G. Mendel, *Verhandl. Naturforsch. Ver. Brunn* **4** (1865); English translation *in* "Classic Papers in Genetics" (J. A. Peters, ed.), p. 1. Prentice-Hall, Englewood Cliffs, New Jersey, 1961.
2. W. S. Sutton, *Biol. Bull.* **4,** 231 (1903).
3. H. J. Muller, *Am. Naturalist* **56,** 32 (1922).
4. H. J. Muller, *Genetics* **13,** 279 (1928).
5. C. Auerbach, J. M. Robson, and J. C. Carr, *Science* **105,** 243 (1947).

CHAPTER 2

Nature of Genetic Material

I. Genes and DNA

To understand why certain chemicals have a potential for producing changes within the hereditary material of a living organism while others do not, we must concern ourselves with the chemical foundation of heredity. It is necessary but insufficient to know that heredity is determined and mediated by individual units called genes which are organized, at stages in the cell cycle, into microscopically visible structures called chromosomes which contain (in higher organisms) thousands of such units; that each gene possesses the necessary information for specifying the formation of a particular enzyme or other protein; and that the spectrum of these enzymes and other proteins are primarily responsible for the differences among organisms and species.

Evidence of an indirect and direct nature has accumulated so that it is now universally accepted that deoxyribonucleic acid (DNA) is the material of which genes are composed (with the exception of certain ribonucleic acid viruses which do not contain DNA). Chemical mutagens have been instrumental in providing some of this evidence, for as C. Auerbach stated over 20 years ago, "If we assume a mutation is a chemical process, then knowledge of the reagents capable of initiating this process should throw light not only on the reaction itself but also on the nature of the gene, the other partner in the reaction" (1). And so it has been, although direct and indirect evidence has been derived from many other sources.

It is beyond the scope of this review to consider all the evidence on which the concept is based [see reviews by Peacocke and Drysdale (2) and Sadgopal (3)] but we will deal with certain key aspects.

What indirect evidence exists is based on a correlation between the properties exhibited by DNA and that required or expected of genetic information. Obviously, if genes are composed of DNA, the distribution of the latter should parallel that of the genes in terms of cellular location and quantity. This is the case: DNA is common to all those entities that carry genetic information, i.e., germ cells, nuclei, chromosomes, mitochondria, and viruses (again, with the exception of ribonucleic acid viruses). As originally demonstrated by Vendrely and Vendrely (4), the DNA content of cells of higher organisms is fairly constant from one tissue to another except for germ cells which contain only half the number of chromosomes (haploid set of chromosomes) and half the quantity of DNA. Just as impressive are the observations showing a direct correlation between the quantity of DNA in yeast cells and the ploidy of the organism (number of chromosomal sets per cell) (5).

Another characteristic of genetic material is its metabolic stability. Once formed, genetic material persists until the cell dies. This feature is satisfied by DNA whose metabolic stability has been demonstrated in many tissues and organisms (6, 7). There are, however, some notable exceptions (8, 9), but these are not relevant to our discussion.

Assuming all cells of an organism contain the same genetic information, the chemical composition of the genetic material should be identical from one tissue to another within the same organism. In terms of this concept, the relative proportion of the purine and pyrimidine bases of which DNA is composed should be constant for all tissue. In fact, no significant differences have been found among DNA samples of different organs of the same organism [with the exception of certain germ cells which amass large quantities of special genes (10)]. This constancy is particularly meaningful when we consider how variable the base composition of DNA is from species to species (11).

By definition, genetic material must be able to self-replicate, that is, synthesize a copy of itself so that all the genetic information of a parental cell may be conferred on daughter cells. Recently, this has been demonstrated by Kornberg and colleagues (12) who synthesized the infectious DNA of a bacteriophage virus *in vitro* from the chemical constituents of DNA, polymerizing and rejoining enzymes, and a viral DNA template which was prepared *in vitro*. The demonstration constitutes both indirect and direct evidence for the relation between DNA and genetic material. Indirect, in the sense that it demonstrates the ability of DNA to direct its own synthesis; direct, in the sense that the DNA prepared *in vitro* was infectious, possessing all the genetic information necessary for making new virus.

A series of experiments have shown how DNA could serve as genetic information and how it can carry the "code" for specifying the spectrum of enzymes and other proteins which characterize an organism (13, 14). This will be considered more fully.

The observation that isolated, highly purified DNA can impart (transform) new genetic character to bacterial cells (15, 16) and cells of higher organisms (17) is taken as direct evidence that genes are composed of DNA, as are the observations which show that viral DNA alone is capable of producing new viral particles (18) whereas other viral constituents are not (again, with the exception of certain ribonucleic acid viruses).

The validity of the concept is further consolidated by the congruity between the structure and composition of DNA (18) and what genetic analyses (i.e., genetic maps) demand of the structure of the genetic material. For example, certain bacteriophages and bacteria were shown, by genetic analyses, to possess a circular chromosome, while subsequent investigation showed that their DNA was circular (19).

II. Chemical Composition and Structure

DNA is composed of heterocyclic bases (purines and pyrimidines) which are attached to the C-1 position of 2-deoxy-D-ribose through a β-glycosidic bond. The base-sugar units are termed nucleosides. Upon phosphorylation of the ribose ring they are converted to nucleotides (Fig. 2.1). DNA is simply

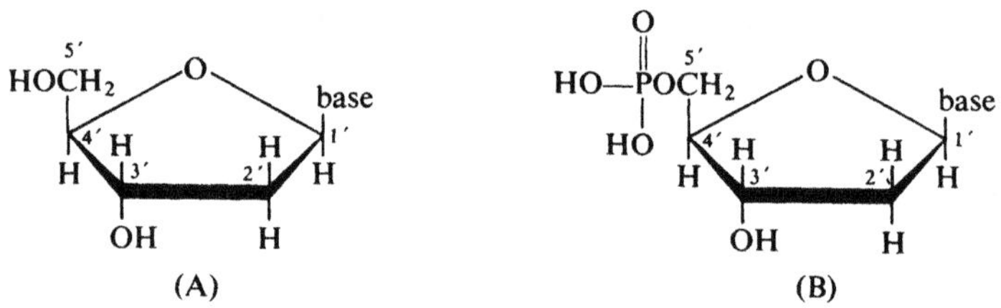

FIG. 2.1. Structure of (A) nucleoside and (B) nucleotide.

a long phosphodiester polymer of nucleotides in which the phosphate attached to the 5′-sugar position is further esterified to the 3′-sugar position of its nearest neighbor (Fig. 2.2).

Depending upon the organism and the quantity of genetic material, there may be many thousands of nucleotides in a single DNA molecule. For instance, the genome (total genetic information) of *E. coli* appears to consist of a single DNA molecule (20) containing an estimated 10^7 nucleotides (21). In higher organisms, it is not clear whether each chromosome represents one or many molecules of DNA.

The commonly occurring bases are adenine (A), guanine (G), cytosine (C), and thymine (T) (Fig. 2.3). With some exceptions, these bases predominate in the majority of DNA's, though 5-methylcytosine has been found in both

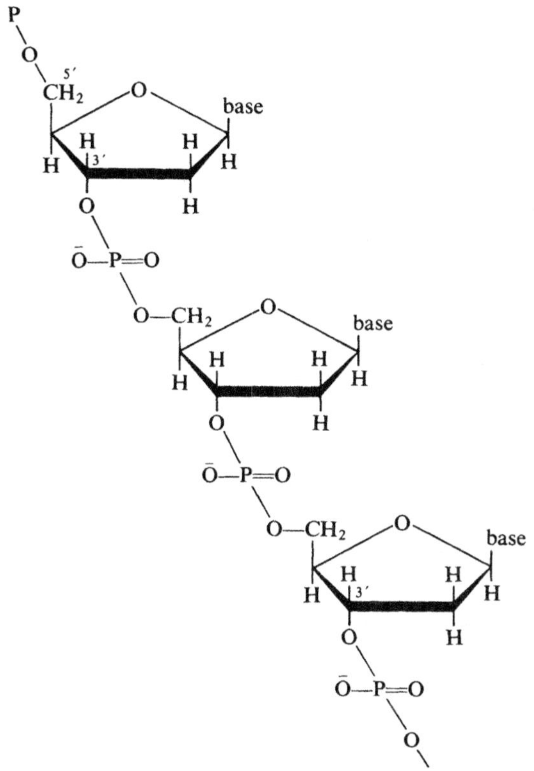

FIG. 2.2. Polydeoxyribonucleotide.

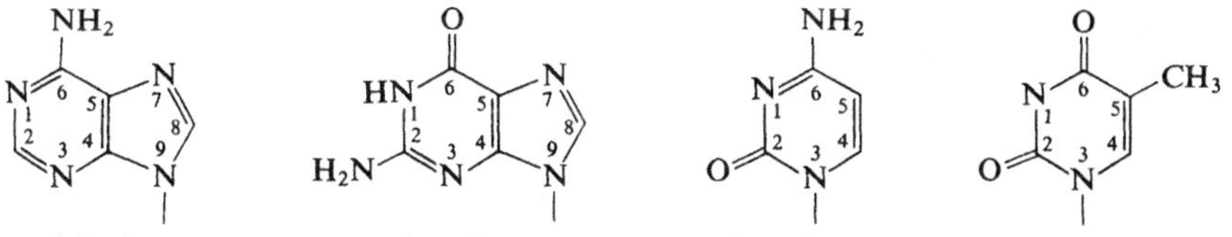

FIG. 2.3. The four most common bases found in DNA.

plant (22) and animal (23) DNA, and T-even bacteriophages contain 5-hydroxymethylcytosine instead of cytosine. There are numerous examples of organisms which utilize, in their DNA, bases other than just A, G, C, and T.

What is important is how these bases (A, G, C, and T) are arranged within a nucleic acid molecule and how this arrangement indirectly or directly accounts for many of the properties required of genetic material. Clearly, the work of Chargaff and co-workers was very important in this regard, for as early as 1949 (24) they demonstrated that, despite large differences in chemical composition, the DNA's of different species had certain regularities in their composition. The molar proportion of A was always equal to that of T, and the molar proportion of G was always equal to that of C among a wide range of organisms which differed greatly in terms of overall base composition (25). In fact, among bacteria, the molar ratios of (G + C) to (A + T) varies from 0.25 to 0.75 from one species to another (26), although between

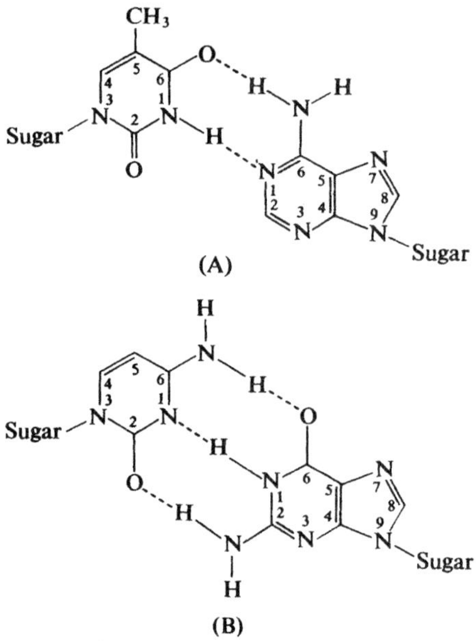

FIG. 2.4. Hydrogen bonding between complementary bases. Base-pairs (A) thymine–adenine and (B) cytosine–guanine.

taxonomically related bacteria the (G + C) to (A + T) proportion is usually similar (27).

It is now understood that A pairs specifically with T and G with C through two hydrogen bonds per pair in the former and three in the latter (28) (Fig. 2.4).

Hence, the hypothetical structure first proposed by Watson and Crick (29) envisaged A, T, G, and C of one polynucleotide strand pairing with T, A, C, and G, respectively, in the other strand. The two polynucleotide strands are complementary to each other and are coiled around a common axis with the sugar-phosphate moieties on the outside forming the backbone of the double helix while the bases are inside where they pair with their complement by hydrogen bonding (A=T, G≡C).

Both strands spiral about a common axis as right-handed helices, with one running from 5′-phosphate to 3′-phosphate and the other (the complementary strand) from 3′ to 5′ (Fig. 2.5). This configuration is referred to as antiparallel (29).

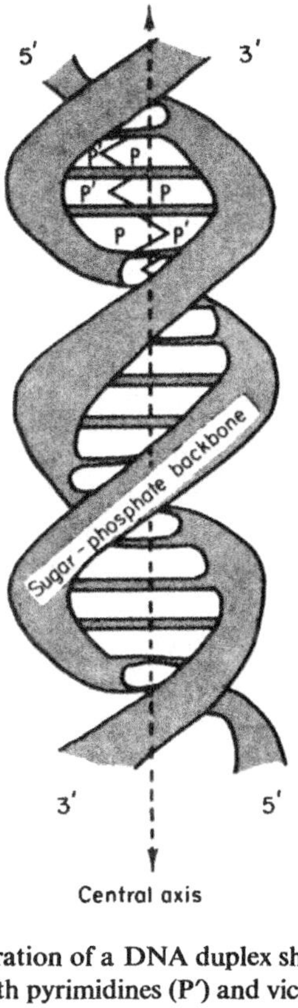

FIG. 2.5. Diagrammatic illustration of a DNA duplex showing 1½ complete turns of the helix. Purines (P) always pair with pyrimidines (P′) and vice versa.

In the paracrystalline B form, which is the form occurring inside the cell, there are 10 base-pairs (each lying perpendicular to the main axis of the helix) for each complete turn of the double helix. One complete turn measures 33.6 Å along the axis indicating that, on average, only 3.4 Å separates one base-pair from another. Appropriately, they are said to be stacked within the DNA molecule giving the appearance, perhaps, of a "pile of pennies" (30).

The great attractiveness of the model (which is now universally accepted as correct) is that it embodies those properties necessary for self-replication and translation of genetic information. The model provides an obvious mechanism by which genetic material can divide (by strand separation) so that each half (each strand) retains the information necessary for specifying its missing half (through specific base-pairing). The ultimate result is a reduplication (doubling) of genetic information prior to cell division.

III. Replication and Gene Expression

Evidence that new DNA is generated by the means alluded to above has been provided by a great number of experiments. Among the most notable are the *in vitro* experiments of Kornberg and his colleagues (31) who demonstrated that new DNA is synthesized according to a molecular pattern predetermined by that of the DNA already present. Presumably, the preexisting DNA strands serve as a template for the synthesis of new strands and specify both composition as well as nucleotide sequence. Such can be envisaged as occurring through specific base-pairing of A with T, T with A, G with C, and C with G. Further, the *in vivo* experiments of Meselson and Stahl (32) and Chun and Littlefield (33) show that the two strands of a DNA double helix separate during synthesis and eventually yield two molecules each consisting of one new strand and one original (preexisting) strand. The process is referred to as "semiconservative replication of DNA," implying that each one of the two original strands is retained (conserved) within each of the two duplexes produced.

The question of how genetic information is expressed can be understood in the same terms as DNA replication. Studies utilizing bacteriophage viruses probably provided the initial clue. Volkin and Astrachan (34) found that, following infection of a bacterium by a host-specific DNA virus, the nucleotide composition of the ribonucleic acid (RNA) being synthesized underwent a drastic change. Before infection, the composition of this RNA resembled that of the host DNA, whereas after infection it more closely resembled the composition of the viral DNA. The immediate implication was that DNA not only serves as a template for self-replication but specifies the composition of RNA as well.

At about the same time, *in vitro* experiments revealed that DNA could, indeed, serve as a template for the production of RNA (35) and that the RNA produced was complementary in composition to the particular DNA strand used (36), except that U replaced T. Artificial hybridization of RNA to DNA has been used to demonstrate that *in situ* DNA specifies the nucleotide sequence of RNA (37, 38).

The relevance of these observations became clear after it was demonstrated that the nucleotide sequence of an RNA molecule is capable of specifying a definite amino acid sequence for a polypeptide molecule (39). Hence, DNA can specify the nucleotide sequence of RNA which then carries the genetic message to the cytoplasm where the apparatus for protein synthesis exists.

Precisely how the nucleotide sequence of an RNA molecule determines an amino acid sequence became the subject of one of the most exciting periods of research in biology. We now understand the "code" and, for the most part, how it functions (40). Each amino acid is coded according to its particular sequence of three ribonucleotides (a combination of adenine, cytosine, guanine, or uracil). The protein-synthesizing apparatus of the cell recognizes this triplet of nucleotides which is called a codon (41) and adds to the growing polypeptide chain that particular amino acid which the triplet has specified.

Codons are nonoverlapping: a given nucleotide used in one codon is not used a second time in an adjacent codon. Hence, a protein consisting of 100 amino acids will have been coded for by an RNA molecule possessing 300 nucleotides. Since there are four nucleotides (A, G, T, and U) in an RNA molecule, there are 4^3 (64) ways in which they can be ordered into groups of three. There are, however, only (approximately) 20 amino acids which raises the question of whether only 20 of the 64 triplets specify amino acids with the rest representing nonsense codons or, alternatively, whether the code is "degenerate" in that certain amino acids are coded for by two or more different types of codons.

It is now well known that the code is "degenerate." The amino acid, alanine for instance, can be specified by either GCT, GCC, GCA, or GCG. Hence, with some amino acids (including alanine) any one of a whole series of code words (codons) can be used to code for a particular amino acid, but the reverse is not true; a single codon does not usually specify more than one particular amino acid.* Nonsense codons also exist, as evidenced by the so-called amber and ochre mutants which are represented by UAG and UAA. Since these codons do not specify an amino acid (in a nonpermissive or restrictive host), the growth of a polypeptide chain is terminated at the point where the codon is read and an incomplete protein (fragment) is released from the ribosome.

It should be pointed out, because it is relevant to discussion in subsequent chapters, that the code is essentially universal, i.e., with only minor exceptions, codon assignments derived from *in vitro* studies using bacterial systems apply to other organisms as well.

*Actually, this is an oversimplification; certain ambiguities do exist in the code, e.g., UUU codes for phenylalanine as well as small but significant amounts of leucine.

REFERENCES

1. C. Auerbach, J. M. Robson, and J. C. Carr, *Science* **105**, 243 (1947).
2. A. R. Peacocke and R. B. Drysdale, "The Molecular Basis of Heredity." Butterworth, London and Washington, D.C., 1967.
3. A. Sadgopal, *Advan. Genet.* **14**, 326 (1968).
4. R. Vendrely and C. Vendrely, *Experientia* **4**, 434 (1948).
5. M. Ogur, S. Minkler, G. Lindegren, and C. C. Lindegren, *Arch. Biochem. Biophys.* **40**, 175 (1952).
6. R. M. S. Smellie, *in* "The Nucleic Acids" (E. Chargaff and J. N. Davidson, eds.), Vol. 2, p. 393. Academic Press, New York, 1955.
7. R. Y. Thomson, J. Paul, and J. N. Davidson, *Biochem. J.* **69**, 553 (1958).
8. S. Nass, M. M. K. Nass, and U. Hennix, *Biochim. Biophys. Acta* **95**, 426 (1965).
9. W. Flamm, H. Bond, H. Burr, and S. Bond, *Biochim. Biophys. Acta* **123**, 652 (1966).
10. J. G. Gall, *Proc. Natl. Acad. Sci. U.S.* **60**, 553 (1968).
11. C. L. Schildkraut, J. Marmur, and P. Doty, *J. Mol. Biol.* **4**, 430 (1962).
12. M. Goulian, A. Kornberg, and R. L. Sinsheimer, *Proc. Natl. Acad. Sci. U.S.* **58**, 2321 (1967).
13. M. W. Nirenberg, O. W. Jones, P. Leder, B. F. C. Clark, W. S. Sly, and S. Pestka, *Cold Spring Harbor Symp. Quant. Biol.* **28**, 549 (1963).
14. M. Nirenberg and P. Leder, *Science* **145**, 1399 (1964).
15. O. T. Avery, C. M. MacLeod, and M. McCarty, *J. Exptl. Med.* **79**, 137 (1944).
16. J. Spizizen, *Proc. Natl. Acad. Sci. U.S.* **43**, 694 (1957).
17. E. Szybalska and W. Szybalski, *Proc. Natl. Acad. Sci. U.S.* **48**, 2926 (1962).
18. J. S. K. Boyd, *Biol. Rev.* **31**, 71 (1956).
19. R. L. Sinsheimer, *Progr. Nucleic Acid Res. Mol. Biol.* **8**, 115 (1968).
20. J. Cairns, *J. Mol. Biol.* **4**, 407 (1962).
21. J. Cairns, *J. Mol. Biol.* **6**, 208 (1963).
22. V. Doskocil and Z. Sormova, *Biochim. Biophys. Acta* **95**, 513 (1965).
23. G. R. Wyatt, *Biochem. J.* **48**, 584 (1951).
24. E. Chargaff, *Experientia* **6**, 201 (1950).
25. N. Sueoka, *J. Mol. Biol.* **3**, 31, (1961).
26. A. N. Belozersky and A. S. Spirin, *Nature* **182**, 111 (1958).
27. K. Y. Lee, R. Wahl, and E. Borbu, *Am. Inst. Pasteur* **91**, 212 (1956).
28. E. Chargaff, *Progr. Nucleic Acid Res. Mol. Biol.* **8**, 297 (1968).
29. J. D. Watson and F. H. C. Crick, *Nature* **171**, 737 (1953).
30. M. H. S. Wilkins, A. R. Stokes, and H. R. Wilson, *Nature* **171**, 738 (1953).
31. M. J. Bessman, I. R. Lehman, J. Adler, S. B. Zimmerman, E. S. Simms, and A. Kornberg, *Proc. Natl. Acad. Sci. U.S.* **44**, 633 (1958).
32. M. Meselson and F. Stahl, *Proc. Natl. Acad. Sci. U.S.* **44**, 671 (1958).
33. E. H. Chun and J. W. Littlefield, *J. Mol. Biol.* **7**, 245 (1963).
34. E. Volkin and L. Astrachan, *Virology* **2**, 149 (1956).
35. S. Weiss, *Proc. Natl. Acad. Sci. U.S.* **46**, 1020 (1960).
36. M. Chamberlen and P. Berg, *Proc. Natl. Acad. Sci. U.S.* **48**, 81 (1962).
37. B. D. Hall and S. Spiegelman, *Proc. Natl. Acad. Sci. U.S.* **47**, 137 (1961).
38. S. A. Yankofsky and S. Spiegelman, *Proc. Natl. Acad. Sci. U.S.* **48**, 1069 (1962).
39. P. M. B. Walker, *in* "Introduction to Molecular Biology" (G. H. Haggis, ed.), p. 300. Longmans, Green, New York, 1964.
40. C. R. Woese, *Progr. Nucleic Acid Res. Mol. Biol.* **7**, 107 (1967).
41. S. Brenner, *Proc. Natl. Acad. Sci. U.S.* **43**, 687 (1957).

CHAPTER 3

Mode of Action and Types of Mutations Induced by Chemicals

From the preceding chapter, it follows that mutation involves alterations in DNA; an alteration of its shape, structure, or constituent nucleotides. All three types have been discovered and will be discussed here. There are, however, two main categories into which mutations can be grouped. One is referred to as a point mutation implying a small, submicroscopic alteration of DNA affecting only one or a very small number of nucleotides. The other category concerns gross chromosomal alterations which can be seen with the light microscope, i.e., chromosomal breaks, deletions, and translocations.

From the point of view of molecular mechanisms of action, only point mutations are considered understood. The concept that gross mutations are always a consequence of many point mutations or are caused by point mutations is presently without adequate foundation. In fact, the relationship between the two categories of mutations is obscure. One reason is undoubtedly the necessity of using different biological test systems for each of the two categories: point mutations are most readily tested for in microbial systems while gross mutations are generally observed in higher plants and animals.

I. Substitution by Base Analogs

The genetic code depends upon a transcription of DNA into RNA such that the RNA represents a complementary copy of its DNA template (its gene). The RNA produced specifies an amino acid sequence for a particular protein via the protein-synthesizing machinery of the cell which is able to "read" the RNA by translating groups of three nucleotides (codons) into appropriate amino acids. By substituting one nucleotide for another (for instance, A for G) within a given codon, the code can be changed, and hence a different amino acid may be specified or protein synthesis may be terminated at that point. Such changes, produced within an amino acid sequence, can (and frequently do) lead to alterations and inactivations of protein function (particularly mutation to a nonsense codon), the net effect of which is referred to as the mutant phenotype.

A. Halouracil and Uridine Derivatives

One of the best known and most studied compounds which produces point mutations through a mechanism involving nucleotide replacement is 5-bromodeoxyuridine (BUdR). This compound (including certain other halouracil and uridine derivatives) is virtually unique in the sense that it can quantitatively replace all the thymidine in DNA under conditions of DNA synthesis when free thymidine is lacking. The depletion of free thymidine is accomplished either by incubating cells with inhibitors of thymidine synthesis or by employing various mutant strains which are unable to synthesize their own (1, 2).

Under conditions of thymidine depletion, BUdR is incorporated in place of thymidine when new DNA is synthesized. The incorporated BUdR preferentially forms Watson-Crick base-pairs with adenine as does its natural counterpart, thymidine, but tends to make occasional pairing "mistakes." BUdR will, though very infrequently, pair with guanine; should this occur during DNA replication, guanine is inserted into the new DNA strand in place of adenine (3). The ultimate consequence is a transition mutation in which a particular A-T site on the DNA is converted to a G-C base-pair in the third generation (Fig. 3.1).

Transitions are point mutations defined by Freese (4) as alterations which involve the substitution of one pyrimidine by another pyrimidine (C replacing T or T replacing C) or one purine by another (A replacing G or G replacing A). Transition substitutions are far more frequently encountered than transversions which involve the substitution of pyrimidine by purine or purine by pyrimidine.

The occasional pairing error by BUdR is thought to be due to the higher electronegativity of the bromine atom as compared to the CH_3 of thymidine. The greater inductive effect of bromine tends to make the pyrimidine ring poorer in electrons, a condition which favors the enolic form (5) responsible for guanine pairing (Fig. 3.2).

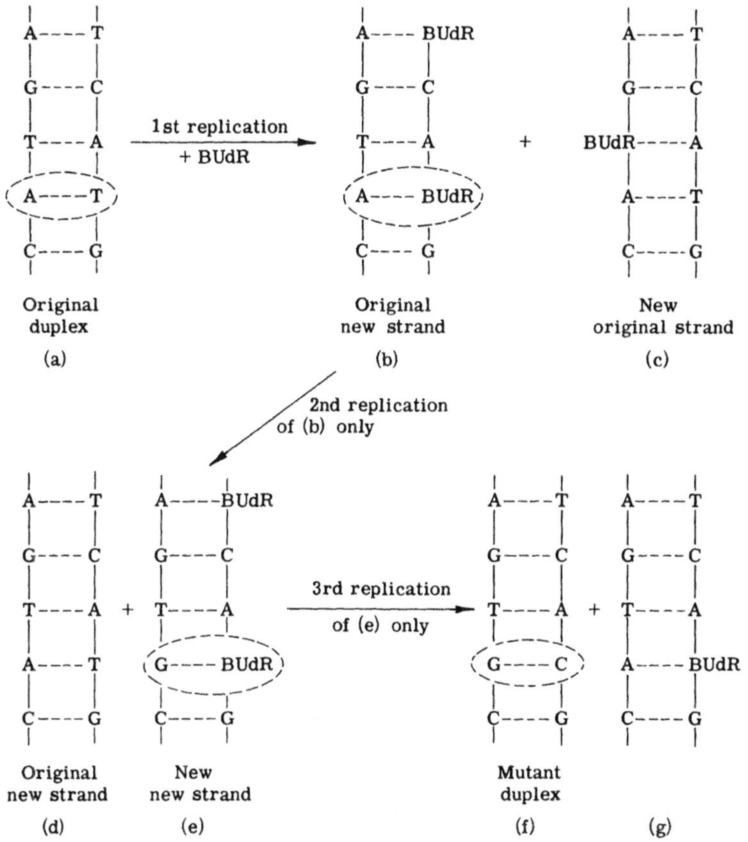

FIG. 3.1. (a)–(f) Induction of an A-T → G-C transition through the erroneous base-pairing of BUdR with G instead of A.

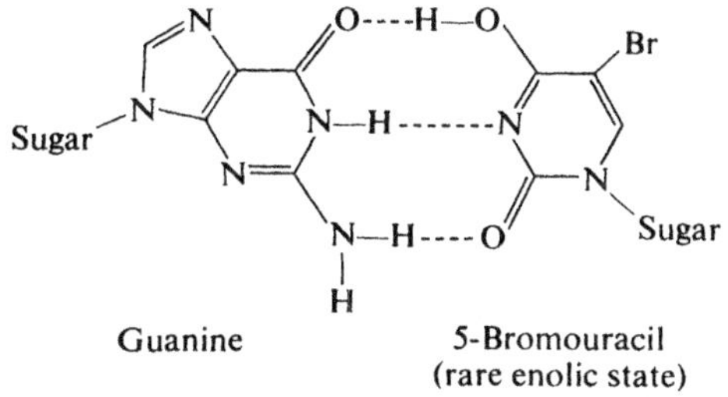

FIG. 3.2. Base-pairing of the enolic form of BUdR to guanine.

In addition to transitions of the A-T to G-C type arising from pairing errors, BUdR can induce G-C to A-T transitions (3). Presumably this occurs when BUdR is incorporated into DNA in place of cytosine, and since it preferentially pairs with adenine (except for the occasional mistake), a G-C to A-T transition is produced (Fig. 3.3). As with the A-T to G-C transition,

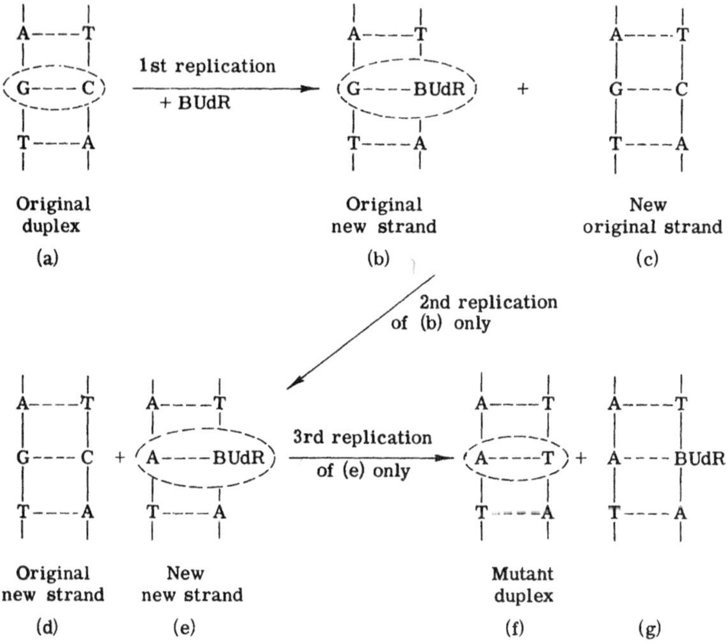

FIG. 3.3. (a)–(f) Induction of a G-C → A-T transition by the erroneous incorporation of BUdR for C.

it occurs very infrequently, but infrequency is a characteristic of all mutations: they are rare events, readily observed only when millions of individual organisms are reproducing in a short time. The attractiveness of microbial organisms for genetic analysis is related to this requirement. The fact that BUdR can induce both A-T to G-C and G-C to A-T transitions suggests that mutations produced by BUdR should also be induced to revert by BUdR. This has been observed (6).

Because of its relevance to the general problem of biological synergism, it should be pointed out that organisms containing BUdR-substituted DNA are very much more sensitive to chemical attack and irradiation by UV, visible light, and x-rays than those containing normal DNA (7–9). Why this should be so is not well understood, and the effect is nearly always lethal and not mutagenic.

5-Bromouracil is enzymically converted to BUdR and will induce the same types of transitions described above, however, somewhat less efficiently. 5-Chlorouracil, 5-iodouracil, and their deoxyribonucleosides will replace thymine in DNA and are also mutagenic (10, 11), but they have been studied in less detail.

B. Aminopurines

The mutagenic properties of 2-aminopurine have been extensively studied. Its effects differ quantitatively from those of the halo derivatives of uracil and uridine (12–14). 2-Aminopurine is incorporated into DNA to such a

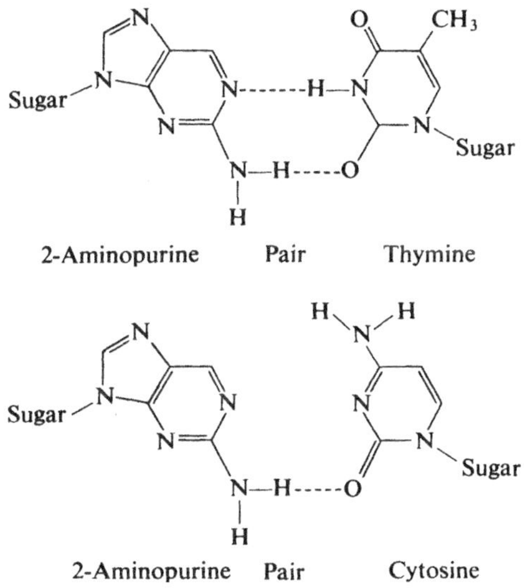

FIG. 3.4. Base-pairing of 2-aminopurine to cytosine and thymine.

limited extent that its presence has been difficult to detect, nor has it been possible to determine chemically the particular base it substitutes. Since 2-aminopurine induces mutations with as great a frequency as the halouracil derivatives, despite its low level of incorporation, it seems likely that it makes many more pairing "mistakes" than BUdR (15).

It is known that 2-aminopurine can pair with T as well as with C through two hydrogen bonds in the former and one in the latter (12) (Fig. 3.4). Though the 2-aminopurine to T base-pair is preferred, its occasional pairing with C would lead to a transition in which an A-T base-paired site is converted to a G-C base-pair (Fig. 3.5A). In support of this possibility Pauling (16) has

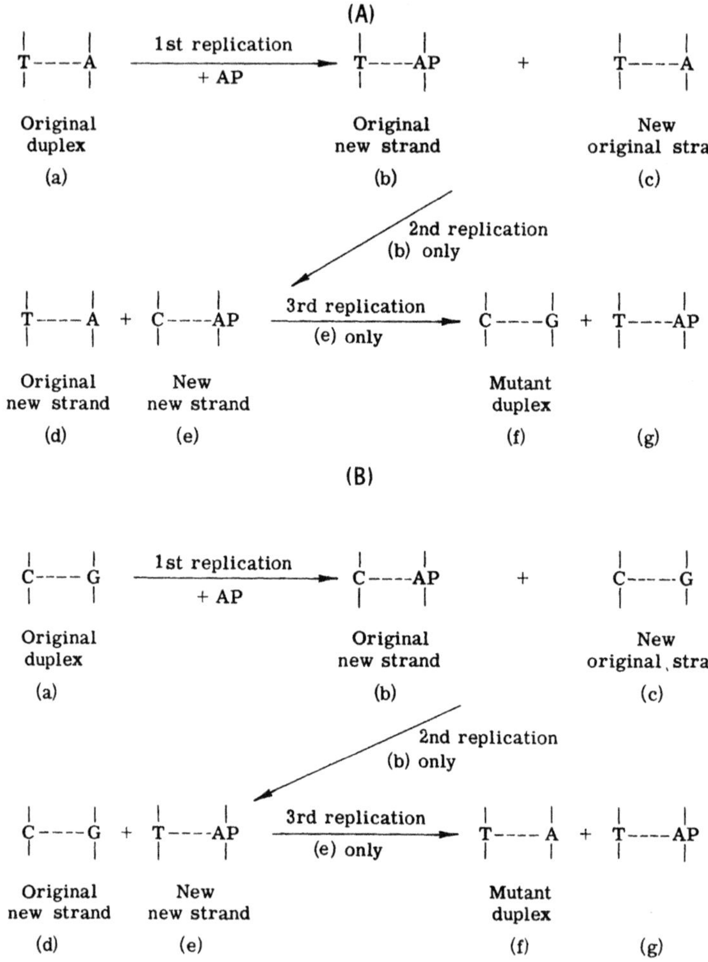

FIG. 3.5. Induction of transition mutants by 2-aminopurine; (A) A-T to G-C transitions through erroneous pairing, (B) G-C to A-T transitions via erroneous incorporation.

demonstrated that thymine-starved bacteria undergo this type of transitional mutation more frequently than normally expected, presumably by increasing the probability of 2-aminopurine pairing with C instead of T.

As with BUdR, 2-aminopurine may induce mutations through errors in incorporation (17). For instance, it may be incorporated instead of G during a period of DNA synthesis, and since 2-aminopurine generally pairs with C, transitions of the G-C to A-T type are produced in the third generation (Fig. 3.5B). The idea is supported by the fact that 2-aminopurine will induce its own mutations to revert (18). Mutations induced by 2-aminopurine which are of the A-T to G-C type are fully reversible by the same agent through G-C- to A-T-induced transitions. The opposite also applies as it does for BUdR. In fact, mutations induced by BUdR can be induced to revert by 2-aminopurine and vice versa.

In addition to 2-aminopurine, 2,6-diaminopurine has been shown to be mutagenic in bacteria although it has not been studied in as much detail (19). It is clear that 2-aminopurine and 2,6-diaminopurine are less inhibitory, that is, they cause fewer inactivations than many other purine and pyrimidine analogs whose effects are often more lethal than mutagenic.

II. Chemical Alterations of DNA and Its Constituent Nucleotides

A. Alkylating Agents

The largest class of "potential mutagens" present in man's environment are represented by the alkylating agents (see Chap. 7). Alkylating agents (nitrogen and sulfur mustards) were among the first chemicals recognized as mutagens (20) in certain organisms. The term "potential mutagen" is used here to apply to those agents shown to induce mutation in at least one of the several, commonly used genetic test systems (see Chap. 5). A "potential mutagen" need not and may not be mutagenic in man for reasons which will be explored later in this chapter.

Alkylating agents can be divided into two classes: (a) those which carry a single reactive alkyl group, i.e., the monofunctional alkylating agents and (b) the bifunctional or polyfunctional agents which carry two or more reactive alkyl groups. The primary difference between them is related to the latter's ability to form covalent cross-links between the two individual strands of a DNA double helix (21, 22). Since interstrand cross-links prevent the DNA duplex from undergoing complete strand separation necessary for proper replication (see Chap. 2), bifunctional and polyfunctional agents produce a higher cytotoxicity than their single-armed counterparts (23). It is not surprising, therefore, than in many test systems, polyfunctional and bifunctional agents have been shown to induce a greater number of lethal events per mutational event than their monofunctional analogs (22).

In addition to cross-linking DNA, bi- and polyfunctional alkylating agents may also mimic the action of monofunctional agents by reacting with only one of the two strands of a DNA duplex. The ratio of cross-links to single-strand reactions presumably depends upon the nature of the alkylating agent as well as upon the DNA involved. The reasons for this are discussed later in Section II, A, 2.

1. *Specificity for the N-7 Atom of Guanine*

The N-7 position of guanine is highly susceptible to alkylation whether within nucleic acids, nucleotides, or nucleosides (24–26). Within the DNA molecule, the N-7 atom of guanine is the most reactive site (of all bases) toward nearly all alkylating agents (27). This is not unexpected in view of wave-mechanical studies (28) and three-dimensional model building experiments (29). The latter showed that the N-7 position of guanine is sterically available since it is situated in the wide groove of the DNA double helix.

Under nonphysiological conditions, in which the nucleic acid is suspended in ether, alkylation by diazomethane involves the N-1 atom of guanine (30), but this does not appear to occur *in vivo*. All three of the unsubstituted ring-nitrogen atoms of adenine can be alkylated in DNA. The N-3 position is the most reactive (31, 32), although it is less reactive than the N-3 of guanine. With most alkylating agents, alkylation of adenine would only amount to 5% to 10% of the total alkylated purines (33). According to Brookes and Lawley (34) the N-1 atom of cytosine is the only reactive site among the pyrimidine moieties of DNA although its reactivity is greatly reduced by the hydrogen bonding of C to G. In RNA, and presumably in single-stranded DNA, the N-1 of cytosine is about one-eighth as reactive toward diazomethane as the N-7 of guanine (32). Little, if any, alkylation of the N-1 of cytosine in DNA is to be expected *in vivo* with the possible exception of single-stranded DNA viruses. For a more complete description of the sites alkylated by different agents the reader is referred to reviews by Ross (35) and Orgel (36).

Alkylating agents react with DNA either by an S_N1 or S_N2 mechanism. The former includes certain "aromatic nitrogen mustards" (35) although recently two such compounds (chlorambucil and L-sarcolysin) have been demonstrated to be of the S_N2 class (37). S_N1 reagents react through a carbonium ion intermediate which rapidly alkylates guanine or any other appropriate nucleophile. Since the formation of the carbonium ion is the slow, rate-determining step, the reaction proceeds with unimolecular kinetics. On the other hand, aliphatic nitrogen mustards and the alkyl alkanesulfonates are of the S_N2 type since they form a transition complex with the nucleophile and thus exhibit second-order kinetics. No differences relative to their sites of reaction within DNA have been substantiated—both S_N1 and S_N2 types appear to react preferentially with the N-7 of guanine.

2. *Cross-Linking of DNA*

The high specificity for guanine is also exemplified by the cross-linking reaction of DNA with bifunctional alkylating agents. Hence, diguaninyl derivatives are the principal dialkylated products obtained from the reaction of sulfur and nitrogen mustards with DNA (21, 38). Examination of the Watson-Crick model for DNA reveals that the interstrand cross-linking of guanines most readily occurs between GpC sequences, located on opposite strands, while the covalent linking of two neighboring guanines, present on the same strand (GpG sequence), is possible only if the alkyl group is folded rather than extended. Since the frequency of GpG sequences depends upon G + C content, the proportion of diguaninyl derivatives among the total alkylated products depends upon the base composition (G-C content) of the DNA employed in the reaction. With DNA of low G-C content (34%) only 13% of the total alkylated product was of the diguaninyl type, whereas 20% and 26% of the total consisted of diguaninyl derivatives when DNA's of 50% and 67% G + C contents were studied (39).

Evidence that interstrand cross-links are formed has been provided by denaturation–renaturation experiments on DNA (40, 41) which show that complete strand separation is prevented after its reaction with bifunctional reagents. Using the nitrogen mustard, HN2, Kohn (42) demonstrated that, on the average, one interstrand cross-link is formed for every nine alkylations of DNA, indicating that roughly one-half of the diguaninyl products originate from interstrand cross-links. The ratio of interstrand to intrastrand cross-links may actually be somewhat higher, since the conditions used to quantitate interstrand cross-links tend to labilize the bonds on which the analysis depends. In addition to other compounds which are obvious bifunctional alkylating agents, the antibiotics mitomycin C and porfiromycin cross-link the DNA duplex following their enzymically mediated reductions (43).

The biological significance of interstrand cross-linking is evident from studies on transforming DNA and bacteriophage viruses (42, 44). One interstrand cross-link is sufficient to cause the inactivation of at least 3000 base-pairs within a DNA molecule, presumably as a consequence of blocking complete strand separation.

3. *Possible Molecular Mechanisms Mediating Genetic Effects*

There are two main hypotheses concerning the mechanisms by which alkylation ultimately leads to mutation. The first (45, 46) involves a mechanism similar to that described for base analogs except that instead of inserting an altered base during DNA duplication, a base is altered by the reaction of mutagen with DNA. The altered base, usually a guanine quaternized at the

N-7 position, is assumed to ionize differently, leading to occasional pairing errors, e.g., alkylated guanine pairing with thymine instead of cytosine (Fig. 3.6). Were this the sole means by which alkylation leads to mutation, only mutants of the transition type would be expected. In contrast, however, Freese (47) and co-workers (24) find that both transitions and transversions are produced by several different alkylating agents. Hence, pairing errors, if involved at all, cannot be exclusively responsible for the changes observed (48).

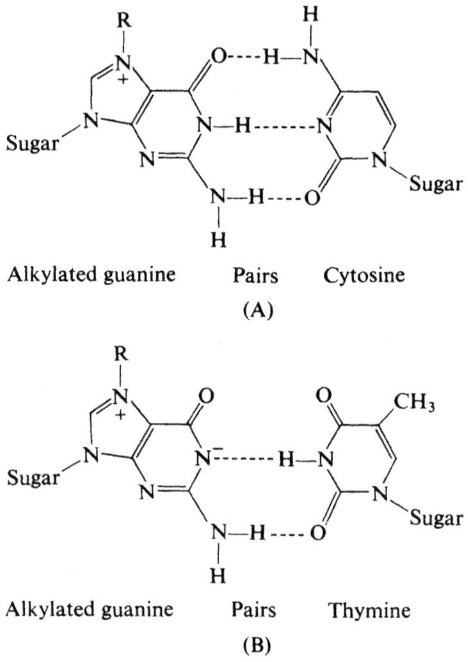

FIG. 3.6. Base-pairing of alkylated guanine to cytosine (A) and of the ionized form (N-1) to thymine (B).

The second hypothesis is based on the observation (49, 50) that the alkyl group on the N-7 of guanine labilizes the β-glycosidic bond resulting in depurination. The loss of this base leaves a gap in the DNA template so that during replication this is either ignored (a deletion) or a base selected at random is inserted into the new strand opposite the deletion. Hence, either a pyrimidine or purine may be inserted opposite the gap and may thus give rise to the combination of transitions and transversions which have been observed. Since acid treatment causes rapid depurination of DNA, it is significant that alkylation simulates acid-induced mutation (47) suggesting depurination as a primary cause of mutagenesis.

Although there are ample reasons for believing that the above modes of action are involved in the induction of mutation, there are also reasons for believing other events may participate as well. It is well established that depurination of guanine leaves an unstable deoxyribose residue which undergoes rapid hydrolysis (51) resulting in cleavage of the polynucleotide chain. Whether this event induces mutational effects or is exclusively lethal in effect has not been decided though this will probably depend upon the type of organism being considered and the extent to which such cleavages can be dealt with by the cell's repair processes (see Chap. 4).

It has been suggested by Alexander *et al.* (52) that the immediate product of DNA alkylation is a phosphotriester (see Fig. 3.7) which, in the case of

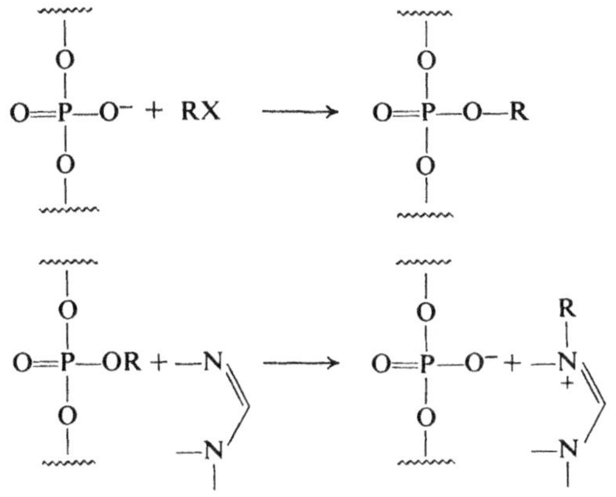

FIG. 3.7. Formation of phosphotriester and its rearrangement.

most alkylating agents, undergoes rearrangement to alkylate the N-7 atom of guanine. According to this study, the presumed triester formed in the reaction between DNA and ethyl methanesulfonate is unusually stable and does not undergo rapid transalkylation as do triesters of methyl methanesulfonate and certain other alkylating agents. Since ethylating agents, according to Alexander *et al.* (52), are, in general, considerably more mutagenic than their methyl counterparts (24, 53), it was suggested that triester formation is more germane to mutagenesis than the alkylation of guanine. However, Lawley, in a recent review (50), discounts this possibility, claiming that phosphotriester formation is not a significant reaction even under conditions yielding a high extent of alkylation. Lawley further suggests that the formation of ionic complexes between basic alkylating agents such as 2-chloroethylamines and DNA are more likely to be observed and may have been confused with phosphotriesters. However, the basic problem considered by Alexander *et al.* (52) remains unresolved.

Since alkylation of guanine followed by depurination labilizes the phosphodiester chain, chromosome breaks might well be expected in higher organisms treated with alkylating agents. This has been observed in many such organisms (54–56), but the question as to whether there is a direct relationship between chromosome breakage and the behavior of alkylated DNA is uncertain. As was pointed out by Lim and Snyder (57), the monofunctional alkylating agent, ethyleneimine, induces chromosome breakage at a faster rate and within a shorter period than that predicted simply on the basis of the rate of hydrolysis of depurinated DNA. Even more surprising is the observation that the frequency of breaks failed to increase upon storage of ethyleneimine-treated spermatozoa in females (58).

Just as perplexing are the observations (57) showing that ethyleneimine fails to induce many small deletions (deficiencies) within the *Drosophila* genome though a relatively high proportion of gross alterations (breaks, translocations) were observed. On the other hand, ethyl methanesulfonate induced far fewer chromosomal breaks and translocations and yet it was approximately as effective in producing sex-linked recessive lethals. Hence, it seems unlikely, based on that report, that chromosome breakage and small deletions necessarily represent different degrees of damage resulting from the same initial reaction of DNA with alkylating agents.

There can be little doubt that mono- and polyfunctional alkylating agents react with the genetic material (DNA) of mammalian cells in the same fashion as reported for bacteria, bacteriophages, and isolated DNA. The products of alkylation are identical, and the appropriate ratios of diguaninyl and monoguaninyl products are obtained from alkylations with bifunctional agents (59). Nevertheless, it is evident that much larger numbers of alkylations are needed to inactivate mammalian cells than that required in microorganisms.

In the case of the bifunctional mustard gas, a mean lethal dose of 2×10^4 alkylations per cell was reported for mouse lymphoma cultures; of these, approximately 2×10^3 consisted of interstrand cross-links. The resistance of mammalian cells to alkylation has been attributed to highly efficient repair processes (60) which remove the alkylated lesion and restore the DNA to normal. Other suggestions concern the growing awareness of the differences that exist between genomes of complex organisms and those of bacteria and viruses. Possibly, the so-called "master-slave" hypothesis (61) which envisages the genome of higher organisms as consisting of master genes which direct and control synthesis of slave genes play an important role in this respect. According to this hypothesis, extensive damage to slave genes could be readily corrected by the resynthesis of "slaves" in which the master would serve as a template (62).

B. Deamination of DNA by Nitrous Acid

Nitrous acid is produced by the reaction of nitrite salts with acid or acidic buffers and is known to oxidatively deaminate adenine, guanine, and cytosine. Hence, free amino groups are lost and replaced by hydroxyl groups converting adenine to hypoxanthine, guanine to xanthine, and cytosine to uracil (63–65). Nitrous acid also reacts with these bases in nucleosides, nucleotides, and nucleic acids.

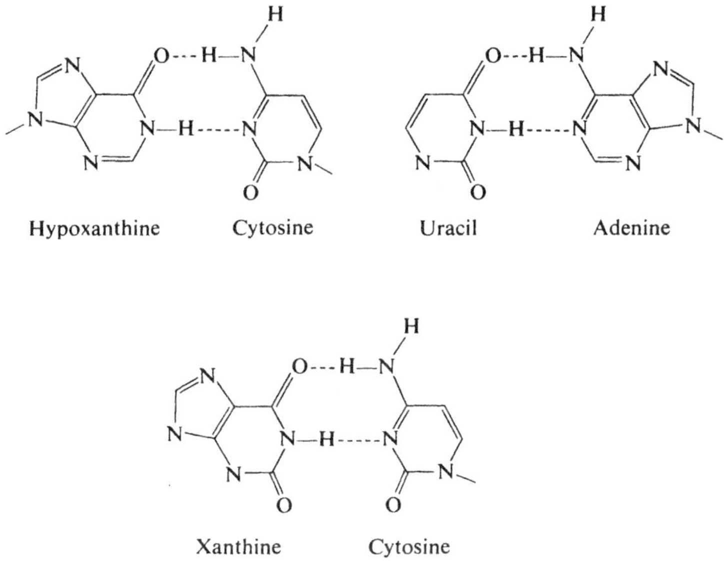

Fig. 3.8. The products of the deamination of adenine, thymine, and guanine and their base-pairing reactions.

The conversion of adenine to hypoxanthine is expected to induce A-T to G-C transitions since hypoxanthine pairs preferentially with cytosine. Similarly, the oxidative deamination of cytosine to uracil should cause transitions of the G-C to A-T type since uracil forms base-pairs much more readily with A than G. On the other hand, it is claimed, and appears structurally feasible, that xanthine would pair preferentially with C as does its amino analog, guanine, and hence would not induce transitions (see Fig. 3.8).

It is known, however, that xanthine does not adequately replace guanine's function in RNA as evidenced by the inability of deaminated polyguanilic acid to support the incorporation of amino acids in a cell-free protein synthesizing system (66). Furthermore, studies with the DNA-bacteriophage, S-13, show that all four bases are implicated in the nitrous acid production of mutants indicating that reactions other than deamination must be involved since thymine cannot undergo deamination (67). Apparently, the reaction with thymine is too small to be detected by the usual chemical means. Conceivably, the lower pK_a of xanthine (68) may lead to occasional pairing mistakes in which a xanthine to thymine base-pair is formed instead of the more frequent xanthine to cytosine pair.

Most of the evidence to date, however, indicates deamination of adenine and cytosine as primarily responsible for the mutagenicity of nitrous acid. By determining what amino acid changes are produced in the proteins of certain TMV mutants (induced by nitrous acid) it has been shown that most codon changes can be accounted for on the basis of the deamination of adenine and cytosine (69). Additional support for this view has been provided from other genetic systems. In one study, the rate of deamination, which differs for each of the three bases, was compared with the rate of mutation and it was concluded that deamination of guanine does not lead to point mutations (65).

In addition to deamination, nitrous acid produces interstrand cross-links within the DNA molecule. Presumably, via the formation of diazonium complexes which then attack free amino groups on nearby bases located on the opposite strand. Geidusheck calculated that the relative frequency of interstrand cross-links is about one per four deaminations (22) but this ratio apparently varies with the source of DNA employed. For reasons unknown, DNA's from different sources differ appreciably in the extent to which their strands will be cross-linked by nitrous acid (70).

As mentioned earlier with regard to polyfunctional alkylating agents, the formation of interstrand cross-links interferes with the replication of DNA and is expected to be more lethal than mutagenic. Nevertheless, it is likely that the induction, by nitrous acid, of deletion mutations is a consequence of cross-linking (38). Presumably, DNA replication is blocked only in the general area where the cross-link is found and that replication at distal regions can continue, thus leading to the formation of a mutant organism whose genome contains a large deletion near the site of the cross-link.

Curiously, and for unaccountable reasons, nitrous acid elicits different biological responses in different test systems. Even among bacterial species such differences have been noted, e.g., between pneumococcus, *B. subtilis*, and hemophilus (71–75). It is tempting, though probably premature, to suggest a causal relationship between the unexplained differences in chemical reactivity of DNA's from different sources and the variability in biological responses.

C. Modification of Pyrimidines in Nucleic Acids

Hydrazine, hydroxylamine, and certain hydroxylamine derivatives (CH_3NHOH and NH_2OCH_3) are mutagenic to different degrees in bacteria and to transforming DNA. Each have in common the ability to interact specifically with pyrimidines under specific conditions including pH, concentration of reagent, and oxygen tension. Their mutagenicity also depends markedly upon the above conditions, although, in general, hydroxylamine and its analogs are appreciably more mutagenic than hydrazine.

1. *Hydroxylamine and Its Derivatives*

At high concentrations and high pH (approximately pH 9), hydroxylamine (0.1 *M* to 1.0 *M*) reacts exclusively with the uracil moieties of nucleic acids.

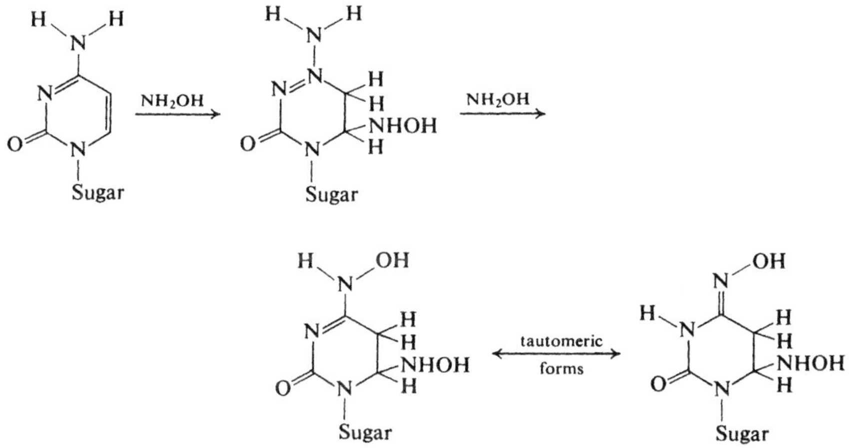

FIG. 3.9. Reaction of hydroxylamine with cytosine.

The pyrimidine ring is opened by the reaction leaving a phosphoribosylurea and splitting out 5-isoxazolone (76). At lower concentrations (10^{-3} *M*) and at pH 9, hydroxylamine reacts with both thymine and guanine as well as with uracil (77). It seems probable, however, that the reaction with thymine and guanine does not involve hydroxylamine per se, but rather a product of its decomposition. The fact that the reaction of thymine and guanine in dilute solutions of hydroxylamine is suppressed by the reduction of oxygen tension or the presence of compounds which chelate heavy metal ions support this view (78).

At low pH (approximately pH 6) and high concentration (0.1 *M* to 1.0 *M*), hydroxylamine reacts exclusively with the cytosine moieties of DNA, aminating only the C-4 atom (78) (see Fig. 3.9). Paradoxically, at lower concentrations of hydroxylamine, the reaction involves all four bases (78, 79), a situation analogous to that reported for higher pH's (see above). Furthermore, it is well known that hydroxylamine is highly toxic at low concentrations (80), where it is only weakly mutagenic (76), and yet, at the higher concentrations at which it is mutagenic, little or no cytotoxicity is observed. The explanation may relate to the degradation products of hydroxylamine rather than to the compound per se. For instance, Bendick *et al.* (81) observed that dilute solutions of hydroxylamine had little effect on DNA when present in a reducing atmosphere or after metal-chelating compounds were introduced to the solution. The possibility arises, therefore, that in an oxygen-containing atmosphere, hyponitrous acid is generated by the oxidation of hydroxylamine and that this product is responsible for both cytotoxicity and the reactivity with G, A, and T.

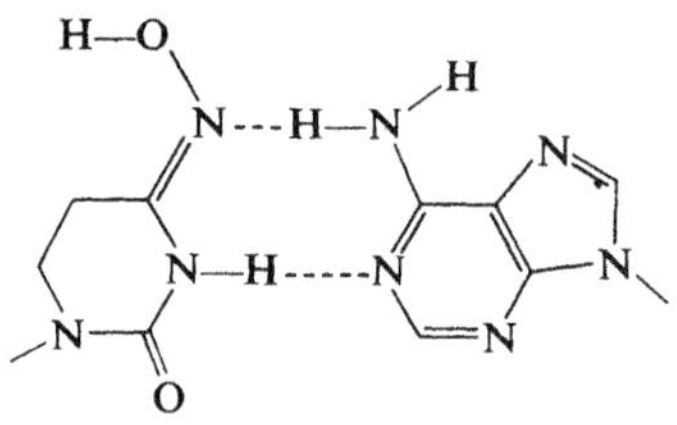

FIG. 3.10. Base-pairing of the proposed reaction product (of hydroxylamine with cytosine) to adenine [see Orgel (36)].

Since none of the effects noted above are caused by strong solutions of hydroxylamine, it has been suggested (15) that the active component (hyponitrous acid, in this case) interacts with hydroxylamine when present at high concentrations and is destroyed. On the other hand, Freese and Freese (77) showed that catalase and peroxidase eliminated the inactivating effects of dilute solutions of hydroxylamine suggesting that peroxide formation or resultant free radicals produced by it are responsible for the side effects. Whether the active substance is a peroxide or hyponitrous acid, it is evident that hydroxylamine-induced mutations arise from the reaction of hydroxylamine with cytosine, while toxicity and the reactions with purines and thymine are the results of the interaction with decomposition products.

Judging from the evidence obtained with single-stranded bacteriophages (80), all hydroxylamine-induced mutations are C to T transitions; or as in the case of double-stranded DNA, G-C to A-T transitions. A mechanism has been proposed by Orgel (36) to explain how a hydroxylamine-altered cytosine might specifically pair with adenine (Fig. 3.10).

2. *Hydrazine*

In contrast to hydroxylamine and certain of its derivatives (*N*-methylhydroxylamine, and *O*-methylhydroxylamine), hydrazine is only very weakly mutagenic even when used at high concentrations (1 *M*) (18). The activity of hydrazine, however, is similar to hydroxylamine in terms of its specificity for pyrimidines and the strong inactivating (cytotoxic) effect it exerts on cells that have been exposed to low concentrations of the substance.

As in the case of hydroxylamine, hydrazine's cytotoxicity is highest at low concentrations, which again indicates an indirect effect in which decomposition products, rather than hydrazine itself, would seem to be involved. In this regard, Freese and Freese (17) have provided evidence implicating peroxides and their free radicals as the actual causes of cytotoxicity. The nature of peroxide and free radical reactions are discussed below.

At very high pH, hydrazine reacts with all three pyrimidines—U, C, and T—while at lower pH (8.5) it reacts mainly with U and T, opening the pyrimidine ring and leaving only urea attached to the sugar-phosphate chain. The new β-glycosidic bond is unstable, leading to the hydrolysis of urea and rendering the sugar-phosphate backbone sensitive to breakage (17). It is not surprising, therefore, that the effects of hydrazine, even at high concentrations are predominantly lethal (78). In fact, when compared at equivalent concentrations (1 *M*), hydrazine produces only one-hundredth as many mutations per lethal event as does hydroxylamine (3, 78).

III. Compounds That Physically Bind to DNA

A wide variety of chemical compounds are known to interact with either DNA or RNA so as to induce certain structural alterations within the nucleic acid molecule but apparently without involving the formation of new covalent bonds. The acridines, which are antibacterial agents, and the phenanthridines, which are important trypanosides, are two of the best known representatives of this class, though a number of other compounds have been implicated. Among these are the polycyclic hydrocarbons, nitrogen-containing heterocyclic carcinogens, certain mycotoxins and antibiotics, and even caffeine.

A. Acridines and Phenanthridines

The ability of acridines to combine with nucleic acids has been known for many years, and this interaction forms the basis of an important histological stain for DNA and RNA. With many compounds of this class it is apparent that an interaction with DNA has occurred, since shifts in their UV absorption spectrum or quenching of their fluorescence have been observed (82). It is also clear from the changes in the properties of the affected nucleic acid molecule that some interaction has occurred. For instance, the nucleic acid molecule tends to sediment more slowly following combination with the drug while its viscosity is increased substantially (83). x-Ray diffraction data (84) show that although the basic 3.4 Å reflexion is retained (this corresponds to the distance by which adjacent base-pairs are separated), the longer range ordering of DNA is lost. Hence, there are interruptions in the stacking of the bases present in the DNA helix.

1. *Nature of the Interaction*

Based on these observations, Lerman (85) proposed a model for the binding of proflavine (an acridine derivative) to DNA. He suggested that the planar

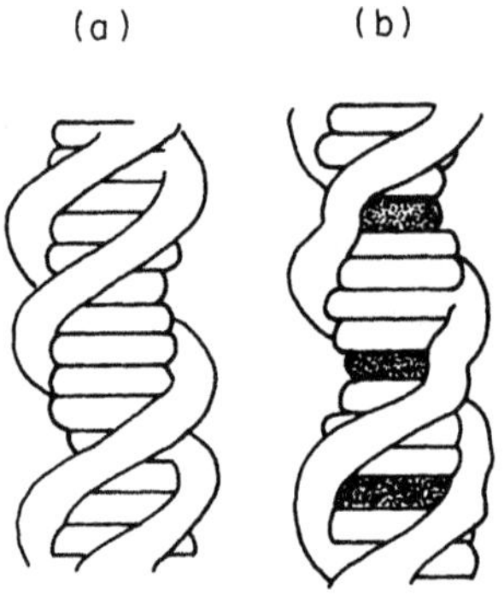

FIG. 3.11. Representation of the secondary structure of normal DNA (a) and DNA containing an intercalated proflavine molecule (b).

triple-ring system of the drug becomes intercalated between adjacent base-pairs of the DNA double helix (Fig. 3.11). This, of course, would lengthen the double helix and give rise to an increased viscosity and decreased sedimentation coefficient. It would also explain the disruption of long-range ordering observed on x-ray diffraction analyses.

Two predictions can be made on the basis of Lerman's intercalation model: first, the length of the DNA molecule should be increased, and second, a certain amount of uncoiling or unwinding of the helix must occur to provide spaces where the intercalating agent is inserted. For the first, direct evidence has been provided by Cairns (86) who demonstrated an actual extension of proflavine-treated DNA on autoradiographs of isotopically labeled DNA. The second prediction has been verified by Waring (87), who followed the changes produced in the sedimentation coefficient of supercoiled DNA as increasing amounts of ethidium bromide (a phenanthridine derivative) were allowed to react with the molecule.

Vinograd and co-workers (88, 89) noticed and took advantage of the fact that less intercalation of DNA is possible when the ends of a double helix are joined in the form of a covalently closed circle than when the DNA is present in the "open-ended" linear form. Since the addition of ethidium molecules to DNA reduces its buoyant density in accordance with the number of ethidium molecules added, linear DNA is made lighter than the closed circles which cannot be as extensively intercalated. This forms the basis by which circular DNA is isolated from mixtures of circular and linear forms. By this method, the mitochondrial DNA of mammalian cells (which is circular) can be isolated directly from preparations of total cellular DNA. This technique has led to the discovery of varying size circles of mitochondrial DNA (called oligomers) as well as the group of interlocked circles of mitochondrial DNA called catenates.

These studies also revealed that at very low concentrations of ethidium bromide more drug is bound to circular, supercoiled DNA than to the linear and other nonsupercoiled forms. The relevance of these observations concerns the possibility that the drug might act selectively on certain small viruses or bacteriophages (e.g., SV40, ϕX174) or kinetoplasts (circular DNA's found outside the main genome) or on mitochondrial DNA.

Although direct experimental evidence is lacking, it seems likely that ethidium bromide and possibly other phenanthridine compounds might exert a reasonably specific antiviral activity in the case of the small circular viruses and bacteriophages.

2. *Genetic Considerations*

It was originally proposed that mutations induced by acridine and its derivatives were of the transversion type (where a purine is changed to a pyrimidine and vice versa) since proflavine (an acridine derivative) failed to revert certain T4 mutants known to be of the transition type (where purine is substituted for purine or a pyrimidine for pyrimidine) (6, 12). Brenner *et al.* (90), however, expressed doubts as to the validity of this proposal, partly because the exact sites of mutation induced by acridine were all independent of those sites mapped for transition mutants. Furthermore, the observation of Lerman (85) that acridines are bound to DNA by sliding between adjacent base-pairs forcing them 6.8 Å apart (twice the normal 3.4 Å separation) raised the possibility that a base might be erroneously added or deleted during replication of DNA. This would be particularly likely if intercalation were limited to only one of the two strands of the DNA duplex. It is now believed that addition and deletion occur during recombination rather than replication.

The deletion or addition of a base-pair generally has a more serious consequence upon the affected gene and its ability to code properly for protein than does a simple substitution of one base for another (transition and transversion mutants) unless, of course, the substitution results in a nonsense codon. As discussed in earlier sections, substitution concerns only the codon in which base substitution has occurred, and hence usually involves only the substitution of one amino acid for another in the finished product of gene expression (a protein molecule). In the case of mutation to a nonsense codon, however, protein synthesis is terminated at the point of mutation.

With a single-base addition or deletion the reading of every codon beyond the point of mutation (addition or deletion) has been changed and is now "incorrect." This is because the code is read by starting at one end of the gene and moving in one direction only. Going from one codon to the next, the bases of one codon never overlap into the next codon. By adding or deleting a base, the new codon either excludes one of the bases originally present (addition) or it includes a base from a neighboring codon (deletion). In either case, all codons transcribed beyond the altered site now differ because reading has either been shifted forward one base (deletion) or back one base (addition) (see Fig. 3.12). Appropriately, such mutations are now referred to as frameshifts mutants and, as predicted, these mutants lead to the production of grossly altered proteins or possibly no protein at all.

Extensive genetic analyses of T4 bacteriophage leave no doubt that acridine-induced mutations arise from frameshifts which may occur in either direction depending upon whether they were mediated by addition or by deletion. Mutants derived from single-base additions are readily reversed by deletion though not by deleting the added base per se but by producing a second mutation at a point near the initially altered site. Hence, when a mutation involves a frameshift in which the reading of codons is shifted back by one base, a deletion mutation can correct it by shifting codon reading forward by one base. The codons lying between the two point mutations (addition and deletion) remain shifted, and hence they still code for the "wrong" amino acids whereas all the other codons, on either side of the double mutation, remain correct or are "restored" and again specify the "correct" amino acids.

Since the coding distance between addition and deletion sites are often very small, only a few amino acids are "wrongly" coded and the resultant protein is but slightly modified. When such gene products have some, but not all, the functional features of the wild-type proteins, the mutant is referred to as leaky.

Addition and deletion mutations are called sign mutations (90) and are denoted by either (+) or (−). In other words, (+) mutants are either all additions or all deletions; we do not know which, but they never represent mixtures of the two. The same is true of (–) mutants; they are either all additions or all deletions, but never both.

(+) Mutants are reverted by (–) mutations and vice versa, though both types, when acting alone, can cause the loss of the function of a gene. The second mutation which is used to revert the first is called a suppressor. Hence, (–) is a suppressor for (+) and (+) is a suppressor for (–). Since the codon is a triplet (three bases in a series), a triple mutant of the form of (+)(+)(+) or (–)(–)(–) should restore frame reading to normal and so yield a functional or

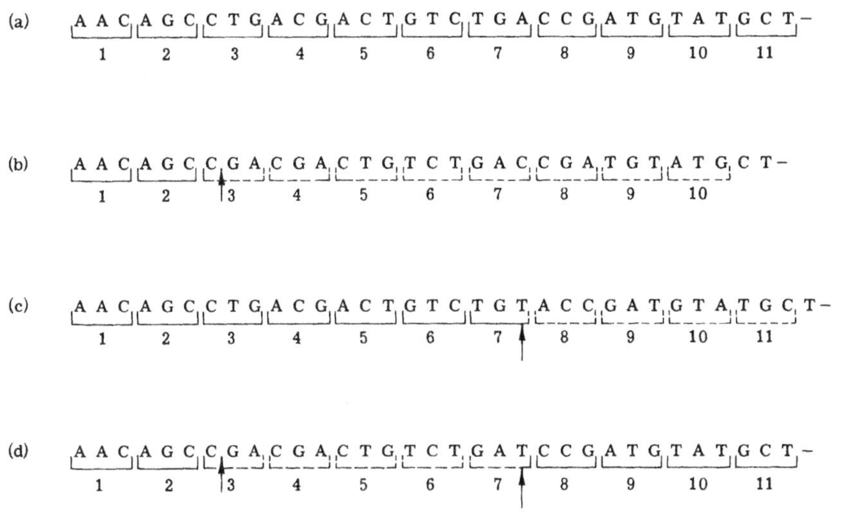

FIG. 3.12. Diagrammatic illustration of frameshift mutations. (a) Part of the nucleotide sequence of a wild-type gene. The arrows indicate the sites at which either deletion (b) or addition (c) has occurred; wild-type codons are indicated by solid bridges, shifted codons by dashed bridges; (d) is a suppressed mutant because the codons 3 through 7 are not essential.

semifunctional (leaky) gene. This has, in fact, been demonstrated with *rII* mutants of T4 bacteriophage (91).

One theory (92) holds that acridine dye-induced frameshifts arise from a recombinationlike event between paired, homologous DNA molecules, provided that one of them contains an intercalated acridine. One feature to its credit is that it explains how acridine mutants are able to form in the absence of DNA replication which formerly was thought to be obligatory. And, it is also consistent with the observation that 5-aminoacridine is mutagenically active in yeast during meiosis but not during mitosis (93). Hence, it appears that the actual mistake (i.e., addition or deletion of a base) occurs when a DNA duplex (in the case of bacteriophage) or a chromosome (in the case of yeast) undergoes recombination or crossing-over with its homologous counterpart which contains an intercalated acridine dye.

Provided mutagenic activity of acridine dyes is entirely dependent upon recombination (as in bacteriophage) or meiotic crossing-over (as in higher organisms) mutagenicity would not be observed in somatic cells (these do not undergo meiosis) but should be limited to germ cells capable of meiotic division.

Although experimental evidence is lacking, it seems probable that the phenanthridines, of which ethidium bromide is a member, would also produce frameshift mutants. They might, however, fail to do so if the distance to which they force adjacent base-pairs apart (as they become intercalated) is not exactly equivalent to that which would be occupied by another base pair.

B. Polynuclear Aromatic Hydrocarbons and Nitrogen-Containing Polycyclic Carcinogens

Although there is appreciable evidence to suggest that certain polycyclic hydrocarbons and nitrogen-containing polycyclic carcinogens form complexes and are intercalated by native DNA in much the same manner as acridines and phenanthridines, the biological significance of such findings remains unclear. Evidence for intercalation is derived primarily from two separate observations: (a) the ability of native DNA to solubilize aromatic hydrocarbons in aqueous media and (b) the increase in the viscosity of native DNA upon its interaction with certain polycyclic hydrocarbons or heterocyclic carcinogens.

1. *Physicochemical Studies*

It is known that purines, their nucleosides, and the nucleotides can solubilize benzo[*a*]pyrene and a number of other hydrocarbons in aqueous media (94, 95), presumably by forming planar purine-to-hydrocarbon complexes (96). The naturally occurring purines in DNA (guanine and adenine) appear less effective in this regard than methylated purines (such as caffeine and tetramethyluric acid) (97) though the base-pairs within DNA seem to exert a greater solubilizing effect than do monomeric purines (98). This is also true of a number of basic dyes which have a far greater affinity for DNA than monomeric nucleotides.

Apparently, electrostatic factors play an important, if not major, role in the interaction of heterocyclic carcinogens with DNA, even though hydrophobic forces (involved in intercalation) probably determine the precise structure of the complex. The fact that the binding of nitrogen-containing polycyclic carcinogens to DNA is pH dependent (99), requiring the presence of the cationic form of the heterocyclic compound, helps to support this contention as does the fact that polycyclic hydrocarbons which are neutral bind DNA to a lesser extent than nitrogen-containing polycyclics (100). For instance, Boyland and Green (101) showed that only 7 mmoles of benzo[*a*]pyrene are bound per mole of DNA nucleotide under the most favorable conditions while 100 or more mmoles of ethidium bromide are bound.

With both the heterocyclic carcinogens and the neutral polycyclic hydrocarbons, binding to native DNA depends upon salt concentration, or more particularly, upon the ionic strength of the medium. The lower the ionic strength, the greater is the number of moles which can be bound to DNA (101). Boyland and Green (101) found seven times as many moles of benzo[*a*]pyrene bound to DNA in water as compared to that found in aqueous 0.1 *M* NaCl solutions of DNA. The explanation apparently relates to the influence of ionic strength on the secondary structure of DNA. At very low ionic strength, a DNA duplex is more loosely coiled than in solutions of higher ionic strength. Intercalation of polycyclics is favored by the former structure for steric reasons.

T'so (102) and Boyland and Green (97) have interpreted these observations to imply that intercalation of polycyclics *in situ* is most likely to occur during either DNA replication or when RNA is being transcribed from a DNA template. It is at these times that DNA is most loosely coiled and hence most available for intercalation. This explanation, however, is not entirely satisfactory in light of recent studies which show that supercoiled DNA has a greater affinity for certain heterocyclics (at low concentrations) than does nonsupercoiled DNA (89).

The viscometric studies of Lerman (103) indicate that only certain polynuclear hydrocarbons (and derivatives) and nitrogen-containing polycyclic carcinogens become intercalated by DNA. Viscosity enhancement, characteristic of the complex formation with acridine dyes, has been observed with pyrene, 9-methylanthracene, dihydropyrene, dibenz[*a,j*]acridine, 3-aminopyrene, 7,9-dimethylbenz[*c*]acridine. No evidence of intercalation, however, was found for estradiol, naphthalene, or any other compound structurally lacking three fused aromatic rings (which is conceivably a structural requirement for intercalation).

2. *Biological Significance*

Aromatic, polycyclic hydrocarbons and the nitrogen-containing polycyclics mentioned above have been extensively studied for their carcinogenicity, though little is known of their mutagenic properties, particularly with regard to the kinds of genetic systems used to study frameshift mutations (see Section III, A). It has been suggested by some workers and virtually assumed by others that carcinogenicity is a manifestation of mutation (somatic mutation); however, one is cautioned by the fact that the correlations between the carcinogenicity and mutagenicity of many chemicals seems insufficient to prove this (104) (see Section V).

Furthermore, there does not seem to be a correlation between carcinogenic activity and the extent of binding of various polycyclic hydrocarbons to DNA *in vitro*. The noncarcinogenic pyrene is more readily solubilized by DNA than the potent carcinogen, 3-methylcholanthrene (105). In fact, it is questionable whether the physical interaction of polycyclic hydrocarbons with DNA observed *in vitro* bears any relationship to the firm binding found *in vivo* (106).

Brookes and Lawley (107) and others (106, 108) have provided evidence indicating that some kind of covalent attachment of hydrocarbons to DNA is effected *in vivo* and that the nature of the bound product is distinguishable from that of the hydrocarbon itself (109). It is significant, in this regard, that T'so and Lu observed a covalent binding of benzo[*a*]pyrene to DNA following irradiation of the physically bound complex at wavelengths absorbed exclusively by the hydrocarbon (110). On the other hand, the metabolic oxidation of hydrocarbons may yield epoxides (111) which might subsequently alkylate DNA *in situ* (at the N-7 position of guanine) and so explain the covalent nature of the complex.

Lerman (103), however, argues that the formation of a physical complex might necessarily precede the supposed covalent attachment of hydrocarbons to DNA *in vivo*. Should this be so [and it has been disputed (112)] physical binding and intercalation might be related to carcinogenic potency. This possibility arises because of the demonstration by Brookes and Lawley (113) that a positive correlation exists between the *in vivo* binding of hydrocarbons to DNA and their respective carcinogenic potency using Iball's index (114).

In an effort to reconcile the conflicting information, Lawley has offered an interesting theory (115). He suggests that the interaction of aromatic hydrocarbons or heterocyclic carcinogens, or their metabolites, whether covalently bound to DNA or not, could be envisaged to interfere with the processes concerned with repair of DNA. This could allow either a greater number of genetic lesions to go unrepaired and hence to be expressed as spontaneous mutations—or to result in defective recombination and so give rise to mutated genomes.

C. Antibiotics and Mycotoxins

1. *Actinomycin D*

Of the known antibiotics, actinomycin D (see Fig. 3.13) has been the most extensively studied chemical in its reactivity toward DNA. The interest in actinomycin, in this regard, is attributable to its ability to interefere with the function of DNA, particularly as it relates to the transcription of RNA from a DNA template (116). At low concentrations, the antibiotic selectively turns off RNA synthesis both *in vitro* and *in vivo* without inhibiting the synthesis of either DNA or protein (117). At higher concentrations (approximately 10- to 20-fold higher) synthesis of DNA may also be blocked (117).

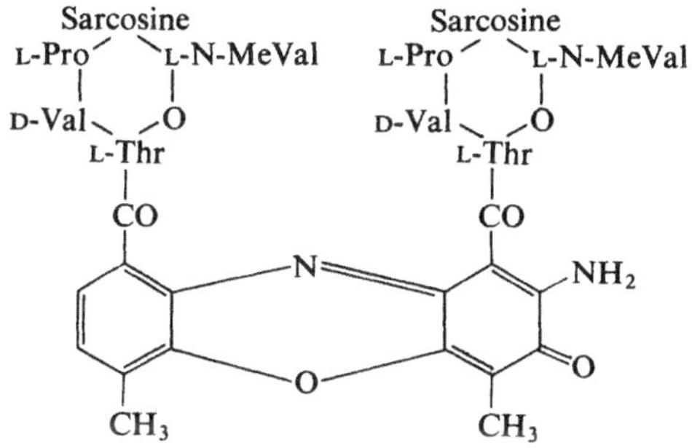

FIG. 3.13. Structure of actinomycin D. (Abbreviations: Pro = proline; Val = valine; Thr = threonine.)

The binding of actinomycin to DNA is known to be dependent on the proportion of guanine present and upon helical structure (118). It is, however, specific for DNA as evidenced by the inability of actinomycin to bind either to natural or synthetic double-helical polyribonucleotides, even to those containing high proportions of guanine (118). Recent studies by Reich (117), however, showed that guanine per se is not essential for actinomycin binding but rather an amino group with the proper steric relationship is required. Hence, a modified poly dAT double helix in which an amino group was substituted at the 2-position of adenine has been found to bind actinomycin very effectively.

Using various chemically modified analogs of actinomycin, other binding studies show that the amino group and quinoidal oxygen of the ring system of the antibiotic are essential for its reaction with DNA (and its biological activity) and that only slight modifications of the peptide side chains are possible without drastically altering both activity and DNA binding (117).

Originally, the binding of actinomycin was thought not to involve an intercalation by adjacent base-pairs of DNA but rather that hydrogen bonds were formed between the quinoidal oxygen and amino group of actinomycin and the 2-amino, N-3 and deoxyribose ring oxygen of deoxyguanosine (119). The peptide chains on actinomycin would then lie within the small groove of the helix where they would form hydrogen bonds with phosphate oxygens. Since these hydrogen bonds are on the strand opposite the guanine complex, the helix is effectively cross-linked which probably accounts for the resultant stabilization of the helix against strand separation (117). This could also explain why DNA synthesis is inhibited by high concentration of actinomycin.

The above views have recently been challenged, however, by Müller and Crothers (120) who believe, from very extensive and involved studies, that actinomycin combines with DNA by an intercalation in which one of the two base-pairs involved is G-C. It is not yet certain which of the above two models are correct, and we must await further developments in this field (121).

Information relative to the mutagenic activity of actinomycin is very sketchy; actinomycin may be slightly mutagenic. If so, it has not been established whether it induces additions, deletions, transitions, or transversions.

2. *Other Antibiotics*

Other antibiotics such as streptothricin and probably the entire class of aminoglycosidic antibiotics (for example, streptomycin, dihydrostreptomycin, neomycin, and kanamycin) combine with DNA, though apparently they are only loosely bound to the negatively charged phosphate groups of DNA by electrostatic interaction (122). It seems unlikely that binding of this type would necessarily lead to mutagenic effects. They will, however, cause phenotypic curing but this is known to function at other than the genetic level.

Antibiotics such as mitomycin C, porfiromycin, and streptonigrin have a more profound effect on DNA, at least on DNA *in situ*. Mitomycin C and porfiromycin are metabolically reduced in the cell to hydroquinone derivatives by a quinone reductase (diaphorase) and these compounds alkylate and extensively cross-link DNA (43, 123). There has been, however, one report suggesting that mitomycin C may have intercalating capabilities as well (124). In any event, their mutagenicity is expected to be low since they effectively block DNA replication and prevent the development of daughter cells. Hence, the ratio of mutagenic to lethal events should be extremely low in most test organisms. Nevertheless, it has been shown that many unusual chromatid exchanges are induced in human leukocyte chromosomes when such cells are treated with mitomycin C at the time of culturing (125) and, further, that streptonigrin induces a relatively high frequency of back-mutations (reverse mutations, see Chap. 5) in a fungal test system (126).

3. *Aflatoxin B_1*

Aflatoxins are metabolic products of the mold *Aspergillus flavus*. The most toxic member of this class (in terms of carcinogenic activity) is aflatoxin B_1 (Fig. 3.14) which combines with DNA and forms complexes, indicative of base-pair intercalation (127, 128). Sufficient information, however, is not yet available to establish this, though binding has been shown to be entirely reversible (127), indicating that covalent bonds are not involved.

Furthermore, aflatoxin, like actinomycin, is a powerful inhibitor of RNA synthesis. It also has the interesting property of drastically reducing the amount of nuclear RNA in liver tissue derived from rats treated with approximately 3 mg of aflatoxin (127).

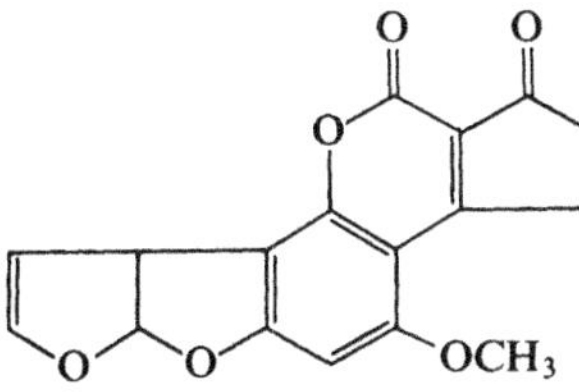

FIG. 3.14. Structure of aflatoxin B_1.

Evidence for the mutagenicity of aflatoxin has been obtained in plant cells of *Vicia faba* (129) and in human cells grown in tissue culture (129). In both cases, the evidence is based on abnormalities produced in mitotic chromosomes; both chromosomal fragments and bridges were observed. In the mouse, evidence for the mutagenicity of aflatoxin has been provided by the dominant lethal test (130) which may be related to the chromosome-breaking effect (see Chap. 5).

D. CAFFEINE (1,3,7-TRIMETHYLXANTHINE)

1. *The Controversy*

Of the compounds that have been investigated for mutagenic activity none have been so cloaked in controversy or afflicted with dispute as caffeine. The heat of the controversy is sustained by the importance of the subject: enormous quantities of caffeine are consumed in coffee, tea, cocoa, and soft drinks by a vast number of individuals of all age groups. Should caffeine prove to be mutagenic in a variety of biological test systems, particularly those which are taxonomically or physiologically related to man, the levels of intake considered safe will need to be reassessed.

Evidence for caffeine's disputed mutagenicity has been reported from studies involving bacteria (131–133), *Drosophila melanogaster* (134, 135), fungi (136, 137), plants (138, 139), mice (140), and human tissue culture cells (140, 141). In bacteria (142), mutagenicity is claimed to depend upon DNA synthesis or cell division; in nondividing bacteria caffeine is reported to be antimutagenic (143). Induction of recessive lethal mutations in *Drosophila* has been reported by some investigators (134, 144) and refuted by others (145–147). There is also disagreement whether caffeine induces dominant lethal mutations in mice, though in this case the positive demonstration was based on chronic feeding (140) while the negative result was not (130).

2. *Genetic Observations*

It seems well established that caffeine is at least weakly mutagenic in the bacterium *E. coli* since it induces back mutations to phage resistance and streptomycin nondependence (131, 133). However, there is no agreement as to how the effect is mediated. Shimada and Takagi (148) claim that caffeine inhibits the initial phases of the process responsible for the repair of UV-irradiated DNA in *E. coli* and present evidence indicating that strand excision (of the DNA) is extensively inhibited by caffeine. Grigg (149), on the other hand, has reported that it is repair replication, not strand excision, which is interfered with by caffeine.

The problem appears even more conflicting when the reported findings in *Drosophila* are examined. According to Andrew (134) and, more recently, others (135, 144), caffeine is weakly mutagenic to *Drosophila*, inducing sex-linked recessive lethals and certain chromosome aberrations. Other investigators find no induction of sex-linked recessive lethals (145–147) and a very low incidence of chromosome loss (147). In fact, Clark and Clark (147) were unable to repeat the experiments of Andrew (134), despite the fact they had been carried out some years before in Clark's laboratory and utilized the same strain of flies (Canton-S).

Even those investigators who have reported positive results have commented on the lack of reproducibility between replicate experiments. This has also been noted in the experiments dealing with chromosome loss and nondisjunction in *Drosophila* (147). Clark and Clark summarized the problem best when they wrote in 1967, "If it were not for the significance of caffeine as a potential mutagen in man, further efforts to study its genetic effects in higher organisms would scarcely be justified" (147).

Lyon *et al.* (150) undertook a rather extensive analysis of the genetic effects of caffeine in mice. They investigated the frequency of dominant lethals in the progeny of caffeine-fed males and the frequency of dominant visible mutations and specific locus recessive mutations (Russell test) (151). Their conclusions were that there was no indication whatever of any induction of specific-locus mutations by caffeine treatment; neither was there evidence for the induction of dominant visibles or lethals. Epstein (130) also failed to observe any significant increase in dominant lethals among the progeny of caffeine-fed male mice.

Nevertheless, Kuhlman *et al.* (140) have reported an induction of dominant lethals in mice after feeding caffeine to male animals for 150 days. When both types of dominant lethals were scored (preimplantation losses and resorptions) a nearly linear relationship was found between the frequency of dominant lethal mutations and the dose of caffeine administered. At the highest dose, the frequency of dominant lethals was approximately three times that found for control animals.

Perhaps the above effect is related to the demonstrated ability of caffeine to break plant chromosomes (138, 139) and mammalian chromosomes (140, 141), particularly since the number of chromosomal breaks in HeLa cells is also dose-dependent and nearly linearly related to the dose.

3. *Interference with Repair of Genetic Damage*

One interesting aspect which sheds light on the mode of action is that caffeine-induced chromosome breaks do not rejoin (140). That is, they form few translocations, rings, or other rearrangements which require fusion of broken ends. These observations together with the reports from bacterial studies (143, 149) and those with mouse L cells (152) led Ostertag (140) to suggest that caffeine owes its genetic effect to its interference with one of the steps (specifically, blocking the step within the repair process that deals with strand rejoining) concerned with the repair of genetic damage (see Chap. 4). Since strand rejoining is also necessary for normal DNA replication, Ostertag further suggests that both aspects of caffeine's action (inhibition of repair and induction of mutants during DNA synthesis) can be accounted for by the same basic mechanism.

Were this so, then the aspect of Ostertag's theory which depends upon the extrapolation of genetic effects (based on chromosome breaks in tissue culture induced with extremely high concentrations of caffeine) to what normal coffee drinking would produce in man seems unjustified.

Caffeine is distributed relatively evenly throughout the body, including the testes in man and the germ cells of an embryo, but if caffeine's mutagenicity is exerted via a mechanism involving enzyme inhibition, there ought to be a threshold below which caffeine would be entirely nonmutagenic. This can be explained simply in terms of basic enzyme kinetics. Assuming the rate at which the enzyme or enzymes function to reseal breaks within DNA is normally dependent upon the number of breaks rather than enzyme levels, then caffeine would not begin to exert its effect upon the genome until a sufficient level of enzyme activity has been inhibited.

Hopefully, something will be gleaned from this depressingly uncertain story. If caffeine is exerting a mutagenic effect by inhibiting repair enzymes, it should be possible to investigate this question as easily in mammalian systems as in bacterial ones. In fact, one such study has recently been reported in which human tissue culture cells (153) were used and their ability to repair the genetic damage produced by UV irradiation was monitored. A quantity of caffeine (sufficient to reduce their survival to UV irradiation) was added to the culture medium immediately following the exposure of such cells to UV light; however, no change in repair or the rate of repair was noted. It should be mentioned, however, that only one of the several steps involved in the process (that concerned with nonsemiconservative replication) was investigated. The possibility remains that some later step in the overall process of repair is blocked by caffeine, i.e., the strand-rejoining step.

IV. Depolymerization of DNA

Many chemical and physical agents will induce single- or double-stranded breaks in DNA. Some of these substances have already been referred to in this context, such as the alkylating agents, low pH hydrazine, and hydroxylamine. As mentioned earlier, these agents break DNA strands by an indirect mechanism, i.e., by facilitating depurination or depyrimidation which has the effect of labilizing the sugar-phosphate backbone.

A. Enzymic Mechanisms

There are also other ways by which strand breakage can be effected. Certain base analogs inhibit DNA replication by blocking the synthesis of one of the obligatory mononucleotide precursors to DNA (for instance, FUdR inhibits the formation of thymidylic acid). The mechanism by which these substances induce breaks is probably related to the effect of pyrophosphate which is known to reverse the action of DNA polymerase (154).

Ordinarily, the DNA polymerase, in the presence of a DNA template, catalyzes the incorporation of nucleoside triphosphates into DNA with the concomitant release of pyrophosphate. Since this is an equilibrium reaction, a high concentration of pyrophosphate will drive the polymerase reaction in reverse causing depolymerization and release of nucleoside triphosphates. This has been demonstrated *in vitro* with a bacterial polymerase (154) and evidence for its occurrence *in situ* has been provided by the observation that pyrophosphate causes fragmentation of chromosomes in the plant *Vicia faba* (155). Since the type of chromosome damage produced by inhibitors of nucleoside triphosphate synthesis simulates that induced by pyrophosphate, it has been suggested that these inhibitors may also reverse the direction of the polymerase, presumably by decreasing or eliminating the presence of one of the nucleoside triphosphates.

An interference with repair, particularly that step concerned with the resealing of single-stranded DNA breaks, could lead to chromosome breaks and partial depolymerization of DNA. As mentioned in Section III, D, such a mechanism has been proposed as the basis of caffeine's mutagenic action in mammalian cells.

It is even conceivable that normal noninhibited or unaltered repair could lead to chromosome breaks. Should the resealing of single-stranded DNA breaks be the rate-determining step in the process of repair, it is conceivable that in response to extensive alkylation more excisions are produced (in the process of dealkylation) than can be dealt with by resealing enzymes (ligases). In consequence to DNA replication, such single-stranded breaks might manifest broken chromosomes.

One study (57) has shown that the induction of chromosome breaks by alkylating agents cannot be explained simply on the basis of chemical, nonenzymic depurination because the breaks are effected within 24 hours and undergo virtually no increase following that time. Were chemical, nonenzymic depurination responsible, the breaks would be expected to develop slowly over a period of several days (156). On the other hand, if such chromosome breaks are caused by a repair process in which strands are broken faster than they can be reesterified then the development of such breaks within 24 hours is not surprising. Present evidence suggests that enzymic strand-breakage, in response to alkylation, is a rapid process—over half of all alkyl groups are removed from the HeLa cell genome within a few hours despite the fact that the amount of alkylation, in this case, exceeded the mean lethal number by a factor of 10 (157).

There is, of course, the other possibility that certain chemical or physical agents may activate "latent nucleases" which function to depolymerize DNA.

B. Peroxides and Free Radicals

1. *Hydrogen Peroxide* (H_2O_2)

There is general agreement that the effects of hydrogen peroxide (and hydrogen peroxide-producing agents) on DNA are caused by the free radicals they generate. This seems to be the case, since chelating agents and radical scavengers protect DNA from the damage and inactivating alterations (within transforming DNA) ordinarily produced by peroxide treatment (158).

Hydrogen peroxide (H_2O_2) decomposes into two ·OH radicals in response to UV irradiation or spontaneously at elevated temperatures. It also gives rise to HOO· radicals in the presence of reduced transition metals (e.g., Fe^{++}, Cu^{+}). These radicals can then react with organic molecules to produce relatively more stable organic peroxy radicals and organic peroxides which may later decompose again into free radicals. This process of "radical-exchange" sustains the effectiveness of short-lived radicals such as ·OH, HOO·, and H· and gives them an opportunity to reach the genome where they can exert their effect. This is a particularly important consideration for *in vivo* studies since the chemical effectiveness of highly reactive radicals is preserved thus providing sufficient time for diffusion to carry the radical or potential radical to its site of action.

2. *Organic Free Radicals and Hydrogen Peroxide-Generating Compounds*

Free radicals are produced in many chemical and metabolic reactions, particularly those involving oxidation. Radicals, therefore, are quite ubiquitous within living matter and may contribute significantly to the level or frequency of spontaneous mutation. The presence of radical scavengers (mild reducing agents, e.g., cysteine, Fe^{++}, hydroquinone) and inhibitors of radical production (chelating agents, e.g., citric acid) will tend to moderate this effect.

Hydroxylamine and its aliphatic (*N*-methylhydroxylamine, hydroxyurea, and hydroxyurethan) and aromatic derivatives produce hydrogen peroxide and nitroso compounds in the presence of oxygen via complex radical chain reactions. Hydrazine reacts similarly (though nitroso intermediates are not produced) and, as was mentioned in Section II, C, both types of compounds generate H_2O_2 more rapidly at low than at high concentration. Presumably this is because the rate of H_2O_2 production at high concentrations of hydroxylamine or hydrazine is limited by the supply of oxygen, whereas its rate of elimination is enhanced by high concentrations of hydroxylamine or hydrazine.

Several other compounds appear to involve free-radical reactions, e.g., 8-ethoxycaffeine (159), maleic hydrazide (160), *N*-methylphenylnitrosamine (161), and formaldehyde (162). The evidence is based on the observation that the presence of oxygen is required for these compounds to exhibit their genetic effects (chromosome breaks and recessive lethal mutations).

3. *Effect of Free Radicals on DNA*

Free radicals produced by hydrogen peroxide reduce the UV absorption of DNA and its constituent mononucleotides. Deoxythymidylic and deoxycytidylic acid are the most rapidly destroyed, though deoxyadenylic and deoxyguaninylic acids are also attacked (163). Very few of the degradation products have been characterized in detail but urea derivatives have been isolated from the reaction products of pyrimidine nucleotides, indicating that the ring system has been opened (163). This will, of course, labilize the sugar-phosphate backbone giving rise to strand breakage.

Apparently, the β-glycosidic bond is susceptible to attack also, since unaltered free bases have been detected (164) among the reaction products along with an altered adenine (17). Again, the consequence is one of labilizing the sugar-phosphate backbone though conceivably these radicals can break phosphodiester bonds directly to cause depolymerization and manifest chromosome or chromatid breaks.

From studies with transforming DNA and viruses it would appear that all radical-producing agents are strongly inactivating (interfere with DNA synthesis) and only very weakly mutagenic. In fact, with H_2O_2, less than one mutation for every ten million lethal events was detected (17). This has been interpreted by Freese (17) as indicating that the major effect of radicals is to block DNA replication, though it is not known whether the block is due to an alteration of the bases or to breaks in DNA or whether it is a consequence of base-removal.

V. Somatic Mutations and Carcinogenesis

Though it was not always explicitly stated, most of our discussion of mutagenesis has focused on germinal cells since it is their genetic material that must be affected for the progeny to be altered (165). In the case of unicellular organisms, however, germ cells and somatic cells can be said either to be one and the same or that the distinction does not apply. In either case, there is no doubt that somatic mutations can and do arise in higher organisms as well as in man. Evidence for this is cited later, for instance, in the discussion of those mutagenic test systems which employ somatic cells of the hamster (Chap. 5, Section VII, A) or in the account given of chromosomal translocations and aberrations which are often observed as defects within somatic cells (Chap. 5, Section VI).

Mutations occurring within somatic cells are not expected to be detrimental to an adult organism. The reason for this is very obvious: most organs and specialized tissue are composed of millions of cells. For one or even a large number of these to be altered and rendered nonfunctional is of no consequence since the many other cells are capable of carrying on the organ's function. In general, it is only when the mutant cell has acquired the ability to outgrow the other cells of that organ that a problem arises.

This outgrowth might be accomplished in a variety of ways. For instance, in tissue which is essentially nondividing, the mutant might simply undergo division after division, and soon, even when dividing at a slow pace, become a major contributor to the mass of that tissue. On the other hand, this overgrowth could be accomplished in rapidly dividing tissue (where new cells are constantly being regenerated and old cells are dying) should the mutant cell resist death and continue to proliferate as rapidly as do the others of this particular organ. A third mechanism would be provided if the mutant cell secreted a substance which would be inhibiting the growth of surrounding cells but not of itself, as occurs in fact in lower organisms. No doubt, other mechanisms can be envisaged to account for how a mutant somatic cell might outgrow its normal counterparts and establish itself at their expense but there is little evidence to support any of these ideas. Could these mechanisms, for instance, account for the development of malignant tumors? Is there, in fact, any relationship between chemical mutagenesis and chemical carcinogenesis? (107, 166, 167).

These are difficult questions for which there are as yet no satisfactory answers. Conceptually, the idea that cancer and other tumors might be mediated, in some instances at least, by mutational events is a compelling one since cancer cells do breed true. That is, the daughters or progeny of a cancer cell are also cancerous and so the effect is propagated from one generation of cells to another. Though this is an indication that somatic-cell inheritance is involved, it by no means is certain that the chromosomes and the DNA of which they are composed have been either directly or permanently affected as must be true of a mutation.

Alternatives to this hypothesis are replete in the literature though the only one which concerns us here relates to viral carcinogenesis [or, as in the more general sense, viral oncogenesis (168–171)]. However, this process also leads to the development of cells with an altered genome (168) and hence even these cells can be regarded as being mutants.

If the underlying mechanism by which chemicals cause cancer consisted of special kinds of mutations within susceptible organisms, all proven carcinogens should be mutagens. On the other hand, the hypothesis does not require that all chemical mutagens be carcinogenic since it can be argued that only specific types of mutations may lead to oncogenic processes. However, provided the molecular mechanisms proposed for chemical mutagenesis are correct, predictions of carcinogenicity should be possible within certain selective classes of compounds. For instance, if a particular type of alkylating agent, e.g., a nitrosamine, is proved to be carcinogenic, other closely related nitrosamines should also be effective. In fact, since many alkylating agents appear to be similar in terms of their chemical reactivity toward DNA and ability to induce mutations of the transition type, there is no immediately obvious reason why, if one is carcinogenic, the entire group should not be so. This is, however, not the case, but an argument can be made for different alkylating agents being metabolized and transported differently and being thus, in reality, not comparable (166).

Another, and perhaps, more decisive argument concerns those observations which show that certain closely related chemical mutagens (believed to form identical products with DNA) will consistently give rise to different types of mutations or will be surprisingly different in their mutagenic potency. Hence, at our present level of knowledge, the lack of predictability associated with chemical carcinogenesis cannot be taken as evidence for or against the mutation hypothesis of chemically induced cancer.

Perhaps the reasons for the different biological activities observed among individual but closely related compounds are based on differences in their metabolism as well as in their transport. Certainly, the importance of metabolism and of understanding the nature of the metabolites in relation to carcinogenesis is well established. For instance, it has been shown that many compounds are not carcinogenic per se but must be metabolized to an active form (50). Were the mutagenic test organism used in the assay unable to produce the metabolite, mutagenicity would not be expected. In fact, this points up one of the fundamental difficulties inherent in any attempt to correlate mutagenicity with carcinogenicity. Carcinogen testing requires higher organisms, usually mammals, while the testing of potential mutagens is usually confined to microbial systems. In an attempt to overcome some of these difficulties, "active" metabolites have been isolated and tested in simple genetic systems (e.g., transforming DNA) where the problems associated with permeability barriers and the metabolism of the chemical are avoided.

One such study has involved testing the mutagenicity of esterified *N*-hydroxy metabolites of 2-acetylaminofluorene and *N*-methyl-4-aminoazobenzene (172). Neither of the parent compounds, as such, appear to be mutagenic or carcinogenic unless converted via *N*-hydroxylation in the liver to esters which attack guanine residues in DNA (173, 174). Not surprisingly, in addition to their carcinogenicity in rats, these derivatives have also proved to be highly mutagenic in test systems which utilize transforming DNA (172). The mutations induced were spontaneously revertible and so appeared to be caused by a single base-pair change.

The nitrosamines constitute another class of carcinogens which are not mutagenic per se but are metabolically reduced to diazomethane which decomposes to yield mutagenic carbonium ions (175). Aromatic polycyclic hydrocarbons are still another class of highly active carcinogens, but again, metabolic conversion may be necessary for them to show any mutagenic activity—for instance, epoxide formation (111). Perhaps, though evidence is lacking, all carcinogenic substances would be mutagenic were the active metabolite used and tested in the appropriate system. Where chemical intermediates and metabolic products can be disregarded, the results are in favor of the mutation theory of carcinogenesis, namely that the types of radiation which induce cancer are also very effective mutagens (176–178).

In further support of the idea that direct genomal effects are involved and probably instrumental in the induction or initiation of carcinogenesis are the findings of Cleaver (179) who demonstrated that patients suffering from *Xeroderma pigmentosum* lack the ability to repair genetic damage produced by exposure to ultraviolet light. Since such individuals develop skin cancers following exposures to sunlight, the very strong possibility is raised that unrepaired, damaged DNA is somehow directly related to the initiation of carcinogenesis.

Though there are good biological reasons and a good deal of circumstantial evidence to support the mutation hypothesis of chemical carcinogenesis, a strong antithesis exists as well. That is to say, there are many puzzling phenomena which cause one to doubt the theory. For instance, why should it take so long after a person receives a dose of radiation or is exposed to a carcinogenic agent before a tumor appears? Why should aging play such a dominant role in carcinogenesis?

All of these and other questions lead one to suspect that there is more to carcinogenesis than a single event such as a mutation. In fact, Berenblum and Trainin (180) have provided much evidence to support a two-factor theory of carcinogenesis in which it would appear necessary to have initiation followed by promotion in order to achieve cancer induction. The initiator may also act as a promoter but usually these studies have involved promoters with little or no initiating capabilities. Conceivably, the initiator induces the specific mutation required while the promoter creates the more general conditions needed for the appearance of cancer. It can be argued that aging itself is a kind of promotion. The nature of these general conditions probably depends, for their understanding, upon learning how cells within specialized tissues function and interact together and how the mitotic activity, rate of proliferation, and cell turnover is controlled in the normal cell. In other words, the problem of carcinogenicity includes, but transcends, genetic considerations and very probably involves the basic mechanisms by which many of the functions and activities of the cell are controlled.

The observation that certain oncogenic viruses can, when activated, insert their DNA into a mammalian chromosome so that the DNA not only replicates along with the host's genetic material but also codes for certain virus-specific proteins has suggested the possibility that chemical carcinogens may function indirectly by activating these viruses. As mentioned previously, the very fact that the virus incorporates its DNA into that of its host gives rise to what can be defined as a mutation. Whether, and by what mechanism, chemicals activate potentially oncogenic viruses and thereby encourage them to incorporate their DNA into the host's genome is unknown. It seems, however, unlikely that all carcinogens function via this route in all demonstrated cases of cancer induction. The possibility, however, has not been eliminated.

In conclusion, it must be admitted that the precise relationship between mutagenesis and carcinogenesis is not clear though it can hardly be doubted that it exists. Clearly too, other factors figure prominently in the process but much less is known about these than about the basic mechanics of mutagenesis.

REFERENCES

1. E. Harbers, N. K. Chandhuri, and C. Heidelberger, *J. Biol. Chem.* **234**, 1255 (1959).
2. H. D. Barner and S. S. Cohen, *J. Bacteriol.* **68**, 80 (1954).
3. E. Freese, *in* "Molecular Genetics" (J. H. Taylor, ed.), Part 1, p. 207. Academic Press, New York, 1963.
4. E. Freese, *Brookhaven Symp. Biol.* **12**, 63 (1959).
5. P. D. Lawley and P. Brookes, *J. Mol. Biol.* **4**, 216 (1962).
6. E. Freese, *Proc. Natl. Acad. Sci. U.S.* **45**, 622 (1959).
7. F. Stahl, J. Craseman, L. Okum, E. Fox, and C. Laird, *Virology* **13**, 98 (1961).
8. B. Djordjevic and W. Szybalski, *J. Exptl. Med.* **112**, 509 (1960).
9. E. Freese, E. B. Freese, and E. Bautz, *Proc. Natl. Acad. Sci. U.S.* **47**, 845 (1961).
10. F. Weygand, A. Wacker, and H. Dellweg, *Z. Naturforsch.* **7b**, 19 (1952).
11. D. B. Dunn and J. D. Smith, *Nature* **174**, 304 (1954).
12. E. Freese, *J. Mol. Biol.* **1**, 87 (1959).
13. R. Rudner, *Z. Vererbungslehre* **92**, 336 (1961).
14. A. Wacker, S. Kirschfeld, and S. Trager, *J. Mol. Biol.* **2**, 241 (1960).
15. R. M. Herriot, *Cancer Res.* **26**, 1971 (1966).
16. E. C. Pauling, Doctoral Dissertation, University of Washington (1964).
17. E. Freese and E. B. Freese, *Radiation Res. Suppl.* **6**, 97 (1966).
18. R. Rudner, *Biochem. Biophys. Res. Commun.* **3**, 275 (1960).
19. S. Benzer, *Proc. Natl. Acad. Sci. U.S.* **47**, 302 (1961).
20. C. Auerbach and J. M. Robson, *Nature* **157**, 302 (1946).
21. P. Brookes and P. D. Lawley, *Biochem. J.* **80**, 496 (1961).
22. E. P. Geiduschek, *Proc. Natl. Acad. Sci. U.S.* **47**, 950 (1961).
23. R. J. Goldacre, A. Loveless, and W. C. J. Ross, *Nature* **168**, 667 (1949).
24. E. Bautz and E. Freese, *Proc. Natl. Acad. Sci. U.S.* **46**, 1585 (1960).
25. P. Brookes and P. D. Lawley, *Biochem. J.* **77**, 478 (1960).
26. J. T. Lett, G. M. Parkins, and P. Alexander, *Arch. Biochem. Biophys.* **97**, 80 (1962).
27. T. Bardos, N. Datta-Gupta, P. Hebborn, and D. Triggle, *J. Med. Chem.* **8**, 167 (1965).
28. B. Pullman and A. Pullman, *Bull. Soc. Chim. France* p. 1502 (1958).
29. B. Reiner and S. Zamenhof, *J. Biol. Chem.* **228**, 475 (1957).

30. E. Kriek and P. Emmelot, *Biochemistry* **2**, 733 (1963).
31. P. D. Lawley and P. Brookes, *Biochem. J.* **89**, 127 (1963).
32. E. Kriek and P. Emmelot, *Biochim. Biophys. Acta* **91**, 59 (1964).
33. P. D. Lawley and P. Brookes, *Biochem. J.* **92**, 19C (1964).
34. P. Brookes and P. D. Lawley, *J. Chem. Soc.* p. 1348 (1962).
35. W. C. J. Ross, "Biological Alkylating Agents." Butterworth, London and Washington, D.C., 1962.
36. L. E. Orgel, *Advan. Enzymol.* **27**, 289 (1965).
37. C. E. Williamson and B. Witten, *Cancer Res.* **27**, 33 (1967).
38. P. Brookes and P. D. Lawley, *J. Chem. Soc.* p. 3923 (1961).
39. P. D. Lawley and P. Brookes, *Ann. Rept. Brit. Empire Cancer Campaign* **41** (1963).
40. K. Kohn, N. Steigfigel, and C. Spears, *Proc. Natl. Acad. Sci. U.S.* **53**, 1154 (1965).
41. J. Jones, R. H. Golder, and R. J. Rutman, *Federation Proc.* **22**, 582 (1963).
42. K. Kohn, D. Green, and P. Doty, *Federation Proc.* **22**, 582 (1963).
43. V. N. Iyer and W. Szybalski, *Proc. Natl. Acad. Sci. U.S.* **50**, 355 (1963).
44. E. F. Becker, Jr., B. K. Zimmerman, and E. P. Geiduschek, *J. Mol. Biol.* **8**, 337 (1964).
45. P. D. Lawley and P. Brookes, *Nature* **192**, 1081 (1961).
46. C. Nagata, A. Imamura, H. Saito, and K. Fukui, *Gann* **54**, 109 (1963).
47. E. B. Freese, *Proc. Natl. Acad. Sci. U.S.* **47**, 540 (1961).
48. D. G. Fahmy and M. J. Fahmy, *Ann. Rept. Brit. Empire Cancer Campaign* **37**, 116 (1959).
49. D. R. Krieg, *Genetics* **48**, 561 (1963).
50. P. D. Lawley, *Progr. Nucleic Acid Res. Mol. Biol.* **6**, 89 (1966).
51. D. B. Brown and A. R. Todd, *in* "The Nucleic Acids" (E. Chargaff and J. N. Davidson, eds.), Vol. 1, p. 444. Academic Press, New York, 1955.
52. P. Alexander, J. T. Lett, and G. Parkins, *Biochim. Biophys. Acta* **48**, 523 (1961).
53. A. Loveless and S. Howarth, *Nature* **184**, 1780 (1959).
54. M. J. Bird and O. G. Fahmy, *Proc. Roy. Soc.* **B140**, 556 (1953).
55. O. G. Fahmy and M. J. Bird, *Heredity* **6**, Suppl. 149 (1953).
56. O. G. Fahmy and M. J. Fahmy, *J. Genet.* **54**, 146 (1956).
57. J. K. Lim and L. A. Snyder, *Mutation Res.* **6**, 129 (1968).
58. W. A. F. Watson, *Z. Vererbungslehre* **95**, 374 (1964).
59. A. R. Crathron and J. J. Roberts, *Progr. Biochem. Pharmacol.* **1**, 320 (1964).
60. A. R. Crathron and J. J. Roberts, *Nature* **211**, 150 (1965).
61. H. G. Callan, *J. Cell Sci.* **2**, 1 (1967).
62. J. E. Edström, *Nature* **220**, 1196 (1968).
63. H. Schuster, *Biochem. Biophys. Res. Commun.* **2**, 320 (1960).
64. H. Schuster and R. C. Wilhelm, *Biochim. Biophys. Acta* **68**, 554 (1963).
65. W. Vielmetter and H. Schuster, *Biochem. Biophys. Res. Commun.* **2**, 324 (1960).
66. A. Peacocke and R. B. Drysdale, "The Molecular Basis of Heredity." Butterworth, London and Washington, D.C., 1967.
67. I. Tessman, R. K. Poddar, and S. Kumar, *J. Mol. Biol.* **9**, 352 (1964).
68. P. Brookes and P. D. Lawley, *Brit. Med. Bull.* **20**, 91 (1964).
69. A. Tsugita, *J. Mol. Biol.* **5**, 284 (1962).
70. D. Luzzati, *Biochem. Biophys. Res. Commun.* **9**, 508 (1962).
71. R. Litman and H. Ephrussi-Taylor, *Compt. Rend.* **249**, 838 (1959).
72. C. Anagnostopolous and I. P. Crawford, *Proc. Natl. Acad. Sci. U.S.* **47**, 378 (1961).
73. H. Strack, E. B. Freese, and E. Freese, *Mutation Res.* **1**, 10 (1964).
74. E. E. Horn and R. M. Herriot, *Proc. Natl. Acad. Sci. U.S.* **48**, 1409 (1962).
75. J. Stuy, *Biochem. Biophys. Res. Commun.* **6**, 328 (1961).
76. H. Schuster, *J. Mol. Biol.* **3**, 447 (1961).

77. E. Freese and E. B. Freese, *Biochemistry* **4**, 2419 (1965).
78. E. Freese, E. Bautz, and E. B. Freese, *Proc. Natl. Acad. Sci. U.S.* **47**, 844 (1961).
79. I. Tessman, H. Ishiwa, and S. Kumar, *Science* **148**, 507 (1965).
80. J. D. A. Gray and R. A. Lambert, *Nature* **162**, 733 (1948).
81. A. Bendich, E. Borenfreund, G. Korgold, M. Kirm, and M. Bolis, *in* "Acidi Nucleic e Loro Fungione Biologica," p. 214. Fondazione Baselli, Istituto Lombardo, Pavia, Italy, 1964.
82. M. Waring, *J. Mol. Biol.* **13**, 269 (1965).
83. L Crawford and M Waring, *J. Mol. Biol.* **25**, 23 (1967).
84. L. Lerman, *J. Mol. Biol.* **3**, 18 (1961).
85. L. Lerman, *J. Cellular Comp. Physiol.* **64**, Suppl. 1, 1 (1964).
86. J. Cairns, *Cold Spring Harbor Symp. Quant. Biol.* **27**, 311 (1962).
87. M. Waring, *Nature* **219**, 1320 (1968).
88. B. Hudson and J. Vinograd, *Nature* **216**, 647 (1967).
89. W. Bauer and J. Vinograd, *J. Mol. Biol.* **33**, 141 (1968).
90. S. Brenner, L. Barnett, F. H. C. Crick, and A. Orgel, *J. Mol. Biol.* **3**, 121 (1961).
91. L. Barnett, S. Brenner, F. Crick, R. Shulman, and R. Watts-Tobin, *Phil. Trans. Roy. Soc. London* **B252**, 487 (1967).
92. L. S. Lerman, *Proc. Natl. Acad. Sci. U.S.* **49**, 97 (1963).
93. G. E. Magni, R. C. von Borstel, and S. Sora, *Mutation Res.* **1**, 227 (1964).
94. H. Weil-Malherbe, *Biochem. J.* **40**, 351 (1946).
95. E. Boyland and B. Green, *Brit. J. Cancer* **16**, 347 (1962).
96. F. de Santis, E. Giglio, A. M. Liquori, and A. Ripamonti, *Nature* **191**, 900 (1961).
97. E. Boyland and B. Green, *Brit. J. Cancer* **16**, 507 (1962).
98. J. K. Ball, J. A. McCarter, and M. F. Smith, *Biochim. Biophys. Acta* **103**, 275 (1965).
99. J. Booth and E. Boyland, *Biochim. Biophys. Acta* **12**, 75 (1953).
100. E. Boyland and B. Green, *J. Mol. Biol.* **9**, 589 (1964).
101. E. Boyland and B. Green, *Biochem. J.* **92**, 4C (1964).
102. P. O. P. T'so, *in* "The Nucleohistones" (J. Bonner and P. O. P. T'so, eds.), p. 149. Holden-Day, San Francisco, California, 1963.
103. L. S. Lerman, *5th Natl. Cancer Conf., Proc., Philadelphia, 1964* p. 39. Lippincott, Philadelphia, Pennsylvania, 1965.
104. P. Kotin and H. L. Falk, *Radiation Res.* Suppl. 3, 193 (1963).
105. E. Boyland and B. Green, *Biochem. J.* **96**, 15P (1965).
106. C. Heidelberger and G. K. Davenport, *Acta, Unio Intern. Contra Cancrum* **17**, 55 (1961).
107. P. Brookes and P. D. Lawley, *J. Cellular Comp. Physiol.* **64**, Suppl. 1, 111 (1964).
108. J. J. Roberts and G. P. Warwick, *Intern. J. Cancer* **1**, 179 (1966).
109. P. Brookes and P. D. Lawley, *Ann. Rept. Brit. Empire Cancer Campaign* **42**, 35 (1964).
110. P. O. P. T'so and P. Lu, *Proc. Natl. Acad. Sci. U.S.* **51**, 272 (1964).
111. E. Boyland, *Biochem. Soc. Symp. (Cambridge, Engl.)* **5**, 40 (1950).
112. G. Reske and J. Stauff, *Z. Naturforsch.* **20b**, 15 (1965).
113. P. Brookes and P. D. Lawley, *Nature* **202**, 781 (1964).
114. J. Iball, *Am. J. Cancer* **35**, 188 (1939).
115. P. D. Lawley, *17th Mosbacher Symp., Dental Ges. Physiol. Chem.* (1966).
116. J. M. Kirk, *Biochim. Biophys. Acta* **42**, 167 (1960).
117. E. Reich, *Symp. Soc. Gen. Microbiol.* **16**, 266 (1966).
118. E. Reich, A. Cerami, and D. C. Ward, *in* "Antibiotic" (D. Gottlieb and P. D. Shaw, eds.), Vol. 1, p. 714. Springer, Berlin, 1967.
119. L. Hamilton, W. Fuller, and E. Reich, *Nature* **198**, 538 (1963).
120. W. Müller and D. M. Crothers, *J. Mol. Biol.* **35**, 251 (1968).

121. News and Views, *Nature* **219**, 679 (1968).
122. Ch. Zimmer, H. Triebel, and H. Thrum, *Biochim. Biophys. Acta* **145**, 742 (1967).
123. E. Reich, A. J. Shatkin, and E. L. Tatum, *Biochim. Biophys. Acta* **53**, 132 (1961).
124. M. Kodama, *J. Biochem.* (*Tokyo*) **61**, 162 (1967).
125. P. C. Nowell, *Exptl. Cell Res.* **33**, 445 (1964).
126. G. Zetterberg and B. A. Kihlman, *Mutation Res.* **2**, 471 (1965).
127. M. B. Sporn, C. W. Dingman, H. L. Phelps, and G. N. Wogan, *Science* **151**, 1539 (1966).
128. J. I. Clifford and K. R. Ress, *Nature* **209**, 312 (1966).
129. L. J. Lilly, *Nature* **207**, 433 (1965).
130. S. S. Epstein and H. Shafner, *Nature* **219**, 385 (1967).
131. M. Demerec, G. Bertani, and J. Flint, *Am. Naturalist* **85**, 119 (1951).
132. M. Demerec, E. M. Witkin, B. W. Catlin, J. Flint, W. L. Belser, G. Dissosway, F. L. Kinnedy, N. C. Meyer, and A. Schwartz, *Carnegie Inst. Wash. Publ.* **49**, 144 (1950).
133. K. Gezelius and N. Fries, *Hereditas* **38**, 112 (1952).
134. L. E. Andrew, *Am. Naturalist* **43**, 135 (1959).
135. S. Mittler, J. E. Mittler, and S. L. Owens, *Nature* **214**, 424 (1967).
136. N. Fries, *Hereditas* **36**, 134 (1950).
137. G. Zetterberg, *Hereditas* **46**, 279 (1959).
138. B. A. Kihlman, *Hereditas* **38**, 115 (1952).
139. B. A. Kihlman and A. Levan, *Hereditas* **35**, 109 (1949).
140. W. Kuhlman, H. Fromme, E. Heege, and W. Ostertag, *Cancer Res.* **28**, 2375 (1968).
141. W. Ostertag, E. Dvisberg, and M. Stürmann, *Mutation Res.* **2**, 293 (1965).
142. E. A. Gläss and A. Novick, *J. Bacteriol.* **77**, 10 (1958).
143. G. W. Grigg and J. Stuckey, *Genetics* **53**, 823 (1966).
144. W. Ostertag and J. Haake, *Z. Vererbungslehre* **98**, 299 (1966).
145. A. F. Yanders and R. K. Seaton, *Am. Naturalist* **96**, 272 (1962).
146. T. Alderson and A. H. Khan, *Nature* **215**, 1080 (1967).
147. A. M. Clark and E. G. Clark, *Mutation Res.* **6**, 227 (1968).
148. K. Shimada and Y. Takagi, *Biochim. Biophys. Acta* **145**, 763 (1967).
149. G. W. Grigg, *Mutation Res.* **4**, 553 (1967).
150. M. L. Lyon, J. S. R. Phillips, and A. G. Searle, *Z. Vererbungslehre* **93**, 7 (1962).
151. W. L. Russell, *Cold Spring Harbor Symp. Quant. Biol.* **16**, 327 (1951).
152. A. M. Routh, *Radiation Res.* **31**, 121 (1967).
153. J. E. Cleaver, *Radiation Res.* **37**, 334 (1969).
154. C. Richardson, C. L. Schildkraut, H. V. Aposhian, and A. Kornberg, *J. Biol. Chem.* **239**, 222 (1964).
155. G. Ahnstrom and A. T. Natarajan, *Hereditas* **54**, 379 (1966).
156. C. C. Price, G. M. Gancher, P. Koneru, R. Shibakawa, J. R. Sowa, and M. Yamaguelic, *Biochim. Biophys, Acta* **166**, 327 (1968).
157. A. R. Crathorn and J. J. Roberts, *Nature* **211**, 150 (1966).
158. E. B. Freese, J. Gerson, H. Taber, H. Rhaese, and E. Freese, *Mutation Res.* **4**, 517 (1967).
159. B. A. Kihlman, *Hereditas* **41**, 384 (1955).
160. B. A. Kihlman, *J. Biophys. Biochem. Cytol.* **2**, 543 (1956).
161. B. A. Kihlman, *Radiation Res.* Suppl. 3, 171 (1963).
162. F. H. Sobels, *Radiation Res.* Suppl. 3, 171 (1963).
163. H. Pries and W. Zillig, *Z. Physiol. Chem.* **342**, 73 (1965).
164. J. A. V. Butler and B. E. Conway, *Proc. Roy. Soc.* **B141**, 562 (1953).
165. C. Auerbach, "Mutation: An Introduction to Research on Mutagenesis," Part I. Methods, pp. 1–9. Oliver & Boyd, Edinburgh and London, 1962.
166. H. Marquardt, *Ger. Med. Monthly* **10**, 107 (1965).

167. H. J. Curtis, *Cancer Res.* **25**, 1305 (1965).
168. J. L. Melnick, *in* "Viruses Inducing Cancer" (W. J. Burdette, ed.), p. 169. Univ. of Utah Press, Salt Lake City, Utah, 1966.
169. R. T. Prehn, *in* "Viruses Inducing Cancer" (W. J. Burdette, ed.), p. 169. Univ. of Utah Press, Salt Lake City, Utah, 1966.
170. R. W. Atchison, *in* "Viruses Inducing Cancer" (W. J. Burdette, ed.), p. 198. Univ. of Utah Press, Salt Lake City, Utah, 1966.
171. J. T. Trentin, *in* "Viruses Inducing Cancer" (W. J. Burdette, ed.), p. 203. Univ. of Utah Press, Salt Lake City, Utah, 1966.
172. Sr. V. M. Maher, E. C. Miller, J. A. Miller, and W. Szybalski, *Mol. Pharmacol.* **4**, 441 (1968).
173. E. C. Miller, U. Johl, and J. A. Miller, *Science* **153**, 1125 (1966).
174. E. Kirk, J. A. Miller, U. Johl, and E. C. Miller, *Biochemistry* **6**, 177 (1966).
175. H. Marquardt, F. Zimmermann, and R. Schwarer, *Z. Vererbungslehre* **95**, 82 (1965).
176. A. C. Upton, *Anais Acad. Brasil. Cienc.* 39, Suppl., 129 (1967).
177. M. A. Bender, *Arch. Environ. Health* **16**, 556 (1968).
178. A. C. Upton, *Proc. Intern. Conf. Leukemia-Lymphoma*, p. 55. Lea & Febiger, Philadelphia, Pennsylvania, 1968.
179. J. E. Cleaver, *Nature* **218**, 652 (1968).
180. I. Berenblum and N. Trainin, *in* "Cellular Basis and Aetiology of Late Somatic Effects of Ionizing Radiation" (R. J. C. Harris, ed.), pp. 41–56. Academic Press, New York, 1963.

CHAPTER 4

Repair of Genetic Damage

I. Variation among Species

In the preceding chapter the question of how a chemical mutagen alters the genome of a living organism was discussed as was the problem of how different types of mutations arise from such alterations. Briefly discussed was the importance of the cell's repair mechanism since interference with this process can presumably lead to mutation or inactivation of the genome (blocking the replication of DNA) (Chap. 3, Sections III and IV). It is also possible that the frequency of "spontaneous mutations" (for instance, those induced by the metabolic production of free radicals) might well be enhanced when repair processes are blocked.

An aspect of gene repair which makes it of central importance in deciding how best to test for mutagenicity in man focuses on the fact that repair is highly variable in different species, i.e., not all organisms are able to repair the same kinds of genetic damage. There is a strong possibility that there is greater variability in how and whether different species repair certain kinds of genetic damage than there is variability in the actual types of damage produced by a given mutagen.

Stated in general terms, DNA need not differ from one species to another except in terms of base sequence. (That is, a microorganism having the same DNA base composition as the human genome could be chosen for genetic analysis.) However, metabolic and enzymic processes do tend to differ enormously not only from species to species (man to microbes) but even among the different tissues of an individual organism.

The products formed by the reaction of a given mutagen with DNA are likely to be virtually identical regardless of the source of DNA employed.

This is less true of *in vivo* reactions than *in vitro* ones; however, it does in general apply to both.

Once the chemicals penetrate the cell or attack the genome, differences among species become evident. In one organism the mutagen may rapidly be rendered nonmutagenic by metabolic processes which convert the compound to an inactive substance while in a different species the same compound may not be metabolized at all or a nonmutagen may be converted to a mutagenic substance in one organism while other species may metabolize it by different pathways or not at all so that the activation does not occur.

In the same sense, repair of genetic damage may differ in different species. The formation of UV-induced pyrimidine dimers is readily dealt with (by dimer excision) in most microorganisms (1, 2), but certain mammalian cells lack this ability (3). Hence, these particular lesions should lead to permanent genetic effects in the latter system.

Another example of "between species differences" is apparent from experiments which show that *E. coli* will preferentially remove or repair alkylated sites on the DNA which involve the cross-linking of two guanines (4, 5) while in contrast, certain mammalian cells will not (6).

In fact, large differences are found even among closely related organisms and among bacteria from the same genus and species. *Escherichia coli* of the strain B/r are able to withstand 100 UV-induced lesions within their genome (approximately, the mean lethal dose) whereas the mutant, *E. coli* B_{S-1}, which lacks a repair enzyme, is inactivated by only a few such lesions. Both coli strains, however, are equally competent in repairing genomal damage due to x-rays (7, 8).

It is well known that one of the basic steps concerned with repair differs fundamentally among different species. The rejoining of DNA strands, which is catalyzed by a polynucleotide ligase, utilizes ATP as a cofactor in T4 phage (9) and mammalian cells (10) whereas in *E. coli* diphosphopyridine nucleotide serves as the cofactor (11).

Differences of a quantitative and qualitative nature are also evident among mammalian cells. Crathorn and Roberts (12) have shown that different cell lines of mouse lymphoma cells differ by a factor of 2.5 in their lethal sensitivity to sulfur mustard gas despite the fact they react equally with the mustard.

It seems likely, therefore, that the major reason lethal sensitivities and mutational susceptibilities vary among different species of related organisms (or for that matter, distantly related organisms) is because of metabolic variation (including repair) rather than because of the differences in the reactivity of the mutagen with the genome. The size and ploidy of the genome are important factors as well but these do not differ significantly among normal mammalian cells.

II. Types of Reparable Damage and Mechanism of Repair

A variety of genetic lesions, whether induced by chemical or physical means, are metabolically reparable in microbial and higher organisms. Several recent reviews have dealt with the subject (13–15) which will only be touched upon here. It is clear that various kinds of damaged bases, pyrimidine dimers, certain structural defects, mismatched bases, interstrand cross-links, and single-stranded breaks are reparable in many organisms.

A. Pyrimidine Dimers and Cross-Links

1. *Bacterial Studies*

The first genetic lesion recognized as being reparable by metabolic processes was the pyrimidine dimers which can be induced by ultraviolet irradiation of whole cells (16). Of initial importance to understanding the process was the fortuitous isolation of the UV-sensitive mutants of *E. coli* B_{S-1} (17). Not only was the genome in this mutant unusually sensitive to UV irradiation but when T1 or T3 bacteriophages were irradiated with UV light and then grown in UV-sensitive mutants (*E. coli* B_{S-1}), the phage proved to be highly sensitive also (18). The interpretation was that the sensitive mutant lacked the ability to repair the damage produced by UV exposures.

Genetic analysis of other UV-sensitive mutants revealed that UV resistance is determined by three different genes which map at widely spaced loci, on the *E. coli* chromosome (19–21). These sites are *UVr*, *A*, *B*, and *C*. Mutants defective in either *A*, *B*, or *C* have served as controls in biochemical studies which showed that the excision of pyrimidine dimers (pyrimidine dimers constitute the primary genetic lesion in UV-irradiated cells) could be correlated with the ability of cells to recover from the effects of UV irradiation. The *UVr*, *A*, *B*, and *C* genes were thus shown to be concerned with the excision of thymine to thymine and thymine to cytosine dimers (15) which is obligatory for cell recovery.

Escherichia coli mutants defective in either one or more of the *UVr*, *A*, *B*, and *C* genes have been shown to be more sensitive to certain chemical agents than wild-type strains, particularly to those agents capable of producing interstrand cross-links such as nitrogen mustard, mitomycin C, and nitrous acid (22–24). Hence, in *E. coli*, the enzymes responsible for excising pyrimidine dimers are also capable of recognizing and repairing other kinds of defects. The type of defect recognized is not simply an alkylated guanine because monofunctional alkylating agents are no imore lethal to UV-sensitive coli than they are to UV-resistant strains. Instead, the defect is probably a cross-link which upon removal allows the genome to undergo DNA replication.

It is of interest to consider that although UV-resistant *E. coli* are spared a "would be" lethal effect by the action of an excising enzyme (endonuclease), this does not preclude the possibility that the monoalkylated sites might give rise to point mutations. In other words, excising enzymes could actually increase the mutagenicity of difunctional alkylating agents by preferentially repairing lethal damage, thereby allowing nonlethal alterations to be expressed as mutants.

In both *E. coli* (25) and mammalian cells (6), monoalkylated sites are repaired; however, in UV-resistant *E. coli*, the diguaninyl products are the first to be removed and only subsequently are the monoalkylated products excised from DNA. In mammalian cells, both types (monoalkylated guanine and diguaninyl derivatives) are removed at about the same rate, though the actual interstrand cross-links are preferentially repaired (26). This, apparently, is accomplished by excising only one of the two guanines found cross-linked to opposite strands of the DNA duplex (26). Hence, the diguaninyl derivative persists even though the cross-link per se has been eliminated.

The first stage in the repair of UV damage and presumably of chemically induced DNA cross-links is the excision, by an endonuclease, of a short oligonucleotide which in the case of the UV dimer lesion terminates in pTpT sequence (27). Excision is, of course, limited to only one of the two strands of the DNA duplex so that double-stranded breaks (nearly always lethal) are avoided during the process. Further degradation of the affected single strand continues following the release of the oligonucleotide unit which contained the original lesion (pyrimidine dimer or other cross-link). In *E. coli*, a total of approximately thirty nucleotides are released for each dimer excised. The process continues by an enzymic filling-in of the single-stranded gap made by the endo- and exonucleases. Presumably, this involves replacing all of the original bases in correct order by base-pairing them (in Watson-Crick fashion) with the bases on the opposite, intact strand. This is operationally termed nonconservation replication, but it is not known whether it is catalyzed by the regular DNA polymerase or by a special repair enzyme (see Section III of this chapter). The final step is the rejoining of the single-stranded break which remains after all the nucleotides have been replaced (see Section II, B of this chapter).

2. *Studies in Mammalian Systems*

There is evidence, particularly from cells grown in tissue culture, that higher organisms, including human cells, possess mechanisms similar to those described above for bacteria. However, it is not clear whether the enzymes used to excise pyrimidine dimers in mammals are the same as those used to excise alkylated lesions of the cross-linked type. Probably, they are not since mouse L cells appear to lack the ability to excise pyrimidine dimers (3) and yet they do possess enzymes which selectively eliminate interstrand cross-links induced by sulfur mustard (26).

There is also evidence that following the removal of lesions induced either by chemical or physical means, mammalian cells undertake nonconservative (repair) replication (28–31) and finally reseal single-stranded interruption within the repaired strand (32). The isolation of such an enzyme (ligase) from a mammalian source has been reported (32).

An indication of the importance of repair in humans is exemplified by the human hereditary disease known as *Xeroderma pigmentosum* which is characterized by an extreme sensitivity of the skin to sunlight (33, 34). It has been shown that such individuals lack the ability to repair UV damage to DNA (35). The particular step of repair which appears to be lacking is that concerned with repair replication though it is conceivable other steps are absent as well (35).

Because patients with *Xeroderma pigmentosum* readily develop fatal skin cancers from exposure to sunlight, Cleaver (35) has stated that the failure of DNA repair in skin must be related to carcinogenic events in that organ.

B. Resealing Single-Stranded Breaks

Since the early work of Sax on *Tradescantia* (36, 37) the concept of localized and reparable lesions on chromosomes has been established. It was also evident that chromosomes can be induced to break and that some of these breaks can undergo restitution but, until 1954, rejoining was thought to be a physical process. Later it became apparent that rejoining was stable in the sense that the rejoined or rearranged chromosomes continued in that form even after several cell divisions (38).

Hence, it has long been realized that breakage and rejoining of the genome (the DNA) occurs, and in fact it is now believed that both replication and recombination depend upon its occurrence (39, 40). The main difficulty had always been the detection of its presence in cell extracts so that the reaction and the enzyme catalyzing the reaction could be fully characterized. Once it was recognized that the enzyme in question rejoined DNA in such a way as to maintain base sequence continuity (by rejoining only at single-stranded breaks which are held in place by hydrogen bonding on the opposite strand) selection of the appropriate substrate was obvious. This difficulty having been overcome, it was relatively simple to assay for and to isolate a rejoining enzyme.

Gellert (41) was among the first to recognize the problem, and he selected λ bacteriophage DNA as a substrate since it is a duplex molecule possessing single-stranded ends that are complementary to each other and will combine to form a circle stabilized by hydrogen bonds which will then contain two single-stranded breaks (see Fig. 4.1). Enzymic esterification of these single-stranded ends (breaks) to the rest of the duplex converts DNA to a fully covalently bounded circle, thus providing the basis of the assay.

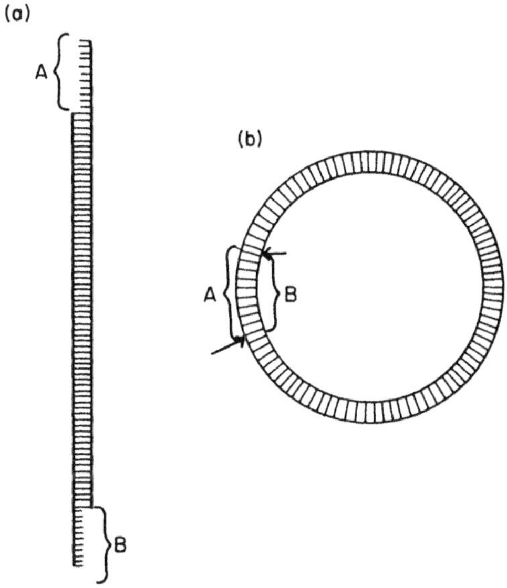

FIG. 4.1. Cyclizing a linear DNA (a) by the hydrogen bonding of its "sticky ends" (where A is complementary to B). Arrows indicate the position of single-stranded breaks in the cyclized molecule (b).

Using either a natural substrate, such as λ, as an assay system or an artificial substrate (see Section III), the rejoining enzyme (referred to as a polynucleotide ligase) has been purified over a 1000-fold from extracts of *E. coli* (42–45). The purified enzyme requires Mg^{++} and diphosphopyridine nucleotide as a cofactor, and catalyzes the joining of 5′-phosphoryl ends to 3′-hydroxy ends to form 3′- to 5′-phosphodiester bonds. Diphosphopyridine nucleotide is cleaved in the presence of the ligase to nicotinamide mononucleotide and 5′-adenosine monophosphate which is bound to the ligase in the form of an activated complex. How "activated" adenosine monophosphate contributes to the ligase reaction is still to be determined, though it is clear that a DNA-adenylate intermediate is formed in the reaction (46).

III. Methods for Studying Repair in Mammalian Cells

As was alluded to elsewhere in the text, the chemical production of mutants involves many factors and steps, most of which are metabolic in nature. First, the mutagen or precursor must enter the cell and reach the genome in a form which is still mutagenic or has become mutagenic; then it must react with DNA or interfere with the enzyme systems concerned with either the replication, recombination, or repair of DNA; and, finally, the alteration must defy the cell's attempts to repair it and persist long enough to influence daughter cells. Following one or two replications of DNA, the original unrepaired lesion has induced a relatively permanent alteration in the genome, one which is no longer reparable. Provided the defect lies in an area of the genome which is not lethal but will effect the phenotype (the physical and biochemical characteristics of an organism) the alteration constitutes a mutation.

Based on these considerations, it is obvious that the scoring of mutants is equivalent to looking at the algebraic sum of a great number of unresolved metabolic events. Mammalian organisms, for reasons discussed in Chap. 5, do not readily lend themselves to this kind of scoring and, for obvious reasons, "scoring results" obtained from microbial systems are not necessarily extrapolatable to mammals or man. It is, however, possible to study many of the individual events that are part of the algebraic sum ordinarily referred to as the mutation frequency.

The study of repair in mammals and mammalian cells and how various environmental agents influence and modify the process is one approach to the problem. The view is that agents which interfere with the process are certain to be either mutagenic or co-mutagenic. ("Co-mutagenic" is defined as any agent which enhances the mutagenicity of mutagenic substances.)

A. Repair Dealkylation

As previously mentioned, different lines of mouse lymphoma cells show differing sensitivity to sulfur mustard even though chemical studies show conclusively that the extent of alkylation is the same in both cell types. A logical explanation for the effect involves repair and the idea that the process is most active in the least sensitive cell line.

To show this, Crathorn and Roberts (6) selected ^{35}S-labeled sulfur mustard as an alkylating agent because it has the advantage of being available at high specific activity (approximately 1 Ci/mmole) and of reacting rapidly in aqueous solutions so that alkylation does not continue beyond 30 minutes.

They then demonstrated that despite the fact only about one alkylation occurred for every 10^5 DNA nucleotides (even at concentrations of mustard gas where extensive killing of cells was observed) sufficient ^{35}S radioactivity was associated with the DNA to permit a study of dealkylation. They were able to estimate, from the specific radioactivity of alkylated DNA, the number of mustard gas molecules attached to the genome, and they could assess the rate at which this number decreased by taking aliquots of cells at various times after mustard treatment. Since the assay for dealkylation depends upon measuring the decrease in the specific radioactivity of ^{35}S-mustard-labeled DNA, it was also necessary to know the extent to which the synthesis of new DNA contributed to this decrease. This was done by labeling the DNA of cells with ^3H-thymidine prior to their alkylation and transfer to fresh medium. Thus, the synthesis of new DNA can be determined by the dilution of the original ^3H-thymidine-labeled DNA.

Their results show that at high mustard concentrations, where cell death is extensive, as many as 50% of the alkylated groups are removed from DNA within a period of time which would be less than a normal generation time. The method, therefore, appears to be a highly sensitive way of assessing enzymic excision (endonuclease activity) *in vivo* and should lend itself to a study of whether certain environmental agents act at this level.

B. Repair Replication

1. *Nonconservative DNA Synthesis*

Normal DNA replication leads to a doubling of the cell's DNA content prior to cell division and is of the semiconservative type (see Chap. 2). That is, the strands of the DNA double helix separate and each strand serves as a template for the synthesis of a new complementary strand (47). If ^3H-BUdR (radioactive bromodeoxyuridine) is substituted for thymidine, then semiconservative replication of DNA will produce a duplex which is composed of one strand that contains about 20% T and no BUdR and a second, complementary strand which consists of about 20% BUdR and no T (see Fig. 4.2). Because incorporation of BUdR into the newly synthesized DNA strand increases the buoyant density of the DNA duplex in CsCl density gradients, the new DNA (consisting of one preexisting strand containing thymidine and one newly synthesized strand containing BUdR) can be centrifuged free of all preexisting duplexes of DNA (see Fig. 4.3).

If radioactive BUdR is used and is incorporated only into the DNA which is undergoing semiconservative replication, then all of the radioactivity should be associated with high-density DNA. On the other hand, if repair replication is taking place in response to short single-stranded gaps or patches produced by endonucleases, then radioactive BUdR will find its way into light-density DNA (Fig. 4.3). This is simply due to the fact that repair replication leaves the product only slightly substituted with BUdR, an amount insufficient to have a measurable effect on buoyant density but which is nevertheless detectable by radioisotope counting procedures.

The analysis, therefore, consists of four steps: (a) first, the cell and its genome are challenged by a mutagenic agent; (b) radioactive BUdR is added to the culture medium and cells are incubated for an appropriate period; (c) DNA is isolated from the cells and separated into heavy- and light-density

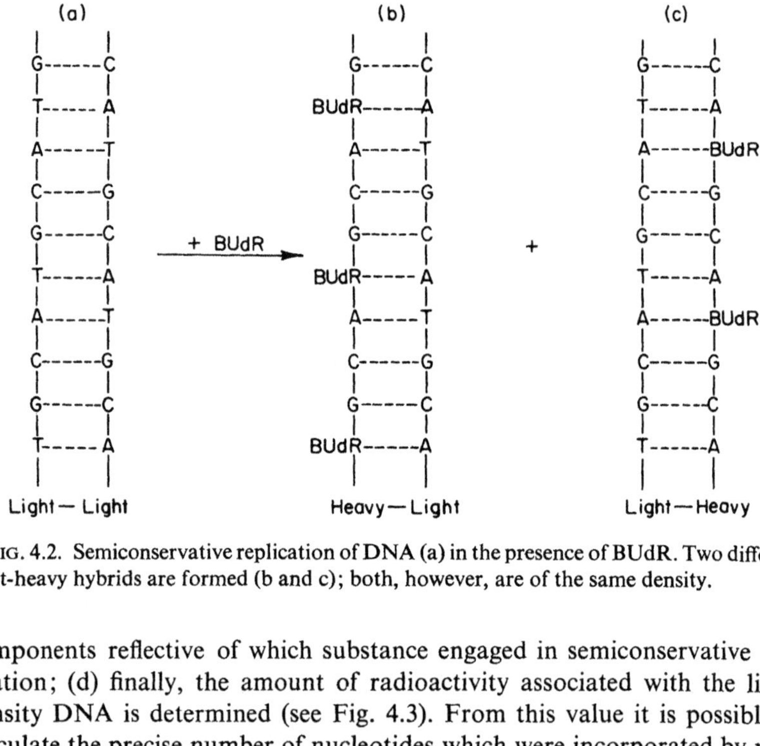

FIG. 4.2. Semiconservative replication of DNA (a) in the presence of BUdR. Two different light-heavy hybrids are formed (b and c); both, however, are of the same density.

components reflective of which substance engaged in semiconservative replication; (d) finally, the amount of radioactivity associated with the light-density DNA is determined (see Fig. 4.3). From this value it is possible to calculate the precise number of nucleotides which were incorporated by nonconservative (repair) replication. Knowledge of the number of gaps that were produced by endonucleases can be derived from dealkylation studies (see Section III, A) so that it should be possible to determine the average number of nucleotides inserted for each gap repaired. In *E. coli*, as was mentioned earlier, thirty nucleotides are used for each lesion but the number remains to be determined for mammalian cells.

The above method was originally developed for bacterial systems (48), but it has since proved to be applicable to mammalian cells grown in tissue culture (28–31). In fact, this method has been used to show that *Xeroderma pigmentosum* is associated with an inability of the patient cells to carry out repair replication. The method has also been used by Cleaver to investigate whether certain drugs and antibiotics interfere with gene repair (31).

2. *Unscheduled DNA Synthesis*

Repair replication can also be detected and quantitated by assaying for unscheduled DNA synthesis (49). This term applies to the event which occurs immediately following UV irradiation of whole cells. Normally, only a fraction

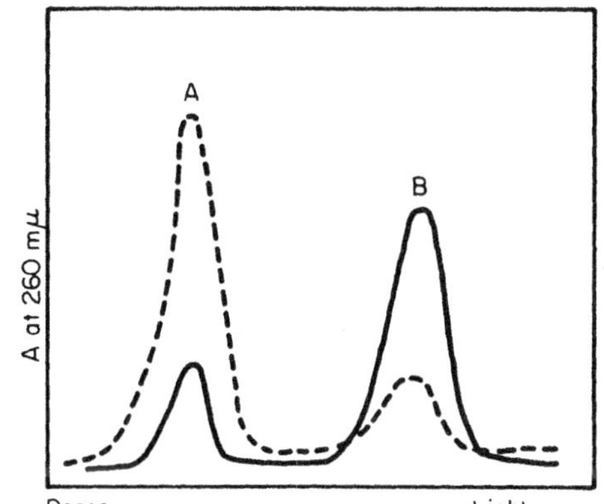

FIG. 4.3. Equilibrium centrifugation of ^{3}H-BUdR-substituted DNA in a CsCl density gradient. Peak A has arisen from the synthesis of heavy-light hybrids depicted in Fig. 4.2. The radioactivity located in peak B arises from repair replication.

of a total cell population would be synthesizing DNA at any given time, and hence only a fraction of the cells would be capable of incorporating isotopically labeled thymidine. After irradiation with UV light, the entire population begins to incorporate thymidine and, presumably, this incorporation constitutes repair replication. In support of this view is the observation that skin fibroblasts from *Xeroderma pigmentosum* patients fail to incorporate thymidine following exposure to UV light (35).

The procedure, as modified by Cleaver (31), involves incubating tissue culture cells for 3 hours in ^{3}H-thymidine prior to irradiation so that those cells which were normally synthesizing DNA can be unambiguously identified by autoradiographs. The cells are then UV irradiated with 100 ergs/mm^2 and incubated for another 3 hours in ^{3}H-thymidine before fixation and autoradiography. The lightly labeled cells are those which have been induced to incorporate thymidine in response to genomal damage. The number of silver grains over these cells constitutes a measure of repair replication. The method has also been used to study the repair of damage induced by alkylating agents (50).

C. Induction of Single-Stranded Breaks and Their Repair

Single-stranded DNA breaks and their repair can be assessed by observing changes in the molecular weight of unsheared single-stranded DNA. It is critical that methods which avoid shear and yield DNA of high molecular

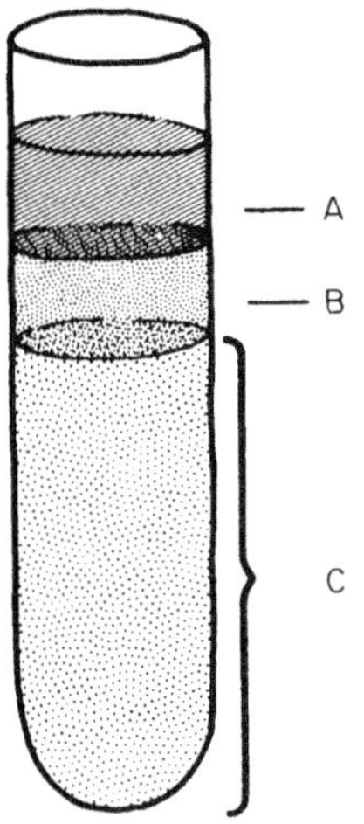

FIG. 4.4. Centrifuge tube containing cell suspension (A), deproteinizing layer (B), and sucrose density gradient (C). Top part of the gradient is approximately 10% sucrose whereas 30% sucrose is present at the bottom of the tube.

weight (10^8 daltons) be used so that a relatively small number of induced breaks can be detected. Usually, this is done by combining two procedures into one, i.e., the extraction procedure with the analytic one from which molecular weight information is derived (51).

The cells are collected and carefully layered on top of a solution which will deproteinize and extract nuclear DNA. Below these two solution layers is a third solution consisting of a sucrose density gradient (see Fig. 4.4). The cells, by being subjected in the uppermost layer to a centrifugal force, move from their original position into the extracting media where the DNA is released and begins to sediment in the sucrose gradient. A gradient of sucrose is used merely to prevent mixing of the DNA as it sediments down the tube.

The pH of the gradient is adjusted to 12.5 so that the DNA duplex dissociates and sediments as single strands.

From the rate at which the dissociated single-strands sediment in alkaline solutions of sucrose an estimate of their molecular weight can be derived (52). High molecular weight molecules will move more rapidly in the centrifugal field and be closer to the bottom of the tube than those of lower molecular weight.

Several reports have appeared in which this technique has been used to observe the induction of single-stranded breaks by ionizing radiation (53–55). Attempts were also made to see if the DNA underwent rejoining reactions. This is carried out by incubating the cells for a longer period of time following their exposure and then analyzing as described above.

REFERENCES

1. R. B. Setlow and W. L. Carrier, *Proc. Natl. Acad. Sci. U.S.* **51**, 226 (1964).
2. R. P. Boyce and P. Howard-Flanders, *Proc. Natl. Acad. Sci. U.S.* **51**, 293 (1964).
3. M. Klimek, *Photochem. Photobiol.* **15**, 603 (1966).
4. P. D. Lawley and P. Brookes, *Nature* **206**, 480 (1965).
5. S. Verritt, *Biochim. Biophys. Acta* **31**, 355 (1968).
6. A. R. Crathorn and J. J. Roberts, *Nature* **211**, 150 (1966).
7. R. B. Setlow, *J. Cellular Comp. Physiol.* **64**, Suppl. 1, 51 (1964).
8. P. T. Emmerson and P. Howard-Flanders, *Biochem. Biophys. Res. Commun.* **18**, 24 (1965).
9. B. Weiss, R. R. Live, and C. C. Richardson, *Federation Proc.* **26**, 395 (1967); A. Becker, N. L. Gefter, and J. Hurwitz, *ibid.* p. 395.
10. T. Lindahl and G. M. Edelman, *Proc. Natl. Acad. Sci. U.S.* **61**, 680 (1968).
11. B. M. Olivera and I. R. Lehman, *Proc. Natl. Acad. Sci. U.S.* **57**, 1700 (1967).
12. A. R. Crathorn and J. J. Roberts, *Progr. Biochem. Pharmacol.* **1**, 320 (1965).
13. W. Ginoza, *Ann. Rev. Microbiol.* **21**, 325 (1967).
14. R. B. Setlow, *Progr. Nucleic Acid Res. Mol. Biol.* **8**, 257 (1968).
15. P. Howard-Flanders, *Ann. Rev. Biochem.* **2**, 666 (1968).
16. E. M. Witkin, *Science* **152**, 1345 (1966).
17. R. F. Hill, *Biochim. Biophys. Acta* **30**, 636 (1958).
18. S. A. Ellison, R. R. Feiner, and R. F. Hill, *Virology* **11**, 294 (1960).
19. P. Howard-Flanders, R. P. Boyce, E. Simson, and L. Theriot, *Proc. Natl. Acad. Sci. U.S.* **48**, 2109 (1962).
20. P. van de Putte and C. A. van Sluis, *Mutation Res.* **2**, 97 (1965).
21. P. Howard-Flanders, R. P. Boyce, and L. Theriot, *Genetics* **53**, 1119 (1966).
22. R. H. Haynes, *Photochem. Photobiol.* **3**, 429 (1964).
23. R. P. Boyce and P. Howard-Flanders, *Z. Vererbungslehre* **95**, 345 (1964).
24. P. Howard-Flanders and R. P. Boyce, *Radiation Res.* Suppl. 6, 156 (1966).
25. P. Brookes and P. D. Lawley, *Biochem. J.* **80**, 496 (1961).

26. B. D. Reid and I. G. Walker, *Biochim. Biophys. Acta* **179**, 179 (1969).
27. P. D. Lawley, *Progr. Nucleic Acid Res. Mol. Biol.* **5**, 89 (1966).
28. J. E. Cleaver and R. B. Painter, *Biochim. Biophys. Acta* **161**, 552 (1968).
29. R. B. Painter and J. E. Cleaver, *Nature* **216**, 369 (1967).
30. R. E. Rasmussen and R. B. Painter, *J. Cell Biol.* **29**, 11 (1966).
31. J. E. Cleaver, *Radiation Res.* **37**, 334 (1969).
32. T. Lindahl, J. A. Gally, and G. M. Edelman, *Proc. Natl. Acad. Sci. U.S.* **62**, 597 (1969).
33. J. H. Seguria, J. T. Ingram, and R. T. Brain, *in* "Diseases of the Skin," p. 73. Macmillan, New York, 1911.
34. V. A. McKusick, "Mendelian Inheritance in Man." Johns Hopkins Press, Baltimore, Maryland, 1966.
35. J. E. Cleaver, *Nature* **218**, 652 (1968).
36. K. Sax, *Genetics* **25**, 41 (1940).
37. K. Sax, *Cold Spring Harbor Symp. Quant. Biol.* **9**, 93 (1941).
38. S. Wolff, *J. Cellular Comp. Physiol.* **48**, Suppl. 1, 151 (1961).
39. R. Okazaki, T. Okazaki, K. Sakabe, and K. Sugimoto, *Japan. J. Med. Sci. Biol.* **20**, 255 (1967).
40. R. Okazaki, T. Okazaki, K. Sakabe, K. Sugimoto, and A. Sugino, *Proc. Natl. Acad. Sci. U.S.* **59**, 598 (1968).
41. M. Gellert, *Proc. Natl. Acad. Sci. U.S.* **47**, 148 (1967).
42. S. Zimmerman, J. W. Little, C. K. Oshimsky, and M. Gellert, *Proc. Natl. Acad. Sci. U.S.* **57**, 1841 (1967).
43. B. M. Olivera and I. R. Lehman, *Proc. Natl. Acad. Sci. U.S.* **57**, 1426 (1967).
44. N. L. Gefter, A. Becker, and J. Hurwitz, *Proc. Natl. Acad. Sci. U.S.* **58**, 240 (1967).
45. J. Hurwitz, A. Becker, N. L. Gefter, and M. Gould, *J. Cellular Physiol.* **70**, Suppl. 1, 181 (1967).
46. B. M. Olivera, Z. Hall, and I. R. Lehman, *Proc. Natl. Acad. Sci. U.S.* **61**, 237 (1968).
47. M. Meselson and F. Stahl, *Proc. Natl. Acad. Sci. U.S.* **44**, 671 (1958).
48. D. Pettijohn and P. C. Hanawalt, *J. Mol. Biol.* **9**, 395 (1964).
49. B. Djordjevic and L. J. Tolmach, *Radiation Res.* **32**, 327 (1967).
50. G. M. Hahn, S. J. Yang, and V. Parker, *Nature* **20**, 1142 (1968).
51. T. Terasima and A. Tsuboi, *Biochim. Biophys. Acta* **174**, 309 (1969).
52. F. W. Studier, *J. Mol. Biol.* **11**, 373 (1965).
53. J. T. Lett, I. Caldwell, C. J. Dean, and P. Alexander, *Nature* **214**, 790 (1967).
54. R. M. Humphrey, D. L. Steward, and B. A. Sedita, *Mutation Res.* **6**, 459 (1968).
55. P. H. M. Lohman, *Mutation Res.* **6**, 449 (1968).

CHAPTER 5

Test System for the Detection and Scoring of Mutants

To determine whether a chemical agent is mutagenic or not, and to assess its relative mutagenicity as compared to other compounds, it is necessary that some kind of biological test system be employed. Fortunately, a variety of such systems have come into being (1); however, these tend to tell us different things about the mutagen and the nature of the mutation, and it should be remembered that most are of little reliability in terms of extrapolating to man.

The simplest organisms generally are the most suitable and give us the greatest amount of information concerning the nature of the mutation. They are also the most sensitive systems; however, taxonomically they are the furthest removed from man. The simple systems are represented by microorganisms and viruses which possess the following advantages for detecting mutagenicity: (1) they are microscopic so that hundreds of millions of organisms can easily be handled in a single experiment; (2) they have short generation times and so give rise to new progeny very rapidly; (3) they have small genomes (the total number of genes in any given microorganism is hundreds to thousands of times less than it is in man) which allows for a higher probability that a given mutant will involve a specific gene (or a specific site). The latter advantage lies in the fact that it is unnecessary to test a great number of genes in order to detect a few mutants.

The difficulty with microbial test systems is that they merely show the potential for mutagenicity in man and do not, as we have discussed elsewhere, prove that such mutagenic hazard exists for man. However, without these microbial systems little would be known of the various types of mutations which have been found (e.g., transitions, transversions, additions, deletions) or how they are induced by different chemical agents. In this chapter we will briefly discuss some of the most frequently used systems and will attempt to single out their advantages and, when appropriate, their disadvantages.

I. DNA Transformation

A. Inactivating Alterations of DNA

As mentioned in Chap. 2, part of the evidence indicating that DNA constitutes the genetic material of bacterial cells is based on experiments which show that "genes" can be added to a bacterium simply by providing it with an appropriate and highly purified DNA (2) (the DNA need not actually be of high purity except in the instance used to show that it is the DNA rather than some contaminant which is responsible for the transformation). For instance, a mutant bacterial cell which is incapable of growing on media lacking tryptophan (tryptophan-requiring mutant) can be converted to tryptophan independence (a cell capable of synthesizing its own tryptophan, a prototroph) by incubating such cells with DNA isolated from wild-type bacteria. The wild-type bacterium possesses the genes which code for and give rise to the enzymes that manufacture tryptophan *in situ*. Hence, wild-type DNA includes these genes which, along with other genes, can be incorporated into recipient bacteria such as the tryptophan-dependent mutant referred to above and does, thereby, transform them to tryptophan independence.

One obvious test system would involve treating isolated wild-type DNA with a potential mutagen to see if it has the power to inactivate the gene which gives rise to tryptophan independence (3). Such events have been termed "inactivating alterations" by Freese (3), who has proposed that the mechanism of inactivation often involves a blockage of DNA synthesis. Compounds known to cause interstrand cross-linking (bifunctional alkylating agents), backbone breakage (hydrazine, peroxides, etc.), or those which prevent base-pairing are the most likely candidates for producing this effect.

B. Scoring Mutants

In addition to being able to detect and score alterations of DNA involving inactivation, the transformation test is used to score mutants (4, 5). This is possible only because the DNA isolated from wild-type cells and used to transform tryptophan-requiring mutants is large enough to include more than a single gene. It would on the average contain about ten genes when isolated by the Marmur method (6).

Genetic mapping studies show that the region adjacent to the gene responsible for the production of tryptophan will, when mutated, allow a blue fluorescent compound to accumulate in the affected cells (7).

The scoring of mutants, therefore, involves counting the number of cells (bacterial colonies) which grow in the absence of tryptophan and at the same time produce a fluorescent colony (Fig. 5.1). The ratio of the number of

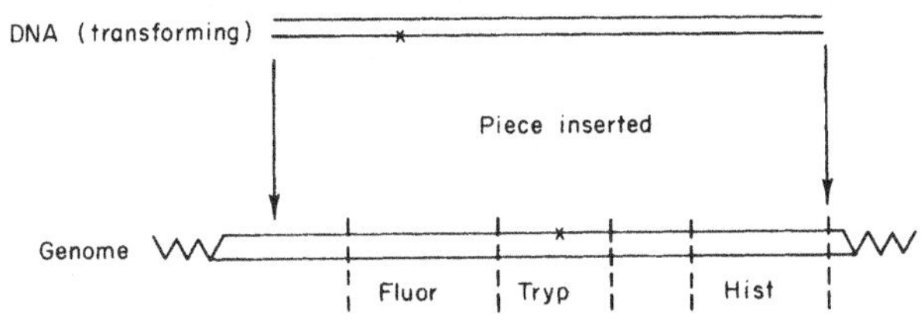

FIG. 5.1. Diagrammatic illustration of how mutations are established within the "genome" of transforming-DNA. Donor DNA transforms recipient cells so that they grow in the absence of tryptophan. Since most of the transformed cells have incorporated other genes which are closely linked to tryp, a mutation (X) in the fluor region of the donor DNA results in colonies which accumulate fluorescent material.

fluorescent cells (colonies) to the total number of transformed cells (colonies) is a measure of the mutation frequency.

C. PROCEDURE

The test is simple and straightforward, and involves methodology which is, for the most part, routine to microbiologists. Once the DNA is isolated from wild-type cells (6), it is treated with the mutagen and then reisolated free of the original compound. The DNA is now ready for testing against the tryptophan-requiring mutant with which it is incubated. Following a short incubation (measured in minutes), the cells are subjected to tryptophan-deficient medium (plated out on an agar-containing medium lacking tryptophan). Those cells which have been transformed to tryptophan independence will, of course, grow on this medium and can be quantitated on the basis of a colony count. The transformed cells which involve an induced mutation accumulate anthranilate derivatives and so can be scored directly on such medium by virtue of their blue fluorescence. The value of this method lies in the fact that the mutagen is tested under *in vitro* conditions wherein cellular metabolism and the necessity to sustain life may be disregarded. It is, therefore, a purely chemical system and yet information concerning the frequency of lethal and mutagenic events can quickly be obtained. Another advantage of the method relates to the fact that the size of transforming DNA corresponds to only about ten genes which in practical terms is equivalent to studying a very small genome. This feature provides very high sensitivity.

It is also possible, in theory at least, to test whether a particular mutagen tends to affect (to mutate) one gene or genetic site more than another. For instance, we have mentioned that the fluorescent gene is adjacent to the tryptophan region, but also adjacent to the later is a gene responsible for histidine synthesis. The effect of the mutagen on either the histidine gene or the fluorescent gene should be approximately the same provided the mutagen shows no particular specificity for special genetic regions apart from the specific bases it attacks. However, were a difference in response observed between the two genetic sites, the possibility is raised that a higher order of specificity exists than can be accounted for by purely chemical procedures. This interpretation, however, is subject to the uncertainties evidenced by "hot spots," which will be discussed in the following section.

II. Bacteriophage

Bacteriophages are highly specific viruses which infect specific types of bacteria. Though some bacteriophage, e.g., the so-called T-even series, have

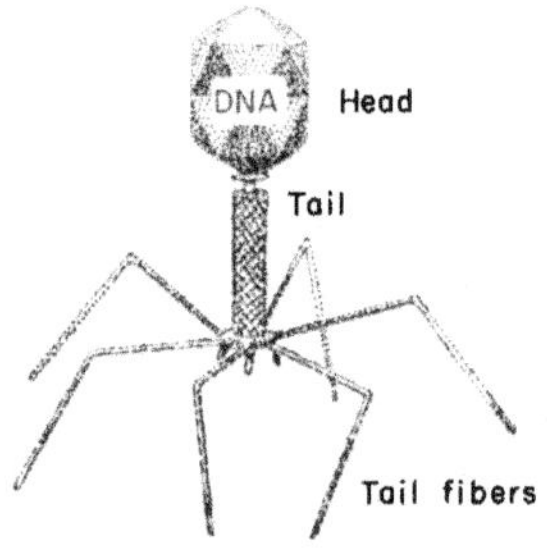

FIG. 5.2. The T-even phage.

a very complex structure (Fig. 5.2), they, like all other viruses, are unable to reproduce outside a living host.

A. T4 Bacteriophage

Much of what is known today about the molecular aspects of mutagenic processes has been learnt from experiments on the phage T4 which is a lytic virus that grows on the bacterium *E. coli.* The virus has a hexagonal head in which its DNA is stored and a long tail to which several fibers are attached (8).

The latter is used for the attachment of the phage to susceptible bacteria. A hole is somehow bored in the bacterial membrane through which the phage DNA is injected to initiate the process of infection (9).

The genetic material of T4 phage consists of double-stranded DNA but unlike the DNA of higher organisms it contains 5-hydroxymethylcytosine (instead of cytosine) which is found to be partially glucosylated (10).

A few minutes following the injection of the phage's genetic material, the synthesis of bacterial components is halted and a number of phage-specific enzymes are made, as well as phage DNA and structural protein used in the phage's coat. After about 30 minutes, several hundred fully formed phage particles are produced and the bacterium undergoes lysis.

1. *Detection and Quantitation of Bacteriophage*

In order to determine the number of bacteriophages present in a given suspension one makes a "lawn" of susceptible *E. coli* (B strain) by mixing the bacteria with warm agar (present in a liquid state at this temperature) which is layered over the surface of a hardened agar plate that contains all the nutrients necessary for bacterial growth. After the top layer has hardened, a suspension, containing phage, is sprayed over the entire surface of the plate. Following an appropriate incubation period at 37°C, the *E. coli* grow and give a hazy appearance to the agar surface. Where a phage has infected a single bacterium, it produces 200 or more progeny and causes lysis of the bacterial cell. The "released" phages infect the surrounding area causing the lysis of many bacteria, giving rise to a clear area on the agar surface known as a plaque (Fig. 5.3). The number of plaques present on the agar plate is equivalent to the number of bacteriophages present in the original suspension.

2. *Detection of Mutants*

One way of obtaining mutant phages is to look for plaques of unusual morphology (11–14), plaques that are larger than normal, or plaques whose edges differ from those usually observed. The wild-type plaque produced by T4 is small and has a characteristic fuzzy edge. Occasionally, however (one out of 10,000 in a good stock) plaques of greater than normal size are found with extremely clear-cut edges. When these are isolated and used to infect fresh bacteria they are found to breed true, i.e., they continue to produce large plaques. The phage is said to be an *r*-gene mutant having mutated from r^+ to r^-.

A second type of mutant was isolated when it was discovered that certain of the r^- mutants failed to grow on the K strain of *E. coli* despite the fact that the T4 wild-type and about half of the r^- mutants would grow as well on the

K strain as on *E. coli* B. The mutants which grow only on B (where they form large round plaques) are classified as *rII* mutants and it is with these that most of the detailed genetic work on mutagens has been done to date.

Because *rII* mutants do not grow on *E. coli* K, any subsequent mutation (whether spontaneous or induced by chemical or physical agents) which causes a few phages to revert to the wild-type can easily be detected simply by plating an appropriate phage suspension on *E. coli* K. These infrequent events (back mutations) are readily detected, since only back-mutated phage will grow on *E. coli* K to produce visible plaques.

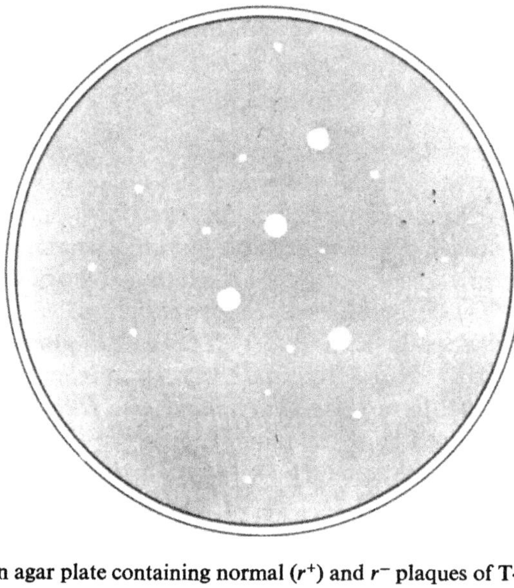

FIG. 5.3. An agar plate containing normal (r^+) and r^- plaques of T-even phage.

Another advantage of this system is that many hundreds of mutants of the *rII* region have been isolated which are, according to genetic mapping experiments, all unique. Hence, hundreds of different mutants of the same gene have been isolated. This is not so surprising when we consider that this gene is composed of about 3000 nucleotide pairs and that a mutation at any of these sites could alter completely the biological activity for which the gene is responsible. (Unfortunately, the molecular function controlled by the *r* gene still remains to be known.)

3. *Mapping the rII Region*

Determining the exact site of mutation within the *rII* region of T4 is done via mapping studies which take advantage of the fact that the DNA of two different mutants will undergo what is called recombination (15). The DNA from two different mutants can intermix (recombine) so as to yield phage with an altered genome—a genome composed partly of the original DNA and partly of "exchanged DNA" contributed by a different mutant. Hence, it is possible through the exchange of DNA (recombination) of two different *rII* mutants to obtain recombinants with a wild-type *r* gene, i.e., phages which grow on *E. coli* K (see Fig. 5.4).

The fact that mixing two independently isolated *rII* mutants produces a few phages which will grow on *E. coli* K attests to the fact that the mutants were altered at different sites within the *r* gene (13, 16). Depending upon the frequency with which successful recombination occurs the two sites are either far apart or close together. The closer together they are, the fewer successful recombinants are seen, whereas if the altered sites are far apart within the *rII* region the probability of recombination is greater, and a higher frequency of

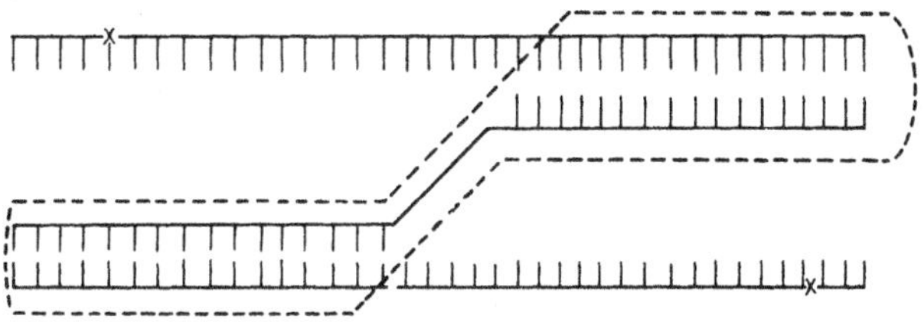

FIG. 5.4. Recombination between two DNA duplexes which contain point mutations (X) at opposite ends. The dotted line indicates the new "recombinant" duplex which is wild-type.

recombinants is observed. The frequency of recombination is used as a mapping unit and relates directly to the distance by which genetic sites are separated. (The method involving complementation between *A* and *B* cistrons rather than recombination is discussed in the section dealing with yeasts and other fungi.)

Since hundreds of "mapped" mutants are now available, the exact intergenic site at which a newly induced *rII* mutant has been altered can be determined from the frequency with which the unknown mutant recombines with "mapped" mutants to produce wild-type (phages which grow on *E. coli* K).

It also is possible, as mentioned in the discussion of acridine dye mutants (Chap. 3, Section III, A, 2), to determine whether the mutation is a consequence of base substitution or of base deletion or addition (sign mutations) (17). This can be done either by recombination with well-mapped stocks or by examining the reversion frequency induced by specific and well-characterized mutagens (16). Usually a combination of both approaches are needed to characterize the new mutant.

4. *Site Specificity*

One of the most significant observations made from the above mapping study is that certain regions of the *rII* gene are much more sensitive to mutagenic change (more mutable) than other sites. In fact, of the 300 or so sites which have been mapped, most have their characteristic mutation frequency. There are certain sites, called "hot spots," where as many as 30% of all spontaneous *rII* mutations occur.

The reason for this high degree of specificity remains obscure but it should serve to caution us against overinterpreting data based on a mutation frequency in only one species where only a single genetic marker has been used. For instance, the marker may have a far greater or far smaller number of "hot spots" than the average gene.

B. Lysogenic Bacteriophage

Certain bacteriophages, for instance lambda, which grow on *E. coli* K, are called lysogenic because they have the ability to insert their DNA (their genetic information) directly into the host cell's genome (18, 19). In the case of lambda, the insertion always occurs at the same site within the coli genome. Its induction is thought to be initiated, in some instances, by a mutagen acting on the viral DNA or on the host genome, possibly by rendering it defective and triggering its release.

There is considerable interest in this event and the mechanism of lysogeny induction because recent advances in the field of viral oncology have shown that certain cancer inducing viruses (e.g., SV40) are incorporated into the genome of their host cell prior to the transformation of the cell to a neoplastic state (20).

C. *In Vitro* and *In Vivo* Observations

One advantage offered by the bacteriophage system relates to the fact that a given chemical mutagen can be tested against exactly the same genome either in the presence of cellular metabolism (*in vivo*) or in the total absence of such activity (*in vitro*). Hence, chemicals which are not themselves mutagenic but must be metabolized to an active intermediate exhibit their mutagenicity *in vivo* but not *in vitro*. For instance, mitomycin C has little effect upon phage treated *in vitro*; however, inside the host cell mitomycin C is enzymically reduced to a compound which is a powerful alkylating agent and has a profound affect upon the phage (21).

On the other hand, there are compounds which appear to be strongly mutagenic *in vitro* but because they are rapidly metabolized inside the bacterium, mutagenicity is not seen *in vivo* unless very high and often lethal doses are used.

III. Bacterial Test Systems

The development of bacterial genetics has contributed significantly to many of the recent advances made in contemporary (molecular) biology (22). It is not surprising that these developments have led to the creation of many excellent and potentially useful test systems, but no attempt will be made to review these comprehensively; instead, only a brief glance at some of the test systems will be given so that the reader gets an idea of the general concepts and techniques.

Bacteria are taxonomically delegated to the class Schizomycetes which are typically unicellular, nonchlorophyll-producing plants that multiply by cell division (23). There are five orders within this class but the one most commonly used for mutagen testing is called Eubacteriales (true bacteria). Among these are thirteen families and nearly a hundred genera, thousands of species and countless varieties and strains. For the most part, the different strains, varieties, and species within a given genus are classified, not on the basis of their morphology (for they are too similar in appearance) but on their metabolic, serological, and chemical characteristics.

The mutants which have been isolated are nearly all metabolic mutants. They differ from the wild-type in terms of their ability or inability to carry out a particular metabolic reaction as catalyzed by one or more enzymes.

A. Forward Mutation

Mutations in which the function of a given gene is lost are referred to as forward mutations.* That is, a gene, previously responsible for the synthesis of an amino acid, nucleoside, or nucleotide (or any other metabolic event), is mutated in the "forward" direction when it fails to direct the synthesis of the enzymes which carry out this process (24). The mutation might arise from a base substitution which, as we have mentioned earlier, could either be a

* Not all geneticists would agree with this definition since originally a forward mutation was defined as any mutation in which the mutant differed phenotypically from the wild-type. The difficulty arises when one tries to decide, within certain species of organisms, which one actually constitutes the wild-type and which the mutant.

transition or a transversion or it might be as a consequence of an addition or deletion. Mutations of the substitution type are the most prone to have an effect on the phenotype (the activity of the enzyme whose synthesis the gene controls) when they give rise to nonsense codons (UAA, UAG) (25). This is so because the synthesis of a growing polypeptide chain is terminated by such triplets and fragments of the enzyme, instead of the whole enzyme, is produced. Additions and deletions are always likely to cause an inactivation of gene function since they cause frameshifts in codon-reading from the point of mutation to the end of the gene (26).

Since forward mutations can occur anywhere within the gene and anywhere within the approximately 3000 genes of the bacterial genome, the incidence of forward mutations is high. Both spontaneous and induced forward mutation frequencies are relatively high (higher than reverse mutations). The difficulty lies in identifying the mutant; in fact, even recovering the mutant can be a problem, since its inability to carry out a particular metabolic function could lead to its elimination in an inappropriate medium. For this reason, attempts to detect and quantitate forward mutants usually involve growing them on medium which has been enriched in a number of nutrients that are not required by the wild-type.

After the cells have grown into visible colonies on enriched medium, each colony can be tested to determine whether it is a mutant or not, simply by reculturing the cells on "nutrient-depleted" medium. The cells which fail to grow on the latter are those derived from mutant colonies. The exact defect can then be studied in greater detail by attempting to grow these same cells in medium which is deficient in only one of the above nutrients. By knowing which medium (deficient in only one of the nutrients) fails to support bacterial growth, the identity of the mutant is obtained.

The above process is obviously very laborious in that it involves testing hundreds of colonies in many different types of growth medium before the mutant can be isolated and identified. There are, however, techniques which can reduce enormously the amount of time and effort involved. One such example has already been referred to in discussing transforming DNA (Chap. 5, Section I, B). The gene which is concerned with the metabolism of anthranilic acid permits a fluorescent product (anthranilate derivatives) to accumulate within the mutant colonies. Hence, a mutation at this locus is rapidly identified and isolated because the colony fluoresces. Certain other mutants give rise to colored colonies or colonies which are morphologically distinguishable from the wild-type and thus lend themselves to rapid identification and isolation. One such example includes a color method for scoring the loss of ability of certain bacteria to ferment lactose or some other sugar. The nonfermenting mutants give rise to white colonies while all others are stained purple when grown on EMB agar medium (eosin and methylene blue containing agar) (27).

Another means of selecting for forward mutations involves the transformation from UV resistance to UV sensitivity. In this case, special advantage is afforded by lytic bacteriophage such as T1 or T3 in *E. coli* or P22 phage in *Salmonella typhimurium* which do not themselves carry UV-repair enzymes but depend upon the host-bacterium for their own repair (28). These phages, when irradiated by UV light, are repaired by their host (host-cell reactivation, HCR) but they "return the favor" by causing the death and lysis of the host cell. However, should the host cell be a mutant, defective in its UV-repair enzymes (HCR^-), the phage cannot reproduce and cannot cause the death or lysis of the bacterium (29). Hence, of the bacteria infected with UV-irradiated phage, only the HCR^- mutants will survive since only these leave the phage in an inactive state.

Forward mutations have also been selected and isolated by making use of "thymineless death." In this case, the cells, which are normal except for their inability to synthesize thymine, die because they continue to metabolize normally despite the fact they are unable to properly replicate their DNA (30). The mutant cells, however, are metabolically arrested because they are unable to synthesize a particular nutrient which is essential for growth. When thymine, along with the necessary nutrient, is added back to the medium, these arrested mutants begin to divide and proliferate and are thus selected from the wild-type.

One advantage of forward mutations is that they can be characterized and understood on the basis of reversion frequencies and the types of mutagenic chemicals which cause them to revert to the wild-type. For instance, were the mutants readily reverted by 5-bromodeoxyuridine or 2-aminopurine, it would be concluded that the original mutation was caused by a base substitution of the transition type (see Chap. 3, Sections I, A and I, B).

B. Back Mutation

Mutations which bring about the restoration of gene function are called back mutations or reverse mutations. These differ fundamentally from those of the forward type by virtue of the fact that back mutations are restricted to only certain highly specific sites within the gene; they must be near (as in addition and deletion types) or at the site of the original mutation (as in substitutions).

Since these mutants will not grow on medium deficient in the particular substance which the mutant alone cannot synthesize, revertants are detected and isolated on the basis of their ability to grow and to give rise to colonies on such medium. Hence, the system is conceptually similar to that described for *rII* mutants of T4 which are unable to grow on *E. coli* K until a second mutation causes it to revert (back-mutate) (Chap. 5, Section II, A). Even though reverse mutations are far less frequent than those of the forward type, they represent a highly sensitive test system for the same reasons as do reversions from *rII* mutants to wild-type T4 phage.

IV. Ascomycete

A. Advantages and Disadvantages

Various species and strains of *Saccharomyces* and of *Neurospora* or *Aspergillus* have successfully served as mutagenic test systems. These organisms share many of the advantages offered by bacteria, i.e., rapid growth, short generation times, small size in terms of cellular dimensions and genome, and growth on chemically defined medium. In fact, it should be noted that the studies which gave birth to the one-gene–one-enzyme concept were performed entirely on *Neurospora* (31).

On the other hand, it must be admitted that yeast and other fungi have the disadvantage of being somewhat larger and having substantially larger genomes (10-fold or greater) than bacteria. Hence, the probability of a mutation occurring at a particular site (somewhere within the marker gene) is less in the former than it would be in bacteria. Also, their generation times are significantly longer, and growing large numbers of these organisms requires considerably more effort than it would for equivalent numbers of bacteria.

However, one advantage held by these nonbacterial yet microbial systems is the fact that they belong to a class of organisms which are eukaryotic, meaning that their genetic information is organized into a discrete nucleus and hence they are more closely related to higher organisms. Another and possibly as important advantage is that these organisms have interesting life cycles which often lend themselves easily to certain types of genetic tests. Certain of these organisms will change from haploid (a single set of chromosomes as carried by human gametes) to diploid (two sets of homologous chromosomes as carried by human somatic cells), and then back to haploid again (1, 32). With others (*Neurospora*), haploid organisms can be made to "fuse" to produce heterokaryons (cells with two different nuclei with a single cell) (1, 33).

Such properties are conductive to genetic recombination (or meiotic crossing-over) and complementation studies (32). Both recombination and complementation are of great value for the construction of genetic maps. Recombination and the mechanism by which it is mediated is of additional significance since certain mutagens (e.g., those causing additions and deletions) seem to function at this level.

Complementation involves the establishment of two "genomes" within a single cell (this can be done via "mixed infections" as in phages or by fusing two nuclei into a heterokaryon or by creating a diploid organism from two haploid cells) in which both "genomes" are of different genetic constitution (though phenotypically they may be and often are identical). To understand complementation let us envisage an organism with two homologous chromosomes both containing a defective *A* gene. Let us assume that in one of the two chromosomes the mutated site is in the first part of the *A* gene, which we will arbitrarily call cistron *a*, while in the homologous chromosome the mutation is located in the last part of *A*, which we will designate *b* cistron. Since each contains one normal cistron within *A*, both mutant chromosomes are capable of coding for a normal polypeptide: a normal *a* from one chromosome and a normal *b* from the other. Under these circumstances the functioning of gene *A* is restored and the process of restoration is known as complementation.

Had the mutation been in the same cistron of both *A* genes (transconfiguration), complementation would not have been observed since one of the two normal products (a polypeptide from either *a* or *b*) would still have been missing.

Complementation tests can be carried out in different ways. With bacteriophages it involves a mixed infection of a bacterium by phages of two different genotypes. In yeast, it might involve the mating of two haploid cells to produce a single diploid cell or, as in the case of *Neurospora crassa*, the fusing of two cells to form a single heterokaryon. In any case, the genetic products from both genotypes mix within a given cell so that each plays its part in restoring normal gene function. A great variety of so-called tester strains of *N. crassa* exist which can be used for rapidly locating the altered site of a given mutant (34).

B. The *ad-3* Region

Because of their convenience, forward mutations within the *ad-3* region (35) are the most frequently studied in both yeast and *Neurospora*. Mutations within this region lead to defects in the synthesis of adenine resulting in the production of a red or reddish-purple pigment. Hence, mutants of the *ad-3* region can be identified and isolated from the midst of perhaps thousands of other normal colonies (that is, white colonies, normal insofar as the *ad-3* region is concerned).

As in bacteria, these mutants can be characterized in great detail on the basis of their reversion rates (back mutations) and the nature of the particular chemical mutagens which induce their reversion. Transitions of both AT to GC and GC to AT types, transversions, additions, and deletions, as well as nonrevertible mutants, have thus been characterized in *N. crassa* (36, 37).

The *Neurospora* system offers a special advantage in that techniques have been developed which permit the unambiguous determination of relative forward frequencies (33). Usually attempts to do this in haploid organisms are frustrated by the fact that any apparent differences in forward mutation rates at different loci might be accounted for by differing viable/lethal ratios. DeSerres and Osterbund (33), however, have developed a system by which they can recover mutants of the lethal type by forming balanced heterokaryons between them and other strains with the appropriate genetic markers.

Their procedure involves producing a heterokaryon between an *ad-3* wild-type and a double mutant (in *ad-3A* and *ad-3B*—these two genes have been shown to be separated by a region known as χ which consists of approximately sixteen genes). The heterokaryon is then treated with the chemical or physical mutagen of interest, and the cells are plated on minimal medium. After a period of growth, mutant colonies (red) are isolated and tested against known strains by complementation. In this way it can be determined whether the mutation occurred in the *ad-3A* or *ad-3B* or both. It is also possible to determine the ratio of viable to lethal homokaryotic mutations among these three different mutant types.

In the case of x-irradiated cells, it was found that 76% of all mutations in *ad-3A*, *ad-3B*, and *ad-3A ad-3B* were of the recessive (homokaryotic) lethal type. Of the *ad-3A ad-3B* double mutants, all were found to be lethal. This, as it turns out, is in good agreement with observations made on other test systems such as mice and *Drosophila*. Since a large proportion of the forward mutations induced within the *ad-3* region are lethal in the haploid-homokaryotic cell, it is evident that accurate estimates of forward-mutation rates depend upon recovering the lethals by methods such as the one outlined above.

V. *Drosophila*

Unlike the other test systems referred to heretofore, *Drosophila* (a fruit fly) is composed of specialized tissues and organs and reproduces by sexual means only. It is, therefore, very much more akin to man than anything we have dealt with in previous sections of this chapter, though there is some question whether those features offer any real advantage. The existence of eyes, legs, a nervous system, etc., do not guarantee that the all-important metabolic events occurring within *Drosophila* simulate those in man. Nor does it suggest that the repair of genetic damage is likely to be handled in a similar way as in mammals. Nevertheless, it must be admitted that *Drosophila* was the first organism from which mutants were isolated (38) and was the test organism used to demonstrate, for the first time, the mutagenicity of certain chemicals (39). It is still widely used for it does offer many unique and virtually unique advantages which will be discussed in turn.

A. Recessive Lethals

1. *Sex-Linked Recessive Lethals*

a. Advantages. As with all higher organisms (including man, of course) the genes of *Drosophila* are organized into multiple chromosomes which are visible under the light microscope at certain stages of the cell cycle (see Section VI). *Drosophila*, like man and other mammals, is a diploid organism; however, it has the advantage of containing only four pairs of chromosomes, one set (haploid set) having been derived paternally, the other, maternally. Again, like man, the female fly is determined by a double set (diploid) of X chromosomes (XX) while the male is determined by an XY constitution. That is, the male is actually haploid in X in the sense that it has no homologous X chromosome with which to pair but instead carries a Y chromosome which appears to be (in *Drosophila*, at least) almost exclusively concerned with the production and maturation of sperm (40). Hence, as with heterokaryons of *Neurospora*, recessive lethals of the X chromosome can be recovered (carried by the female) and subsequently "expressed" by male progeny (actually by the lack of male progeny).

There are many advantages to the use of recessive lethal mutations over those of the visible type. First, there is less bias or virtually no bias in scoring lethal mutations while scoring visible mutants in higher organisms is a subjective matter. Furthermore, there are many more genetic loci which could give rise to a lethal mutation than to a visible mutation (it has been estimated that there are approximately 1000 lethal genes within the X chromosome alone) (1).

The fact that the lethals are a heterogeneous lot is an advantage of a kind when we consider what has been learned from the *rII* phage system (Section II). We refer here to the fact that certain sites within the *rII* gene have been shown to be tens, if not hundreds, of times more mutable than other sites, which raises the question whether certain genes are not far more mutable than other genes. Since the recessive lethals represent a large and heterogeneous assortment of genes, any specific differences existing between individual genes relative to their specific mutability is likely to average out.

Also to the credit of the recessive lethals in *Drosophila* is that they can be unambiguously isolated and the nature of the mutation can be investigated microscopically by examining the giant chromosomes of its salivary gland (41). The examination can reveal whether the new mutant arose from a point mutation (small deficiency), large deletion, or a chromosome rearrangement of some type (42) (see Section V, D).

It would be remiss not to mention that because of the vast amount of genetic work which has already gone into the *Drosophila* system (particularly the species *melanogaster*) and the great number of genetically defined stocks available (43, 44), this system holds advantages offered by no other.

b. Method for Scoring Mutations. Basically, scoring recessive lethals is simple, at least in terms of a general concept. Wild-type male flies are treated with a chemical mutagen (either by feeding, injection, or via an aerosolized spray) then subsequently mated to untreated wild-type females. Were an X chromosome of one of the male gametes (sperms) hit by a recessive lethal mutation, the female descendants would then carry the mutation in a recessive state (unexpressed). This mutant could then be detected in the next generation by crossing the mutant female (X′X) with a wild-type male (XY).

The progeny of this mating would, on the average, be distributed among X′X, XX, X′Y, XY daughters and sons in ratios of 1:1:1:1. However, the mutation in the female is only lethal in the homozygous state (X′X′) while in the male (X′Y) it is always lethal since the male carries only a single X chromosome. Hence, no viable X′Y males are seen. Consequently, on the average, two viable females are born for every viable male.

One difficulty associated with this test is that a large number of progeny must be obtained before the statistical significance of an altered sex ratio (males to females) can be established. This difficulty can be overcome by making use of a tester chromosome which is denotated as X* in Fig. 5.5. Specifically, X* is the "Muller-5" chromosome which is a recessive visible mutant, the nature of which concerns the eyes of the mutant (either in the homozygous female of X*Y male) which are "bar"-shaped and apricot in color (45).

The homozygous "Muller-5" female is mated with mutagen-treated males and the daughters of this mating are tested by crossing them individually with their brothers. Should the daughters contain induced, sex-linked recessive lethals, their progeny will be distributed equally among recessive lethal-bearing females with wild-type eyes (X′X*), nonrecessive lethal-bearing females with mutant eyes (X*X*), and males with mutant eyes (X*Y). No males with wild-type eyes will be seen because these had all carried sex-linked recessive lethals (X′Y) which are nonviable in the male. It should also be noted from Fig. 5.5 that lethal-bearing females (granddaughters of treated males) are recognized by virtue of their wild-type eyes, and hence chromosome examination of salivary gland cells is not limited to daughters of treated males but may involve granddaughters as well.

It should be mentioned that, in addition to the visible marker (bar, apricot eyes), the "Muller-5" chromosome carries an inversion which prevents crossing-over of the treated X chromosome with that of the marker (X*). (Crossing-over is an event which occurs in meiosis of germ cells and involves the intermixing of paternal and maternal genetic information into a single

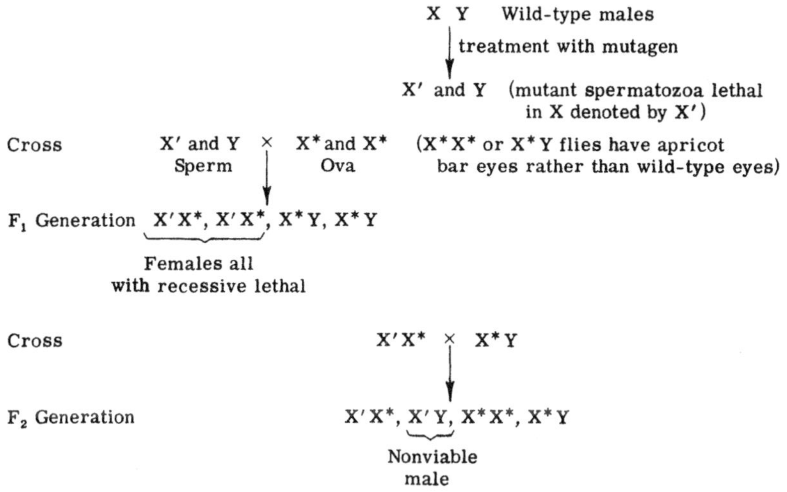

FIG. 5.5. Test for sex-linked recessive lethals in *Drosophila*. The daughters (F_1 generation) of treated wild-type males carry a chemically induced mutation in one of their two X chromosomes. These are detected by individually mating all daughters with X*Y males, and the mutants which produce only X*Y males are scored.

chromosome. It is very much a part of the processes which lead to a mature gamete but in *Drosophila* the male is exempt from crossing-over.) Were crossing-over to occur, wild-type males (grandsons of treated males) could be obtained even though the treated X chromosome had been mutated at a recessive lethal site. However, since crossing-over is prevented with the "Muller-5" chromosome, this problem is avoided and all of the induced mutants are detected.

Other tester chromosomes have been constructed such as the classic C1B which is described in most genetic textbooks. This method, however, has, in recent years, earned disfavor among geneticists, and its disadvantages have led to its nearly complete abandonment.

2. *Autosomal Recessive Lethals*

In addition to the XX of the female and the XY of the male, three other pairs of chromosomes exist in *Drosophila*, and these, because they are not associated with sexual differences, are known as the autosomes. Like sex chromosomes, they too carry recessive lethal genes, and, in fact, their number has been estimated at about 2000 (1).

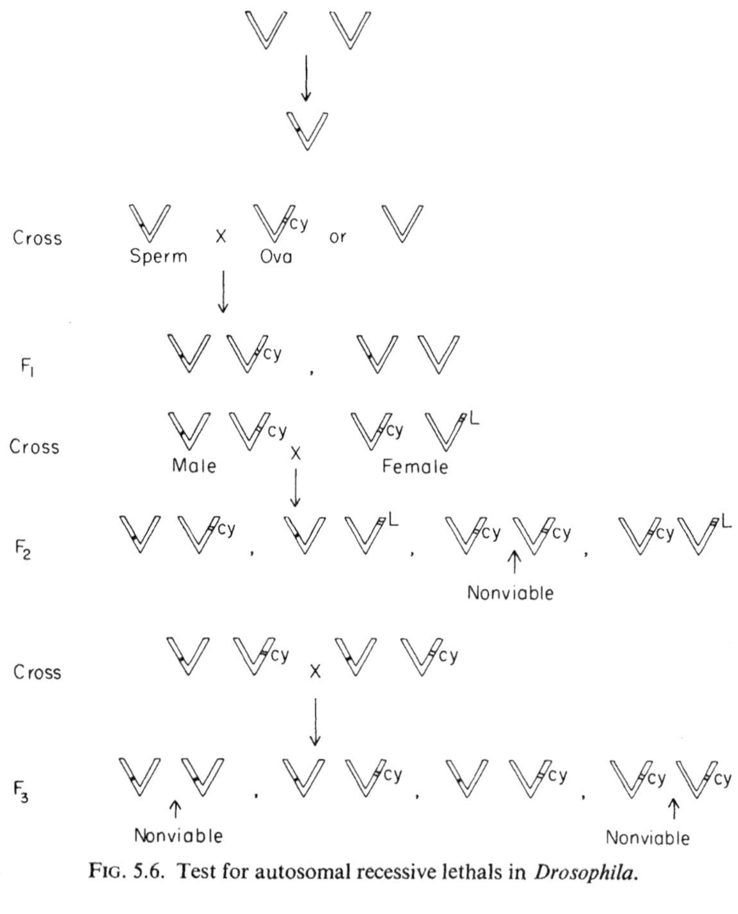

FIG. 5.6. Test for autosomal recessive lethals in *Drosophila*.

The main difference between this test and the one previously described is that the autosomal lethals must be present in a double dose (one on each homologous chromosome) to kill either sex. Hence, an additional generation of inbreeding is required to construct the necessary homozygous state. Obviously, this requires additional time and represents one disadvantage of the method as compared to its sex-linked counterpart. Nevertheless, such methods must be relied upon if information regarding the mutability of autosomal genes is sought.

As in tests for sex-linked lethals, marker genes are required for identification of mutant flies as are inversions to prevent crossing-over. It is necessary, however, that two different markers be used to help compensate for the lack of a sex difference.

A method which has proved extremely valuable is outlined in Fig. 5.6, and involves genes on chromosome II, including the dominant gene *Cy* (curly wings) of that chromosome which is known to be lethal in the homozygous state. Treated males are mated to females which are heterozygous for *Cy*. Curly-winged progeny are all heterozygous for a treated chromosome and the males of this generation are then selected and individually mated to females which are heterozygous for *Cy* as well as heterozygous for another dominant gene (*L*, for lobed eyes) which occurs on the opposite or homologous chromosome. The progeny of this second generation (F_2) consist of flies which are heterozygous for the treated chromosome (II) and those not containing a treated chromosome (II). The heterozygotes with treated chromosomes can be recognized by virtue of their curly wings and wild-type eyes. These are selected and inbred and, should the treated chromosome II contain a recessive lethal mutation, only curly-winged flies will be observed in a third generation since the progeny homozygous in treated chromosomes are nonviable.

B. Dominant Lethals

Technically, the dominant lethal test is by far the easiest to perform (46). Males are treated with a mutagen on test and, at appropriate times, mated to untreated females. From the frequency of unhatched eggs arising from such matings (as compared to the controls) a measure of the mutation frequency is obtained. Though the method is quick and easy to perform, the results obtained are plagued by ambiguities. In the first place, there are a variety of nongenetic causes which might prevent the egg from hatching, i.e., failure to fertilize, failure of syngamy, or damage to the centrosome which the sperm carries into the zygote cell. Hence, even a nonmutagenic compound could conceivably cause a high incidence of unhatched eggs though, of course, it would not be the result of a mutation.

Another drawback of the method is that the dominant lethal cannot be analyzed genetically. In fact, only indirect and imprecise methods are available for their analysis. Furthermore, it is believed that the vast majority of such mutants involve broken or fragmented chromosomes or lethal chromosomal rearrangements rather than point mutations (1). Hence, a mutagen which produces only point mutations would probably go undetected in the dominant lethal test.

C. Visible Mutations

As mentioned above, visible mutations, though of great importance to genetic research and necessary for the construction of special strains, are not very useful for quantitative mutation studies. Personal bias, infrequency of dominant visibles (which are in any case usually associated with chromosome rearrangements), and the need to engage in laborious inbreeding experiments in order to score recessive visible mutations on autosomal genes, are mainly responsible for their limited usefulness. In fact, only the sex-linked recessives can be considered a suitable class of mutations for the quantitative scoring of visible mutations.

The method used to detect sex-linked visibles can be applied to involve the treatment of either females or males. In the former case, sex-linked visibles appear in the sons. It is more convenient, however, to utilize treated males. In this case, normal wild-type males are exposed to a mutagen and then mated to attached X females, i.e., females with an XXY constitution (1, 43). In these females, the two X chromosomes are joined as a consequence of a rearrangement which occurred at some time in the history of the line. Since the gametes of the attached X female are either of XX or Y constitution, an ovum of the first kind gives rise to an attached female when fertilized by a Y-bearing sperm while the Y-bearing ovum gives rise to a male (XY) upon fertilization with a sperm of X constitution. Hence, we have a rather unique situation where the female contributes a Y chromosome and the male an X to their male progeny. The other two possible combinations of gametes yield inviable (YY) or abnormal and sterile (XXX) progeny. Obviously, any recessive visible mutations of the X chromosome will be manifested by F_1 sons of treated males mated with XXY females.

The detection of visible mutations within the first generation of male flies makes large-scale testing possible. Thousands of flies can be rapidly screened though the disadvantages associated with personal bias still exist and represent a serious drawback. The mutants, in general, tend to be less vigorous and are more likely to die or develop later than their wild-type counterparts. Hence, the frequency of visible mutants is likely to be significantly underestimated.

D. Chromosome Rearrangements and Other Effects

As mentioned previously, the salivary glands of *Drosophila* contain giant, or, as they are more often called, polytene, chromosomes. These are "huge" structures which contain hundreds of times more DNA than do the normal

Drosophila chromosomes found in most other tissue. The reason and cause for the existence of polytene chromosomes is not entirely understood.

However, without *Drosophila* and its salivary gland chromosomes the concepts and basic principles which have given rise to modern genetics would have been a much longer time in developing. Salivary gland chromosomes are not only giant in comparison to most other chromosomes but they offer the important advantage of exhibiting a banding pattern which marks and helps the cytologist locate well-defined genetic regions. (Polytene chromosomes are not limited to *Drosophila* nor limited to salivary gland tissue for larva and certain other somatic cells of the adult contain these as well.)

Salivary gland analysis can be used in three ways, each of which serve a somewhat different purpose. As mentioned earlier, sex-linked and autosomal recessive lethals as well as, of course, the sex-linked visibles can be unambiguously identified and the nature of their defect studied by salivary gland chromosome analysis. Generally, this involves classifying the mutant as either constituting a small deficiency or large deletion. The method can also be used to score an unselected sample of progeny from treated flies, except that in this case, only large and medium-sized rearrangements can be expected to be seen, since minute changes (say, a one-band deficiency in a whole chromosome set) would be missed in the screening of hundreds of flies.

A third approach is to use genetic methods for isolating the progeny of parents which have undergone large deletions or translocations (an interchange of segments between nonhomologous chromosomes which has the effect of unlinking certain genetic regions from others). In the case of large deletions, a method has been worked out which involves mating treated males to attached XX females which are homozygous for several sex-linked recessive visible mutations (47). If a sperm with a deleted X chromosome fertilizes an ovum with XX constitution the progeny will have certain of the XX marker genes (recessive visibles) covered by virtue of the X chromosome fragment contributed by the sperm. Hence, the development of females which are phenotypically wild-type for any one of the marker genes used indicates that the treated male was subjected to large deletions within its germinal cells.

VI. Cytological Examination of Chromosomes

As has been mentioned several times before in this review, higher plants and animals exhibit microscopically visible structures known as mitotic chromosomes during the so-called mitotic phase of their cell cycle (Fig. 5.7).

The mitotic phase follows the S period which is characterized as the time during which replication of genetic information (DNA synthesis) occurs and a period known as G_2 wherein the newly replicated DNA is packaged into daughter chromosomes and the apparatus for mitosis is synthesized.

Mitotic chromosomes begin to appear in the initial phase of mitosis known as prophase, and by the metaphase well-defined structures can be seen which are composed of two chromatids (daughter chromosomes) joined at a unit called the centromere (48). It is the vital purpose of mitosis to see to it that the chromatids (daughter chromosomes) are distributed equally among the two

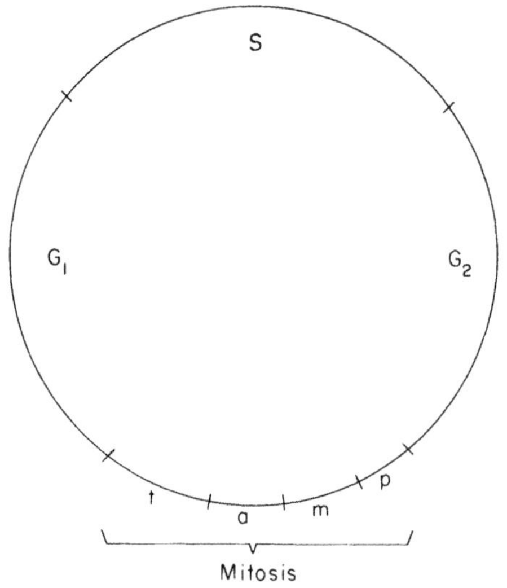

FIG. 5.7. Cell cycle. S phase represents the period of DNA synthesis. Mitosis is divided into prophase (p), metaphase (m), anaphase (a), and telophase (t). G_1 and G_2 may be of varying lengths depending upon the cell and the physiological conditions.

daughter cells which are to arise following telophase (the last step of mitosis). In this way, the daughter cells acquire all the genetic information originally present in the parent. In fact, the entire process of G_2 and mitosis can be looked upon as the cell's way of wrapping up its genetic information into separate packages (chromatids joined by a centromere) which are then distributed equally (by the mitotic spindle apparatus and the division of the centromere) among the two cells which arise from cell fission or cleavage.

Different organisms display different chromosome patterns at metaphase wherein the number, size, and configuration of chromosomes (collectively known as the karyotype) vary from species to species (49–52). For instance, the karyotype of man (Fig. 5.8) is instantly recognizable from that of the

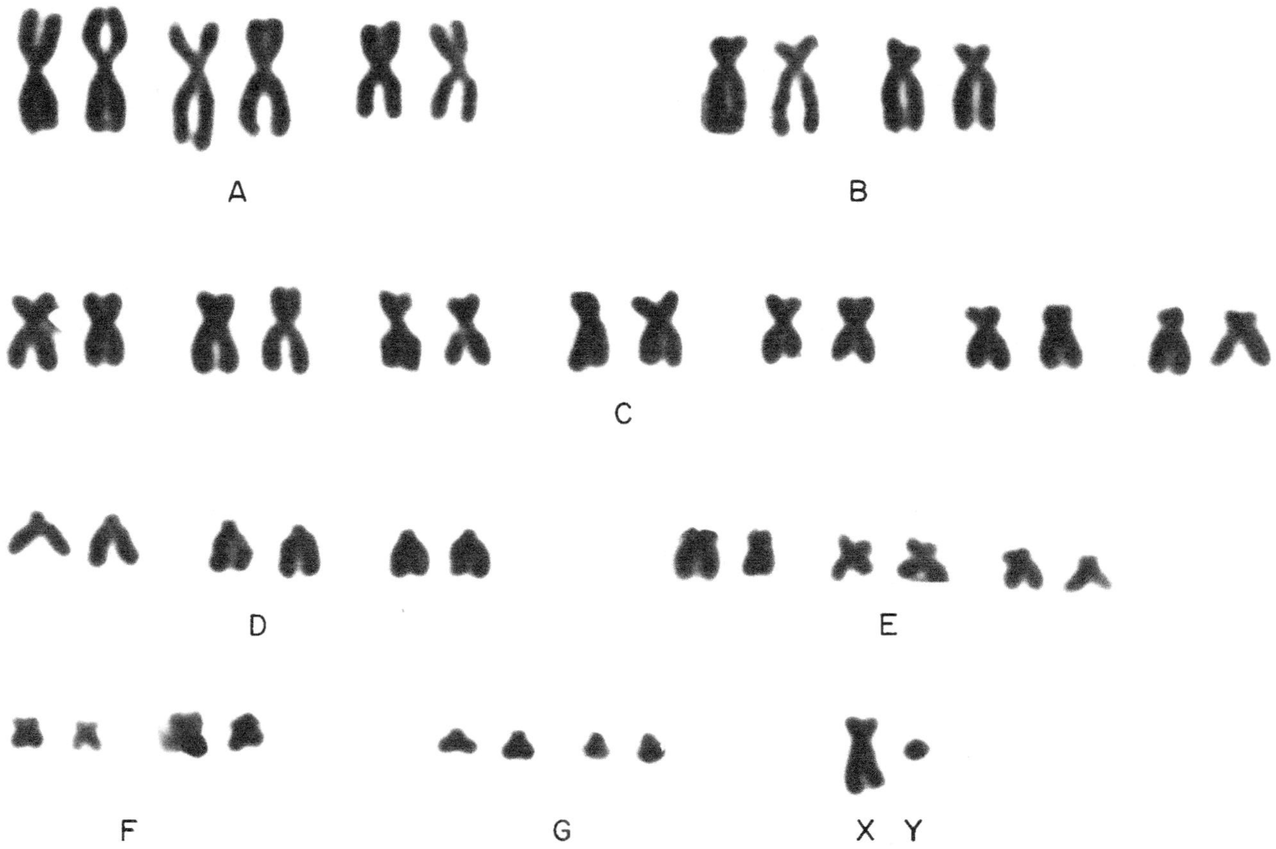

FIG. 5.8. Human chromosomes: diploid normal male. Unpublished photograph provided by M. R. Banerjee (University of Nebraska).

mouse. Since many, if not all, of the metaphase chromosomes of a suitable plant or animal system are individually identifiable, translocations can be recognized as well as a variety of chromosomal breaks and other kinds of rearrangements which presumably are the consequence of double breaks.

It should be realized, however, that certain aberrations give rise to no detectable change in chromosome morphology, e.g., paracentric inversions (essentially equivalent to turning the chromosome upside-down) and other symmetrical interchanges. Furthermore, and it hardly needs stating, point mutation of all types go undetected by this method.

One advantage of the cytological approach is that inviable changes can be detected before their elimination and human cells (cultured leukocytes) can be used about as readily as those of any other organism. The disadvantage relates to ambiguities mentioned above and to the question as to which cytogenetic abnormalities actually constitute a mutation. Some authors are inclined to classify all chromosome breaks as mutations (or to consider all agents which break chromosomes as mutagenic). This surely is an erroneous oversimplification, since not every treatment that leads to visible chromosome breakage is mutagenic; virtually any toxic compound will eventually bring about chromosome fragmentation, even if only in dead and dying cells. There seems, in fact, little doubt that this is a secondary reaction resulting from the activation of certain nucleolytic enzymes (see Chap. 3, Section IV, A) and its consequence is always lethal and cannot, therefore, be considered mutagenic.

It is, however, equally certain that agents which induce typical chromosome rearrangements such as translocations should be called mutagens since these rearrangements will be transmitted from cell to cell as true changes of the genetic material. Should these changes occur within the germ cells, they can be detected by genetic methods.

VII. Mammalian Test Systems

A. Mammalian Somatic Cells Grown in Tissue Culture

One of the great difficulties accompanying the use of mammalian organisms as mutagenic test systems (or for that matter, nearly any higher organism) relates to the obvious fact that it is far easier to handle hundreds of thousands of microscopic organisms than it is to deal with hundreds of thousands of mice or other mammals. Another very great difficulty concerns their long gestation periods which are measured in weeks or months as opposed to minutes or hours in bacteria, yeast, and other fungi. But, like the other eukaryotes discussed, they provide the advantage of being more "like man" than bacteria and bacteriophage.

In order to derive the best of all test systems (that is, to have organisms closely related to man but micro in size and rapidly dividing), attention is turning to mammalian tissue culture cells, which can be grown as single cells in defined media in much the same way as bacterial cells.

Not surprisingly, the work involved is greater, the cells are more difficult to grow (though advances over the past 10 years have solved many of the original problems), and the genome is much larger (1000-fold) than what one finds in bacteria. Nevertheless, a number of mutant mammalian cell lines (either induced or spontaneous) have been isolated over the past decade.

Only recently have attempts been made to quantitate the frequencies of induced and spontaneous gene mutation in tissue cultures of mammalian cells. In this work (53) a Chinese hamster cell line and two different loci within the hamster genome were used. Reverse mutations from L-glutamine dependence (gln^-) to L-glutamine independence (gln^+) and forward mutations from 8-azaguanine sensitive (azg^s) to 8-azaguanine resistance (azg^r) were studied along with reverse mutations from azg^r to azg^s.

Their method involved treating cultured cells for 2 hours with a chemical mutagen (either ethyl methanesulfonate, methyl methanesulfonate, or *N*-methyl-*N'*-nitro-*N*-nitrosoguanidine), after which time the cells were washed to remove all traces of the mutagen and recultured in fresh medium. After various times the appropriate selective chemical was added to the culture medium so that all future growth would only relate to the mutant cells. For instance, in the case of the forward mutation from azg^s to azg^r, only the azg^r mutants would grow following the addition of 8-azaguanine so that afterward when the cells were plated and grew into visible colonies only those cells which underwent a mutation at the *azg* locus were counted. On the other hand, reversions from azg^r to azg^s were scored by adding the selective "THAG" medium (thymidine, hypoxanthine, aminopterin, and glycine) to treated cells prior to plating and counting. ("THAG" medium prevents the growth of azg^r cells so that in this case all colonies were of the azg^s cell type.)

The effects of certain variables on the apparent mutation frequency were studied in detail, i.e., cell concentration (inoculum size); time period between removing the mutagen and the addition of the selecting agent (mutation expression time); and the concentration of the selecting agent. With such information, the investigators were able to maximize the yield of mutants and hence, also, the sensitivity of their assay.

Indications are that the above study, which demonstrated induced forward mutation frequencies of 10^{-3} and reversion frequency of 10^{-5}, will be developed further and may very well help to reveal what agents signal mutagenic dangers to man. However, there are still many questions to be answered, particularly those concerned with how (and why) a diploid organism, such as the cells used here, can exhibit such high forward mutation frequencies. Perhaps part of the genome (in this case, the *azg*-containing portion of the hamster genome) exists in the haploid state but, even so, the mutation frequencies observed appear unexpectedly high. It seems likely, however, that additional work with other cell lines and other genetic loci will eventually bring about a clearer understanding of the situation.

There is hope too that such studies will lead to a deeper understanding of the functioning of the mammalian genome which will improve our position and ability to assess potential genetic hazards.

B. Mouse (Specific Locus Method)

The mouse (*Mus musculus*) is the only mammal that has been used for mutation studies, though, as was pointed out earlier, mammalian cells from many different species (including man) have been used in chromosome investigation and mammalian tissue culture cells appear to be generally applicable for the testing of mutagens.

The mouse, when used as a whole animal, lacks many of the advantages offered by *Drosophila*; for instance, one of its difficulties concerns the fact that the mouse genome is divided into twenty pairs of chromosomes compared to only four in *Drosophila*. Hence, less than 5% of all the genes in the mouse are contained within the X chromosome, whereas over 20% of *Drosophila's* genome is constituted by the X chromosome. Sex-linked recessive lethals are consequently less frequent and form a smaller proportion of the total mutations in the mouse than in *Drosophila*. Even more important, however, is the fact that crossing-over is very much more difficult to prevent in the mouse than in *Drosophila*. In the mouse, crossing-over occurs in both the male and the female. Furthermore, it is not possible to induce inversions which prevent crossing-over (hence, in the mouse, a newly mutated gene within a marker chromosome may cross-over with an unmarked homolog).

Without inversions and the means to prevent crossing-over, recessive lethal mutations can be studied only within small marked regions of the chromosome (closely linked markers) where crossing-over would be most improbable and infrequent. Such techniques have been developed both for sex-linked and autosomal genes; however, the results are unsatisfactory because the frequency of mutation is so low (the low frequency is the expected consequence of focusing on such a minute fraction of the total genome) (54).

The difficulty in using dominant or recessive visibles as tests is essentially the same in mice as it is in *Drosophila* (see Section V).

All of the above-mentioned difficulties are essentially circumvented by the specific locus test developed by Russell (54). In essence, the method is as follows: treated wild-type males or females are mated to partners which are homozygous for seven recessive genes; the progeny are then examined. The question is then asked: Do any of the progeny exhibit a mutant phenotype insofar as the recessive tester genes are concerned? If no mutations were induced, the recessive tester genes would all be heterozygous in the progeny and hence unexpressed. However, with a mutation at one of the tester sites, such progeny would be homozygous for this particular recessive and hence the mutation would be phenotypically evident. The gene loci most frequently used are concerned with coat color and ear and tail morphology.

Though the method is difficult, requiring large numbers (thousands) of inbred and genetically well-defined mice, it nevertheless is explicit and has contributed most significantly to our understanding and appreciation of radiation hazards and the genetic risk of radioactive fallout to mammalian organisms (55, 56).

C. Dominant Lethals in the Mouse

The primary advantage of the dominant lethal test in the mouse over the specific locus method is that fewer animals are required to obtain a statistically significant result (57, 58). On the other hand, the main disadvantage of the method relates to its interpretation and relevance to mutagenesis (57). For instance, it seems likely that this method would be relatively insensitive in assessing the mutagenicity of agents which are only able to act by inducing point mutations (1, 57). In fact, there is, at present, no compelling reason to believe that dominant lethals in the mouse are not (in general) mediated by the same effects which cause lethality in *Drosophila* (1) (i.e., via chromosome breaks which eventually lead to the elimination of a large part of the genome and the death of the animal).

However, in favor of this system it can be said that (when dominant lethal testing involves scoring dead embryos at advanced stages of gestation) scoring is more accurate in mice than in *Drosophila* simply because only fertilized ova are considered in the former case. The method also has certain advantages over "pure" chromosome analysis (Section VI) in that it assesses chromosome damage to germ cells per se and might also provide information relative to precursors of mutagens as well as mutagens.

If a chemical were metabolized, say in the liver, to a mutagenic substance which then attacks the germ cells causing chromosome breaks, such events might well be detected as dominant lethals but would probably go undetected in tissue culture when only chromosome analysis is used. On the other hand, were a chromosome-breaking agent rapidly metabolized or excreted by the body, the frequency of dominant lethals would be minimized whereas this would not necessarily be the case for tissue culture tests of chromosome breaks.

The dominant lethal test usually involves injecting or treating male mice with a suspected mutagen and immediately mating them to one or two virgin females. (Females are never treated for fear that treatment might disturb fertility.) Females are then replaced, either at weekly intervals or some more convenient time and mating is continued with the same groups of treated males. In this way, chromosome-breaking effects can be monitored at all stages of spermatogenesis and it can be established which stage is the most sensitive or susceptible to a given agent. In this connection Ehling *et al.* (57) have investigated the induction of dominant lethals in mice by ethyl methanesulfonate and methyl methanesulfonate. Their evidence indicated that mature sperm and highly developed (late) spermatids are more susceptible to induction of dominant lethals than are developing spermatids. The explanation, they reason, may well relate to one of the molecular changes observed during sperm development. For instance, it might correlate with the exchange of lysine-rich histones for arginine-rich histones which occurs during spermatid maturation.

The actual scoring of dominant lethal mutations in mice involves opening the pregnant female at some advance stage of gestation (2 weeks following conception) and counting the number of live and dead embryos. Hence, only fertilized embryos are considered, which represents an improvement over the *Drosophila* system in which all unhatched eggs are scored. There are, however, various ways of expressing the incidence of dominant lethals. Epstein (59) uses a mutation index which is obtained by dividing the number of dead embryos by the total number of ova fertilized, and multiplying this result by a hundred, while Ehling *et al.* (57) express their data as a frequency of dominant lethal mutations given by

$$100 - \left(\frac{\text{No. embryos (experimental group) per female}}{\text{No. embryos (control group) per female}}\right) \times 100$$

In any case, it must be remembered that we are dealing with a system without progeny and hence genetic tests cannot be carried out, nor can one deduce much about the nature of the effect, since reversions of dominant lethals are not possible or even conceivable. It should also be emphasized that a high incidence of dominant lethals, in response to the treatment of males with a particular agent, does not prove the mutagenicity of the agent since the effect produced may conceivably always result in lethality, never in mutation. Many mutagens will, of course, break chromosomes and presumably give rise to dominant lethals but, as was mentioned earlier, chromosome breaking alone is insufficient to establish mutagenicity.

Based on the results of a comparative study in which methyl methanesulfonate, ethyl methanesulfonate, *N*-methyl-*N'*-nitro-*N*-nitrosoguanidine, and ICR-170 {2-methoxy-6-chloro-9-[3-(ethyl-2-chloroethyl)aminopropylamino]acridine dihydrochloride} were tested, it was concluded by Ehling *et al.* (57) that cross-linkage or depurination of DNA or both events are important to the induction of dominant lethal mutations. And it should be recalled from Chap. 3 that such events lead to single- and double-stranded breaks in DNA which might then be expressed as chromosome or chromatid breaks.

D. Host-Mediated Assay

We have saved for last our discussion of the host-mediated assay, not just because its development is of the most recent vintage, but also because it combines into a single system many of the advantages offered by microbial and mammalian organisms. The purpose of the test is to permit the detection of point mutations which might arise from the action of drugs or other agents and their metabolic products. The host (a mammal) supplies the metabolic activity, i.e., reductions, hydroxylations, etc., and the bacterium (or other microbe) which is injected into the host's peritoneum, provides the genome to be analyzed by genetic methods referred to earlier (Section III).

The method as devised by Legator (60–62) involves the intraperitoneal injection of *Salmonella typhimurium* into a suitable mammalian host. First, however, a suspected mutagen is administered to the host by virtually any route other than the intraperitoneal one so that the animal can activate or detoxify the potential mutagen before it encounters the microbial organisms in the peritoneum. The bacteria used are auxotrophic for histidine (histidine-dependent *Salmonella typhimurium*) and back mutations at this locus are scored 24 hours after mutagen treatment in the manner described (see Section III).

Like all of the methods discussed, it has certain shortcomings, though its disadvantages should not overshadow its value. One obvious shortcoming is the fact that the metabolic activity of the bacterium continues during the test period and this may either inactivate mammalian metabolites or convert them to mutagenic substances. It can also be argued that the effects and consequences of the processes known collectively as repair are likely to be significantly different in microbes as compared to mammals. For example, one might envisage a situation in which a particular type of lesion, induced in the DNA, is lethal and never mutagenic in the bacterium, whereas in the mammalian organism, partial repair of the lesion occurs so that it is no longer lethal but, instead, mutagenic, or the reverse might be true. Or it is conceivable that in one or the other organism an induced mutation arises from faulty repair. In any case, mutations registered in the bacterium will not always apply to the host though these disadvantages should not detract from the fact that the development of the host-mediated assay is a most welcome event which should permit a much better assessment of the genetic hazard due to certain environmental agents. Malling has recently developed a modification of the assay which allows the introduction of microorganisms to specific sites in the mammalian organism, i.e., the liver, kidney, testes, peritoneum, etc., to allow more precise evaluation of persistence, localization, retention, and metabolism of the potential mutagen.

Perhaps, were it possible, and it may be, the host-mediated assay might be improved upon by the use of cultured mammalian cells, injecting these into the host's peritoneum instead of bacteria. For instance, the hamster cell system developed by Chu and Malling (Section VII, A) or a similar system might be used in a hamster host and, after an appropriate treatment with suspected mutagen, removed and tested in the manner already described (Section VII, A).

REFERENCES

1. C. Auerbach, "Mutation: An Introduction to Research on Mutagenesis," Part I. Methods, p. 75. Oliver & Boyd, Edinburgh and London, 1962.
2. O. T. Avery, C. M. MacLeod, and M. McCarty, *J. Exptl. Med.* **79**, 137 (1944); J. Spizizen, *Proc. Natl. Acad. Sci. U.S.* **43**, 694 (1957).
3. E. Freese and E. B. Freese, *Radiation Res. U.S.* Suppl. 6, 97 (1966).
4. C. Anagnostopoulos and T. P. Crawford, *Proc. Natl. Acad. Sci. U.S.* **47**, 378 (1961); E. Nester and J. Lederberg, *ibid.* p. 52.
5. E. Freese and H. B. Stack, *Proc. Natl. Acad. Sci. U.S.* **48**, 1796 (1962).
6. J. Marmur, *J. Mol. Biol.* **3**, 208 (1961).
7. Sr. V. M. Maher, E. Miller, J. Miller, and W. Szybalski, *Mol. Pharmacol.* **4**, 411 (1968).
8. E. E. Horn and R. M. Herriot, *Proc. Natl. Acad. Sci. U.S.* **48**, 1409 (1962).
9. L. E. Orgel, *Advan. Enzymol.* **27**, 289 (1965).
10. R. L. Sinsheimer, *Science* **120**, 551 (1954); E. Volkin, *J. Am. Chem. Soc.* **76**, 5892 (1954)
11. S. Benzer, *Proc. Natl. Acad. Sci. U.S.* **41**, 344 (1955).
12. S. Benzer, *Proc. Natl. Acad. Sci. U.S.* **45**, 1607 (1959).
13. S. Benzer, *Proc. Natl. Acad. Sci. U.S.* **47**, 403 (1961).
14. S. Benzer and S. P. Champe, *Proc. Natl. Acad. Sci. U.S.* **47**, 1025 (1961).
15. C. Thomas, Jr., *Progr. Nucleic Acid Res. Mol. Biol.* **5**, 315 (1966).
16. S. Benzer and E. Freese, *Proc. Natl. Acad. Sci. U.S.* **44**, 112 (1958).
17. S. Brenner, L. Barnett, F. H. C. Crick, and A. Orgel, *J. Mol. Biol.* **3**, 121 (1961).
18. E. Volkin, *Progr. Nucleic Acid Res. Mol. Biol.* **4**, 51 (1965).
19. G. S. Stent, "Molecular Biology of Bacterial Viruses." Freeman, San Francisco, California, 1963.
20. J. L. Melnick, *in* "Viruses Inducing Cancer" (W. J. Burdette, ed.), p. 177. Univ. of Utah Press, Salt Lake City, Utah, 1966.

21. V. N. Iyer and W. Szybalski, *Proc. Natl. Acad. Sci. U.S.* **50**, 355 (1963).
22. F. H. C. Crick, *Progr. Nucleic Acid Res.* **1**, 164 (1963).
23. D. H. Bergey. "Bergey's Manual of Derminature Bacteriology," 6th ed. Williams & Wilkins, Baltimore, 1948.
24. N. Bauman and B. D. Davis, *Science* **126**, 170 (1957).
25. A. Sadgopol, *Advan. Genet.* **14**, 326 (1968).
26. A. R. Peacocke and R. B. Drysdale, "The Molecular Basis of Heredity." Butterworth, London and Washington, D.C., 1967.
27. J. Lederberg, *Methods Med. Res.* **3**, 5 (1950).
28. P. Howard-Flanders, *Ann. Rev. Biochem.* **2**, 666 (1968).
29. G. W. Grigg, *Mutation Res.* **4**, 553 (1967).
30. H. D. Barnes and S. S. Cohen, *J. Bacteriol.* **68**, 80 (1954).
31. G. W. Beadle and E. L. Tatum, *Proc. Natl. Acad. Sci. U.S.* **47**, 499 (1941).
32. D. Michie, *in* "Introduction to Molecular Biology" (G. H. Haggis, ed.), p. 193. Longmans, Green, New York, 1964.
33. F. J. DeSerres and R. S. Osterbund, *Genetics* **47**, 793 (1962).
34. F. J. DeSerres, *Mutation Res.* **8**, 43 (1969).
35. F. J. DeSerres and H. G. Kølmark, *Nature* **182**, 1249 (1958).
36. H. V. Malling, *Mutation Res.* **4**, 559 (1967).
37. H. V. Malling, *Mutation Res.* **4**, 265 (1967).
38. H. J. Muller, *Genetics* **13**, 279 (1928).
39. C. Auerbach, J. M. Robson, and J. C. Carr, *Science* **105**, 243 (1947).
40. W. Hennig, *Chromosoma* **22**, 294 (1967).
41. B. P. Kaufmann, *Radiation Biol.* **1**, Part 2, 627 (1954).
42. J. K. Lim and L. A. Snyder, *Mutation Res.* **6**, 129 (1968).
43. M. Demerec and B. P. Kaufmann, "Drosophila Guide," 7th ed. Carnegie Inst. Washington, Washington, D.C., 1961.
44. H. J. Muller and I. I. Oster, *in* "Methodology in Basic Genetics" (W. J. Burdette, ed.), p. 249. Holden-Day, San Francisco, California, 1964.
45. W. P. Spencer and C. Stern, *Genetics* **33**, 43 (1948).
46. Ø. Strømnaes, *Genetics* **34**, 462 (1949).
47. E. B. Lewis, *Am. Naturalist* **88**, 225 (1954).
48. K. B. Roberts, *in* "Introduction to Molecular Biology" (G. Haggis, ed.), p. 1. Longmans, Green, New York, 1964.
49. T. C. Hsu, W. Schmid, and E. Stubblefield, *in* "The Role of Chromosomes in Development" (M. Locke, ed.), p. 83. Academic Press, New York, 1964.
50. S. Ohno and B. M. Cattanach, *Cytogenetics* (*Basel*) **1**, 129 (1962).
51. H. Harris, J. F. Watkins, G. Cambell, E. P. Evans, and C. E. Ford, *Nature* **207**, 606 (1965).
52. E. Stubblefield, *J. Natl. Cancer Inst.* **37**, 799 (1966).
53. E. H. Y. Chu and H. V. Malling, *Proc. Natl. Acad. Sci. U.S.* **61**, 1306 (1968).
54. W. L. Russell, *Cold Spring Harbor Symp. Quant. Biol.* **16**, 327 (1951).
55. W. L. Russell, *Pediatrics* **41**, 223 (1968).
56. W. L. Russell, *Radiation Biol.* **1**, Part 2, 825 (1954).
57. U. H. Ehling, R. B. Cumming, and H. V. Malling, *Mutation Res.* **5**, 417 (1968).
58. A. J. Bateman, *Nature* **210**, 205 (1966).
59. S. S. Epstein and H. Shafner, *Nature* **219**, 385 (1968).
60. M. G. Gabridge and M. S. Legator, *Proc. Soc. Exptl. Biol. Med.* **130**, 831 (1969).
61. M. G. Gabridge, E. J. Oswald, and M. S. Legator, *Mutation Res.* **7**, 117 (1969).
62. M. G. Gabridge, A. Denunzio, and M. S. Legator, *Nature* **221**, 68 (1969).

CHAPTER 6

Tabular Summaries of Chemical Mutagens

This chapter consists of tabular condensations of the chemical mutagens arranged in terms of chemical or common name, Chemical Abstracts' registry number (where known), structure, and literature citations regarding point mutation in transforming-DNA, phages, other microorganisms or insects, and chromosome aberrations in plant, insect, mammal, or mammalian cell.

The compounds are arranged in the order in which they will appear in the subsequent three chapters and are divided into twelve tabular headings:

1. Aziridines and triazines
2. Mustards (nitrogen, sulfur, and oxygen) and related derivatives
3. Nitrosamines, nitrosamides, and related derivatives
4. Epoxides, aldehydes, lactones, and related derivatives
5. Alkyl sulfates and alkane sulfonic esters
6. Acridines and acridinium salts
7. Antibiotics
8. Miscellaneous drugs and food additives (and related degradation products
9. Pesticides
10. Nitrogen derivatives (hydrazines, hydroxylamines, and *N*-hydroxy derivatives)
11. Hydrogen peroxide and organic peroxides
12. Miscellaneous mutagens

References in italic type in parentheses relate to a nonmutagenic literature citation; all other references shown are positive.

TABLE 6.1

Aziridines and Triazines

Chemical or common name (C.A. registry number)	Structure	Point mutation: Phage	Point mutation: Other micro-organisms	Point mutation: Insects	Chromosome aberrations: Plant	Chromosome aberrations: Insect	Chromosome aberrations: Mammalian cells, mammals
Ethylenimine (151564)	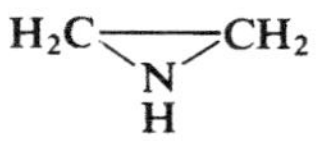		7, 8	1–6	9, 10		9, 10
Triethylenemelamine TEM, tretamine (51183)	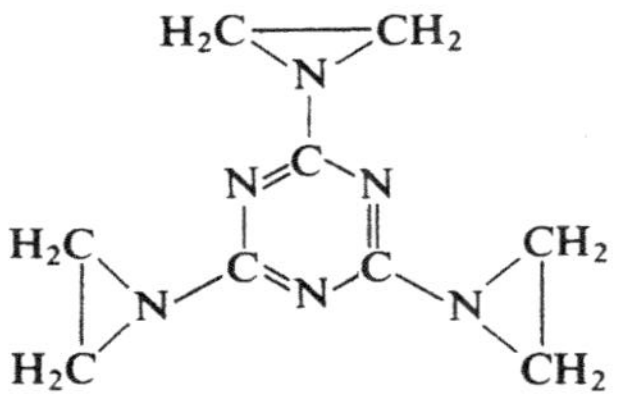		7	6, 18–25	31–36	18, 22	11–17, 26–30
2,3,5-Trisethylenimino-1,4-benzoquinone Trenimon (68768)	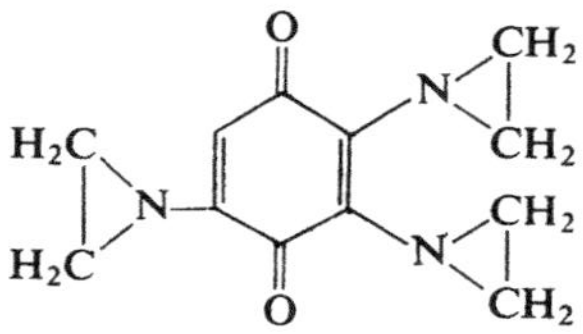			2	31		11–13, 37, 38

TABLE 6.1 (cont.)

Chemical or common name (C.A. registry number)	Structure	Point mutation: Phage	Point mutation: Other micro-organisms	Point mutation: Insects	Chromosome aberrations: Plant	Chromosome aberrations: Insect	Chromosome aberrations: Mammalian cells, mammals
Tris(1-aziridinyl)phosphine oxide TEPA, APO, aphoxide (545551)	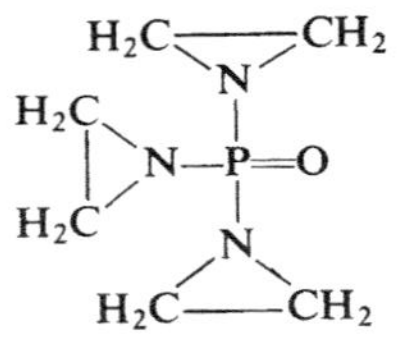	42	40, 41	39			15
Tris[1-(2-methyl aziridinyl)-phosphine oxide] Metepa, mapo, methaphoxide (157396)	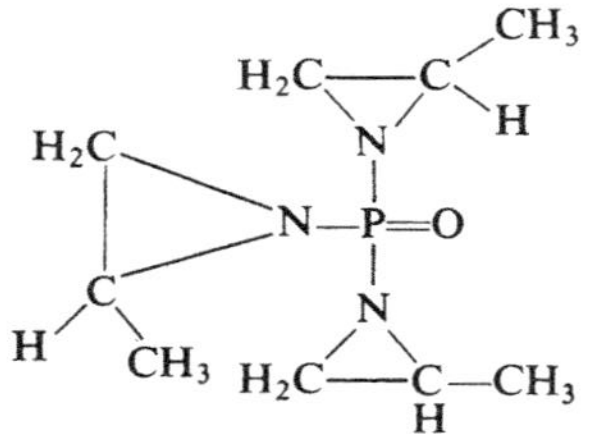	42	40		44		15, 43

2,2,4,4,6,6-Hexakis(1-aziridinyl)-2,2,4,4,6,6-hexahydro-1,3,5,2,4,6-triazatriphosphorine Apholate (52460)	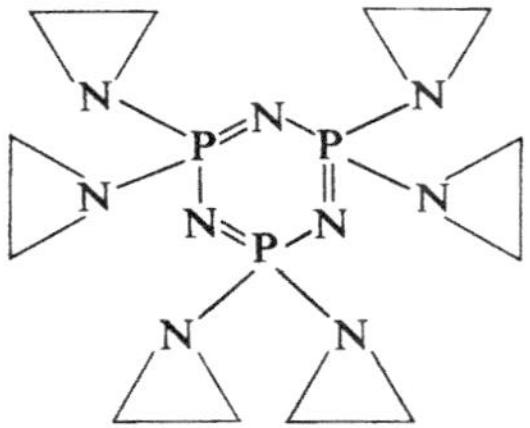	40	45	32	
Tris(1-aziridinyl)phosphine sulfide Triethylene thiophosphoramide, thio-TEPA (52244)	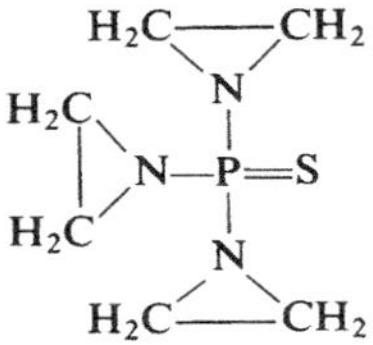		46		15, 47, 48
Trimethylolmelamine 2,4,6-Trioxymethylamino-1,3,5-triazine, TMM	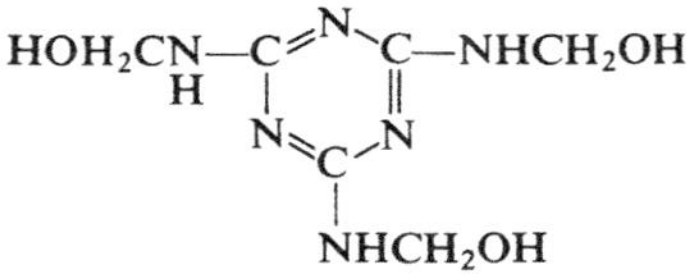		49, 50		

TABLE 6.2

Mustards (Nitrogen, Sulfur, and Oxygen) and Related Derivatives

		Point mutation			Chromosome aberrations		
Chemical or common name (C.A. registry number)	Structure	Phage	Other micro-organisms	Insects	Plant	Insect	Mammalian cells, mammals
2,2′-Dichloroethylamine Nor-HN2	$(ClCH_2CH_2)_2NH$		41, 51, (52)				
Tris(2-chloroethyl)amine HN3 (555771)	$(ClCH_2CH_2)_3N$			53–55			
Methyl di(2-chloroethyl)-amine Nitrogen mustard, HN_2, mechlorethamine (51752)	$(ClCH_2CH_2)_2NCH_3$	71, 72 (83, 84)	41, 51, 54, 55, 65–70, 73–82	57–64	10, 85–88		56
Methyl bis(2-chloroethyl)-amine *N*-oxide Nitromin, *N*-oxide mustard (126852)	$(ClCH_2CH_2)_2N{-}CH_3$ with ↓ O on N			53, 89			

Bis(2-chloroethyl)sulfide Sulfur mustard, mustard gas (505602)	$S(CH_2CH_2Cl)(CH_2CH_2Cl)$	101	51, 55, 98–100	54, 55, 90–97	102–105	94	
(2-Chloroethyl-2-hydroxyethyl) sulfide Half-sulfur mustard, HSM (542814)	$S(CH_2CH_2Cl)(CH_2CH_2OH)$			57			
2,2′-Dichloroethyl ether (111444)	$(ClCH_2CH_2)_2O$			96			
N,*N*-Bis(2-chloroethyl)-*N*,*O*-propylenephosphoric acid ester diamide Cyclophosphamide, endoxan (50180)	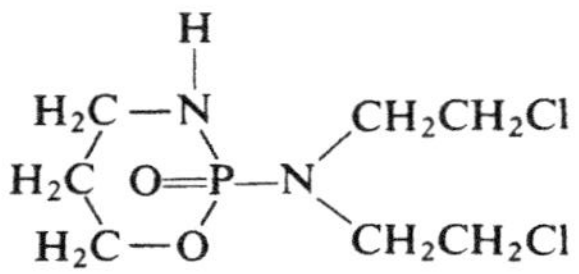			107–109	114		13, 106, 110–113

TABLE 6.3

Nitrosamines, Nitrosamides, and Related Derivatives

Chemical or common name (C.A. registry number)	Structure	Point mutation: Phage	Point mutation: Other micro-organisms	Point mutation: Insects	Chromosome aberrations: Plant	Chromosome aberrations: Insect	Chromosome aberrations: Mammalian cells, mammals
Dimethylnitrosamine (62759)	$(H_3C)_2N—NO$		(120–123), 124	115–119			
N-Methyl-*N*-nitrosourea	$CH_3N(NO)—C(=O)—NH_2$	131	122, 128–130	115, 117, 125–127	132–134		
N-Methyl-*N*-nitrosourethan MNU	$CH_3—N(NO)—C(=O)—OC_2H_5$		123, 129, 137–146	115, 135, 136,	147–150		
N-Methyl-*N*′-nitro-*N*-nitroso-guanidine MNNG	$CH_3—N(NO)—C(=NH)—N(H)—NO_2$	151	122, 128, 137, 139, 145, 152–163, 167	164, 165	166		

N-Methyl-*N*-nitroso-*p*-toluenesulfonamide Diazald, MNTS (80115)	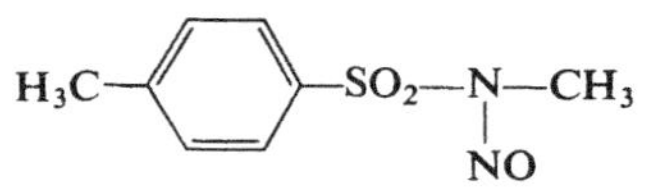	(*122*), 168		
1-Nitroso-2-imidazolidone-2 NIL	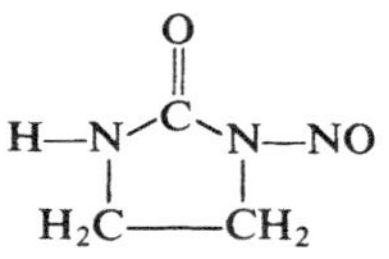	122, 123, 129, 139, 169		134
Diazomethane (334883)	$CH_2{=}\overset{\oplus}{N}{=}\overset{\ominus}{N}$	52, 122, 123, 135, 170, 171	135	
Cycasin and its aglycon methylazoxymethanol MAM (14901087)	$CH_3{-}\underset{\downarrow\atop O}{N}{=}N{-}CH_2O{-}\beta$-D-glucopyranosyl $CH_3{-}\underset{\downarrow\atop O}{N}{=}NCH_2OH$	173, 174	172	175

TABLE 6.3 (cont.)

Chemical or common name (C.A. registry number)	Structure	Point mutation: Phage	Point mutation: Other micro-organisms	Point mutation: Insects	Chromosome aberrations: Plant	Chromosome aberrations: Insect	Chromosome aberrations: Mammalian cells, mammals
Pyrrolizidine alkaloids							
a. Heliotrine	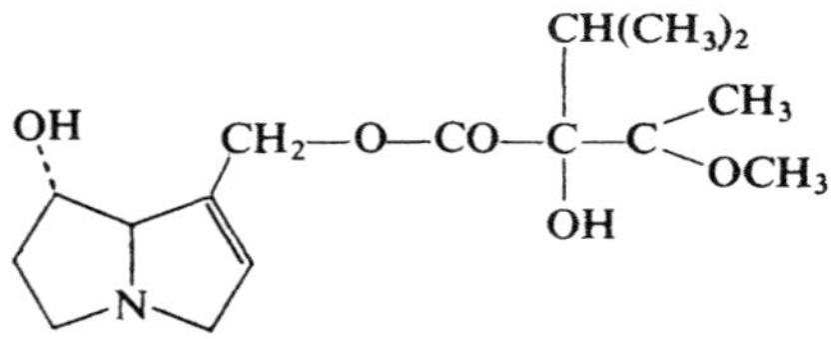		182	176–181	183	178, 181, 183	
b. Lasiocarpine (303344)	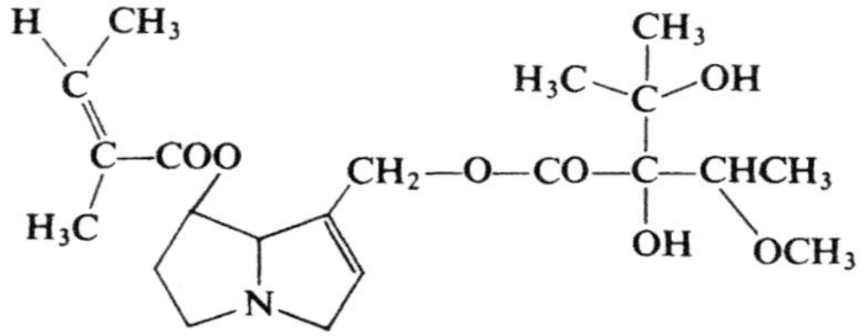		182	176, 177	183	183	
c. Monocrotaline	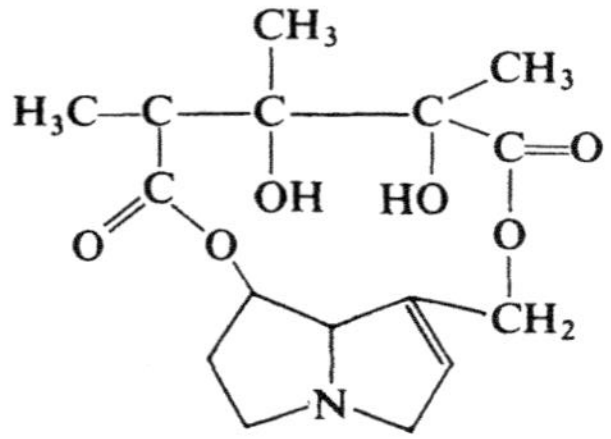			176–178			
Bracken fern				184			184

TABLE 6.4

Epoxides, Aldehydes, Lactones, and Related Derivatives

Chemical or common name (C.A. registry number)	Structure	Point mutation: Phage	Point mutation: Other micro-organisms	Point mutation: Insects	Chromosome aberrations: Plant	Chromosome aberrations: Insect	Chromosome aberrations: Mammalian cells, mammals
Ethylene oxide (75218)	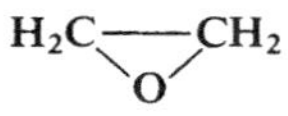		187–190	185, 186	191–193		
Propylene oxide (75569)	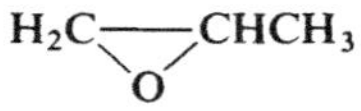			185, 194			
Ethylenechlorohydrin 2-Chloroethanol (107093)	$ClCH_2CH_2OH$		195				
Glycidol 2,3-Epoxy-1-propanol (556525)	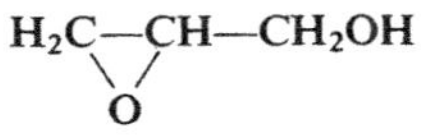		196	185			10
Epichlorohydrin 1-Chloro-2,3-epoxypropane (106898)	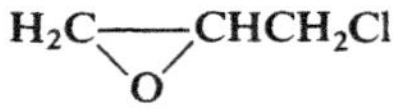		196–198	185			

TABLE 6.4 (cont.)

Chemical or common name (C.A. registry number)	Structure	Point mutation: Phage	Point mutation: Other micro-organisms	Point mutation: Insects	Chromosome aberrations: Plant	Chromosome aberrations: Insect	Chromosome aberrations: Mammalian cells, mammals
1,2:3,4-Diepoxybutane Butadiene diepoxide, DEB (1464535)	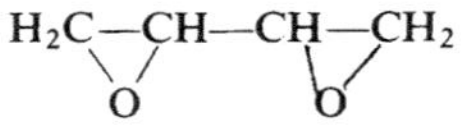		187, 190, 213–220	45, 186, 208–212	86, 91, 220–223	210	26, 224
Di(2,3-epoxypropyl) ether Diglycidyl ether	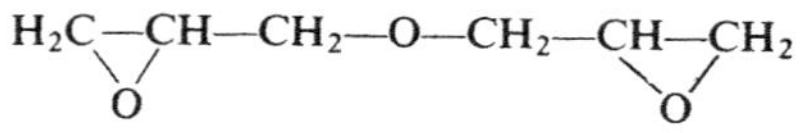	199	190, 196		86, 200–207		
Formaldehyde (50000)	HCHO		52, 239, 240	126, 225–238			
Acetaldehyde (75070)	CH_3CHO			241			
Acrolein Acrylic aldehyde, 2-propenal (107028)	$CH_2{=}CHCHO$			241			

Chloral hydrate 2,2,2-Trichloro-1,1-ethanediol, trichloroacetaldehyde monohydrate	$Cl_3C—CHOH$ with OH on the CH carbon	243	242	
Citronellal (106230)	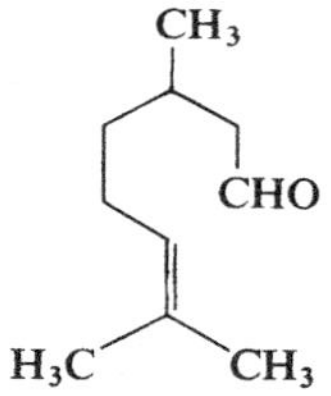		241	
Ketene Carbomethene (463514)	$CH_2{=}C{=}O$	*(52)*	244	
β-Propiolactone Hydracrylic acid lactone, BPL	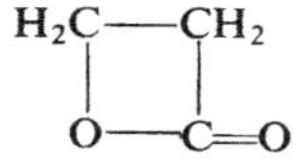	245–247		246, 248, 249
Ethylene sulfide	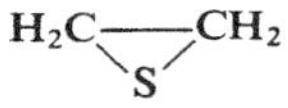	250		

TABLE 6.5

Alkyl Sulfates and Alkane Sulfonic Esters

		Point mutation				Chromosome aberrations	
Chemical or common name (C.A. registry number)	Structure	Trans. DNA	Phage	Other micro-organisms	Insects	Plant	Mammalian cells, mammals
Dimethyl sulfate Sulfuric acid dimethyl ester, DMS (77781)	$CH_3OSO_2OCH_3$			251, 258, 260	251, 253,	261	
Diethyl sulfate Sulfuric acid diethyl ester, DES (64675)	$C_2H_5OSO_2OC_2H_5$	257	83	182, 198, 251, 256–259	251–255		

Methyl methanesulfonate Methanesulfonic acid methyl ester, MMS	$CH_3SO_2OCH_3$	83, 262	263		264, 267	15, 43, 265, 266
Ethyl methanesulfonate Methanesulfonic acid ethyl ester, EMS (62500)	$CH_3SO_2OC_2H_5$	83, 279–281	7, 182, 197, 259, 260, 263, 274–278, 282–287	5, 64, 126, 251, 268–273	291–295	9, 265, 288–290
Myleran Methanesulfonic acid, tetramethylene ester, tetramethylene bis-(methanesulfonate), busulfan	$CH_3SO_2O(CH_2)_4OSO_2CH_3$			208, 272, 296–298	264, 301–304	26, 265, 299, 300
L-Threitolbismethane sulfonate TBMS	$X—CH_2—CHOH—CHOH—CH_2—X$ $X = CH_2—SO_2O—$				223, 305–308	

TABLE 6.6

Acridines and Acridinium Salts

Chemical or common name (C.A. registry number)	Structure	Point mutation			Chromosome aberrations	
		Phage	Other micro-organisms	Insects	Plant	Mammalian cells, mammals
Proflavine 3,6-Diaminoacridine (92626)	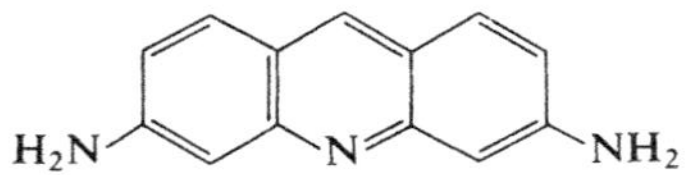	310–314				309
Acriflavine A mixture of 3,6-diamino-10-methylacridinium chloride and 3,6-diaminoacridine (86408)	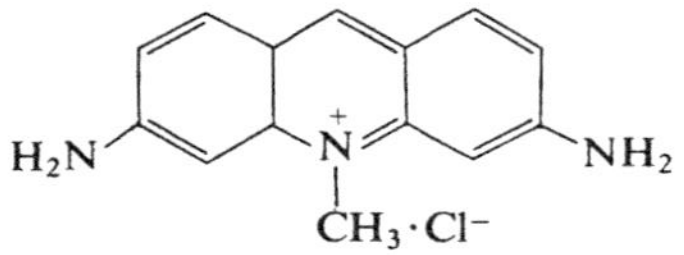 + Proflavine	318	240, 315–317	315		

Acridine orange (494382)	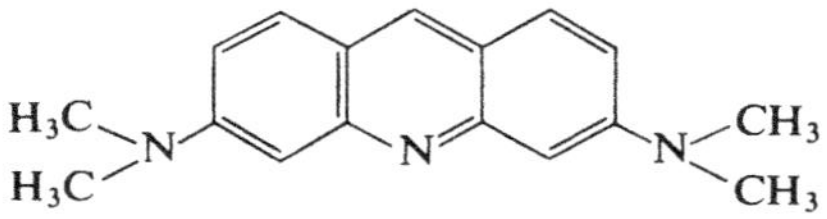		313, 318		
5-Aminoacridine (90459)	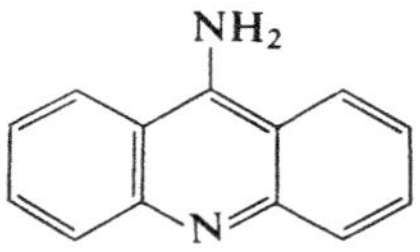	313	320, 321	319	
Acridine mustard 2-Methoxy-6-chloro-9-[3-(ethyl-2-chloroethyl)aminopropylamino]acridine dihydrochloride ICR-170	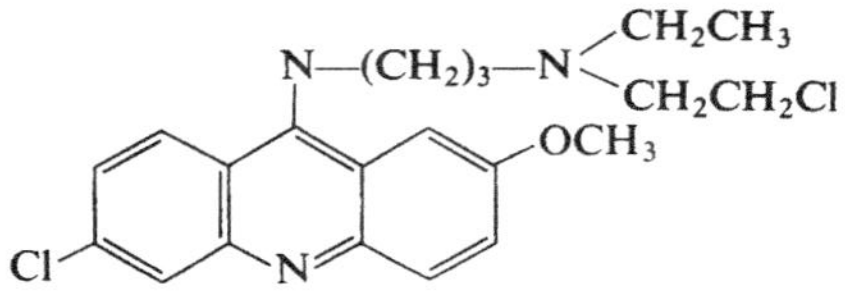		325–328	20, 61, 322–324	329

TABLE 6.7

Antibiotics

Chemical or common name (C.A. registry number)	Structure	Point mutation: Phage	Point mutation: Other micro-organisms	Point mutation: Insects	Chromosome aberrations: Plant	Chromosome aberrations: Mammalian cells, mammals
Mitomycin C (50077)	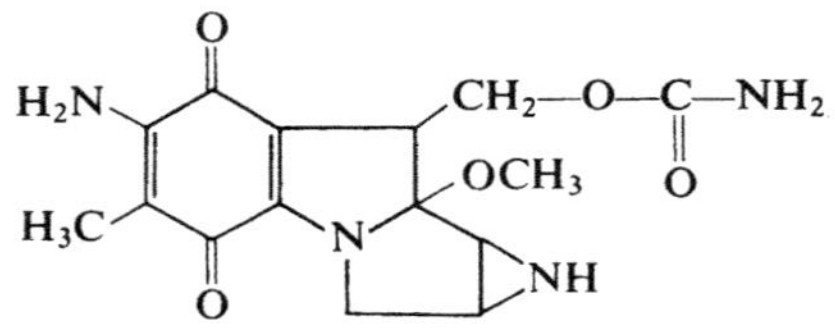			330–332	336	333–335
Streptonigrin (3930196)	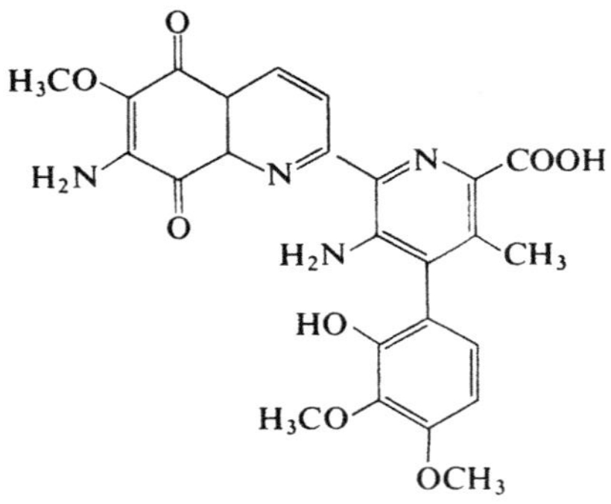		336		339, 340	337, 338

Streptozotocin (11006807)	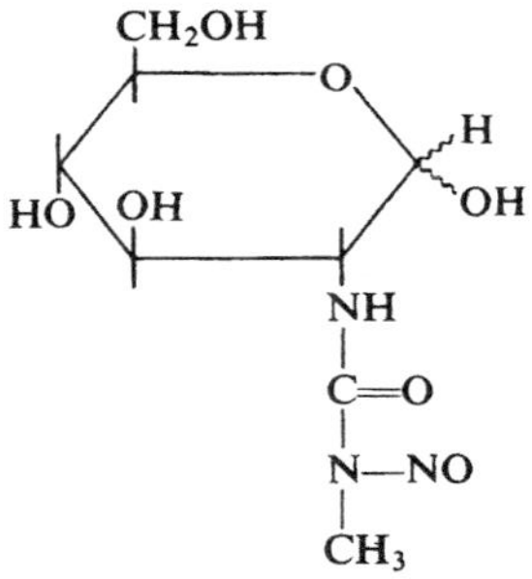		174, 341, 341a		
Patulin 4-Hydroxy-4*H*-furo[3,2-*c*]pyran-2(6*H*)-one, clavicin (149291)	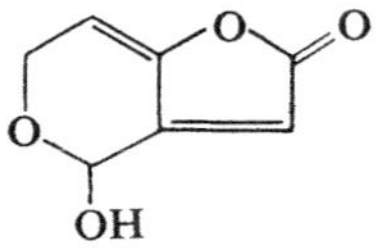		342		343, 344
Phleomycin (11006330)				355, 356	351–354
Azaserine *O*-Diazoacetyl-L-serine, serine diazoacetate (115026)	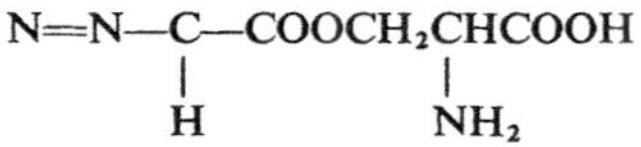	347	41, 66, 240, 345, 346	339, 348–350	
Daunomycin (1407154)					357–361

TABLE 6.8

Miscellaneous Drugs and Food Additives (and Related Degradation Products)

Chemical or common name (C.A. registry number)	Structure	Point mutation: Trans. DNA	Point mutation: Phage	Point mutation: Other micro-organisms	Point mutation: Insects	Point mutation: Mice	Chromosome aberrations: Plant	Chromosome aberrations: Insect	Chromosome aberrations: Mammalian cells, mammals
Ethidium bromide (1239458)	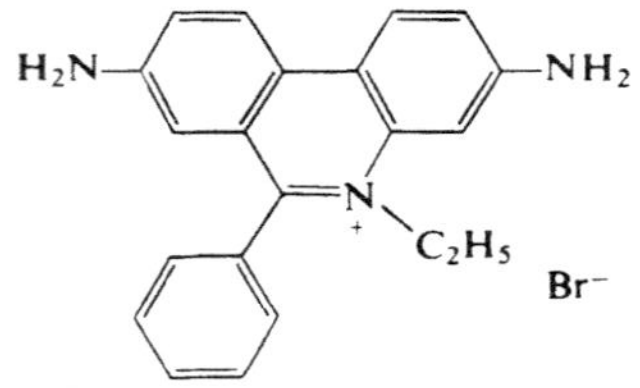			362					
8-Hydroxyquinoline 8-Quinolinol, oxine (148243)	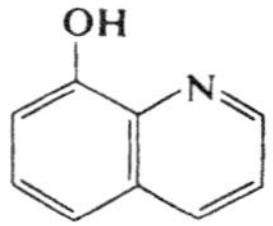			363, 364			365		
Dithranol 1,8,9-Trihydroxyanthracene, anthralin (480228)	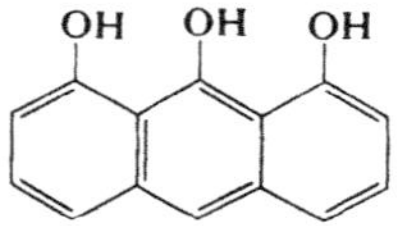			366					
Miracil D (479505)	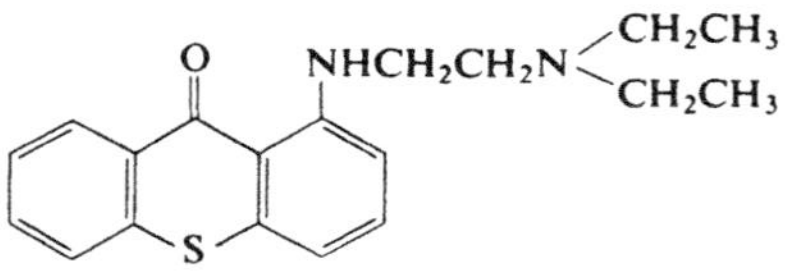				367				368

Steroid diamines a. Malouetine 5α-Pregnan-3β,20α-ylene-bis(trimethylammonium iodide)	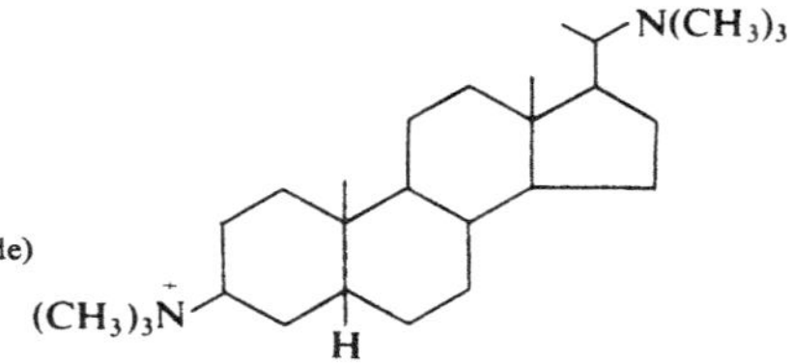	369					
b. Irehdiamine Pregn-5-ene-3β,20α-diamine	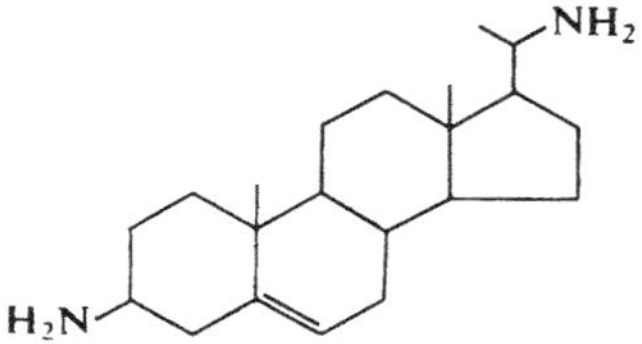	369					
Nitrofurazone 5-Nitro-2-furaldehyde semicarbazone (59870)	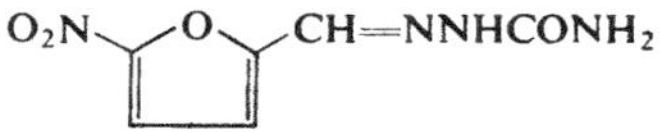		69, (*41*)				
Lysergic acid diethylamide LSD, lysergide (50373)	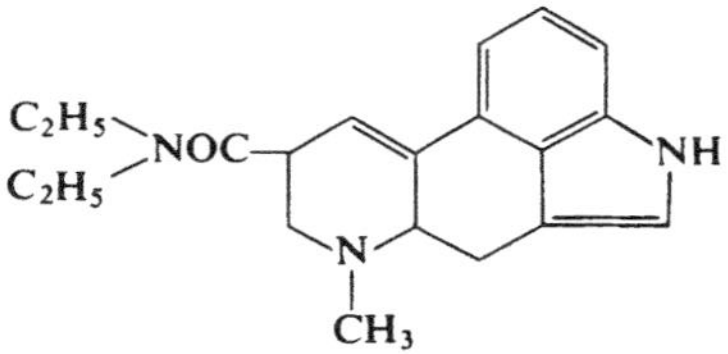		(*383*)	381, (*382*)			370–376, (*377–380*), (*382*)
Caffeine 1,3,7-Trimethylxanthine, methyltheobromine (58082)	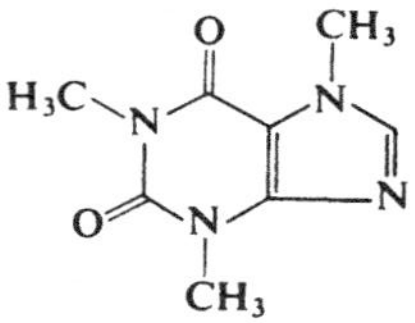	391	240, 389–393	319, 384–388, (*387*)	(*339–340*) (399, 400)	384, 387	385, 394–398

TABLE 6.8 (cont.)

Chemical or common name (C.A. registry number)	Structure	Point mutation: Trans. DNA	Point mutation: Phage	Point mutation: Other micro-organisms	Point mutation: Insects	Point mutation: Mice	Chromosome aberrations: Plant	Chromosome aberrations: Insect	Chromosome aberrations: Mammalian cells, mammals
8-Ethoxycaffeine 8-Ethoxy-1,3,7-trimethylxanthine (577662)	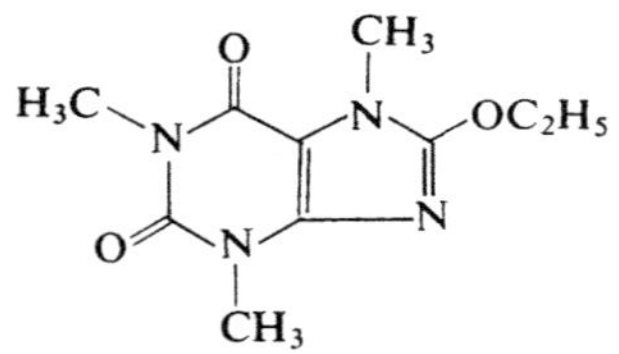						205, 340 402–404		401
Cyclamate Cyclohexanesulfamic acid, sodium salt, sodium cyclamate (139509)	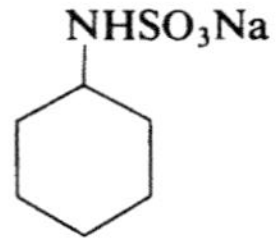						405		406, 407
Cyclohexylamine (108918)	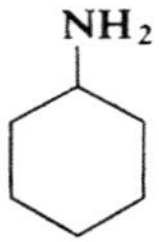								408–409
EDTA Ethylenediaminetetraacetic acid, Versene (60004)	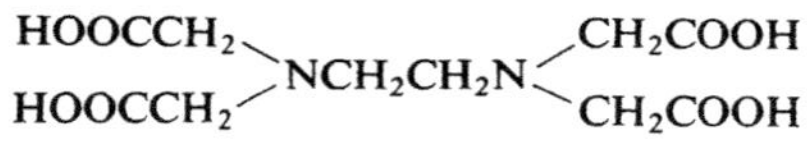						410–415	416, 417	

Allylisothiocyanate Volatile oil of mustard (57067)	$CH_2{=}CHCH_2NCS$			81	91, 418–420		421	418
Sodium nitrite and nitrous acid (7782776)	$NaNO_2$, HNO_2	422–425	426, 436–439	128, 139, 169, 276, 328, 426–435				

TABLE 6.9

Pesticides

Chemical or common name (C.A. registry number)	Structure	Point mutation: Other micro-organisms	Point mutation: Insects	Chromosome aberrations: Plant	Chromosome aberrations: Mammalian cells, mammals
Maleic hydrazide 1,2-Dihydro-3,6-pyridazinedione, MH (12331)	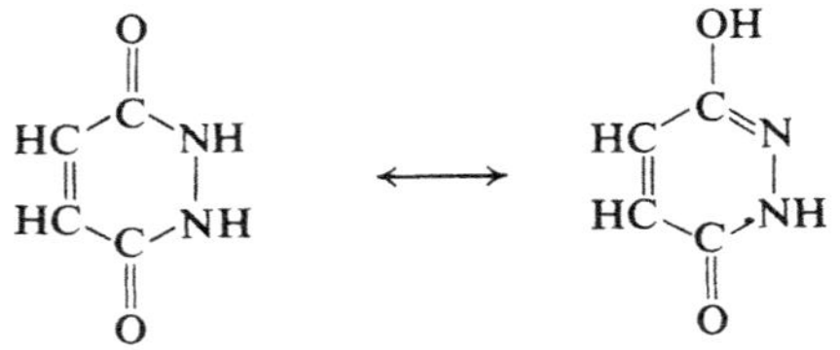		440, 441	202–205, 404, 442–449	
Captan *N*-(Trichloromethylthio)-4-cyclohexene-1,2-dicarboximide (133062)	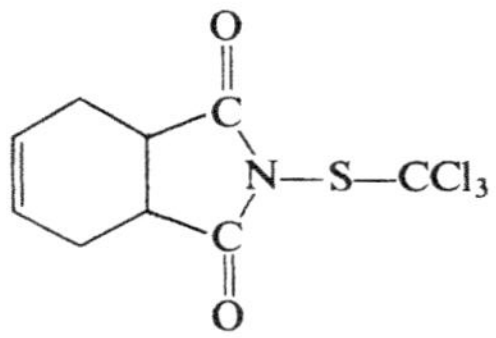	450			450

Hemel 2,4,6-Trisdimethylamino-1-triazine (645056)	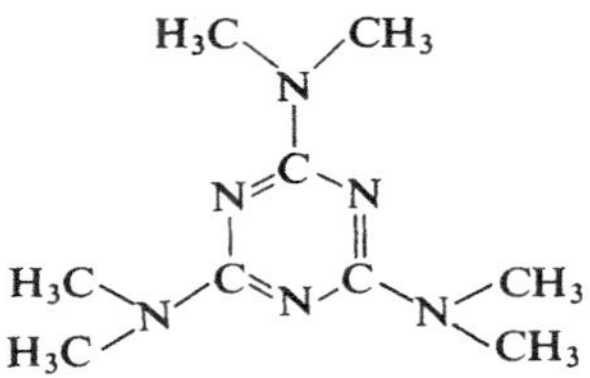	24
Hempa Hexamethyl phosphoric triamide (680319)	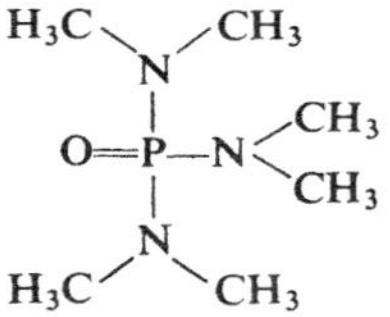	39

TABLE 6.10

Nitrogen Derivatives (Hydrazines, Hydroxylamines, and N-Hydroxy Derivatives)

Chemical or common name (C.A. registry number)	Structure	Point mutation: Trans. DNA	Point mutation: Phage	Point mutation: Other micro-organisms	Point mutation: Insects	Point mutation: Mice	Chromosome aberrations: Plant	Chromosome aberrations: Mammalian cells, mammals
Hydrazine and derivatives a. Hydrazine (10217524)	H_2NNH_2			451–453				454
b. *Unsym*-1,1-dimethylhydrazine UDMH (57147)	$(H_3C)_2N—NH_2$			451–453				
c. *Sym*-1,2-dimethylhydrazine (540738)	$CH_3—NH—NH—CH_3$			451–453, 455				
Isonicotinoylhydrazide Isoniazid, INH (54853)	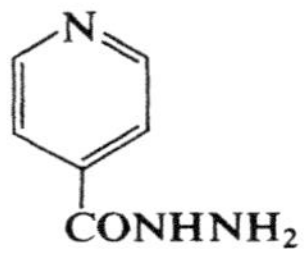	456						
Hydroxylamine (7803498)	NH_2OH	425, 457–459	437, 460, 461	434, 462			293	463–465

N-Methylhydroxylamine	CH_3—N(H)—OH	457	466				465
O-Methylhydroxylamine (67629)	H_2N—OCH_3	457	466				465
N-Hydroxyurea (127071)	H_2N—C(=O)—N(H)—OH					470	467–469
N-Hydroxy-1-naphthylamine	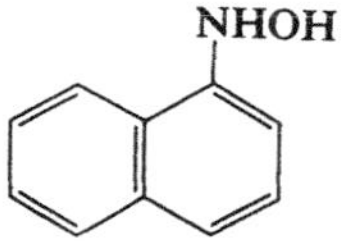		245, 471–473				
N-Hydroxy-2-naphthylamine	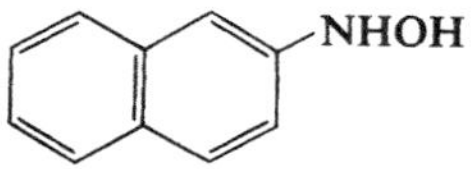		245, 471–473				
Urethan Ethyl carbamate (51796)	H_2N—C(=O)—OC_2H_5		(*52*), 240, 476, 477	474, 475	(*478*)	479	480

TABLE 6.10 (cont.)

Chemical or common name (C.A. registry number)	Structure	Point mutation: Trans. DNA	Point mutation: Phage	Point mutation: Other micro-organisms	Point mutation: Insects	Point mutation: Mice	Chromosome aberrations: Plant	Chromosome aberrations: Mammalian cells, mammals
N-Hydroxyurethan (589413)	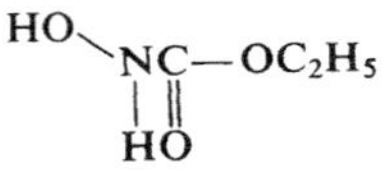	481, 482					480	465, 468, 480
4-Nitroquinoline-1-oxide (56575)	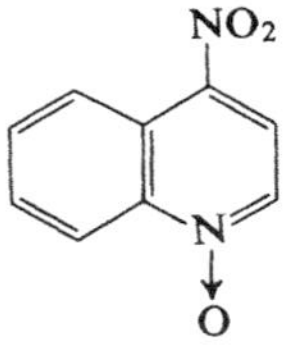		487	41, 483–486				488–490
4-Hydroxylaminoquinoline-1-oxide	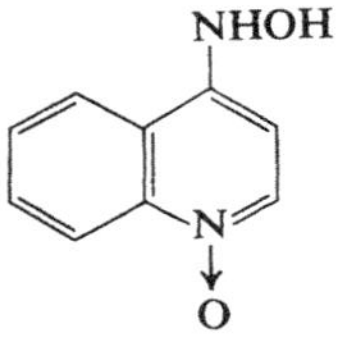			491				

TABLE 6.11

Hydrogen Peroxide and Organic Peroxides

Chemical or common name (C.A. registry number)	Structure	Point mutation: Trans. DNA	Point mutation: Phage	Point mutation: Other micro-organisms	Point mutation: Insects	Chromosome aberrations: Plant
Hydrogen peroxide Hydroperoxide (7722841)	H_2O_2	482, 496–498		52, 56, 239, 240, 492–495, (*499*)		
tert-Butyl hydroperoxide (75912)	$(CH_3)_3COOH$			239, 502, 524	500, 501	86, 502, 503
Cumene hydroperoxide (80159)	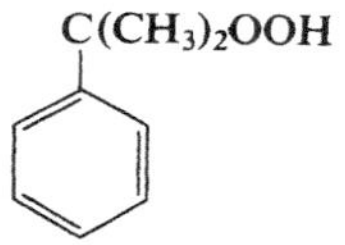			52, 498, 502, 524		
Succinic acid peroxide	$HOOC—CH_2CH_2—C(=O)—O—OH$	496	504	502, 505, 524		

TABLE 6.11 (cont.)

Chemical or common name (C.A. registry number)	Structure	Point mutation: Trans. DNA	Point mutation: Phage	Point mutation: Other micro-organisms	Point mutation: Insects	Chromosome aberrations: Plant
Disuccinyl peroxide (123239)	$COCH_2CH_2COOH$ — O — O — $COCH_2CH_2COOH$	482, 496, 498, 504				
Dihydroxydimethyl peroxide	$HO—CH_2—O—O—CH_2OH$	*(482)*		52, 504	238, 506, 507	
Di-*tert*-butyl peroxide	$(CH_3)_3C—O—O—C(CH_3)_3$	*(498)*		52		

TABLE 6.12

Miscellaneous Mutagens

Chemical or common name (C.A. registry number)	Structure	Point mutation: Other micro-organisms	Point mutation: Insects	Chromosome aberrations: Plant	Chromosome aberrations: Mammalian cells, mammals
Aflatoxin B_1 (1402682)	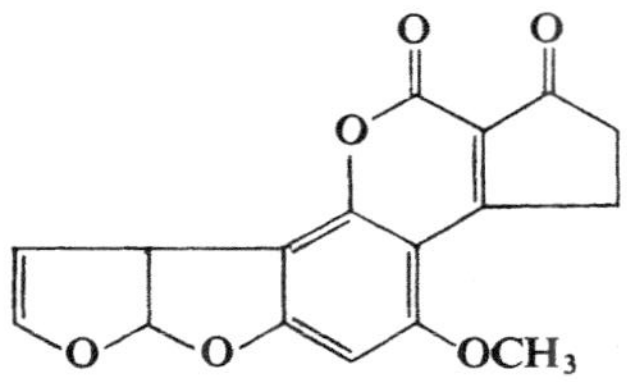				15
Polynuclear aromatic hydrocarbons a. 3,4-Benzypyrene Benzo[*a*]pyrene (50328)	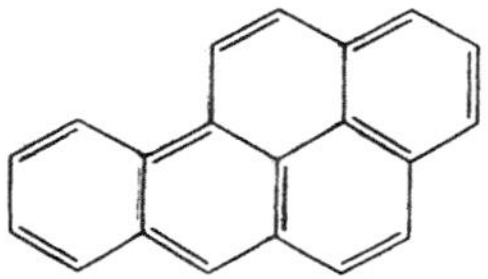	509, (*516*)	508		15
b. 1,2,5,6-Dibenzanthracene Dibenz[*a*,*h*]anthracene (53703)	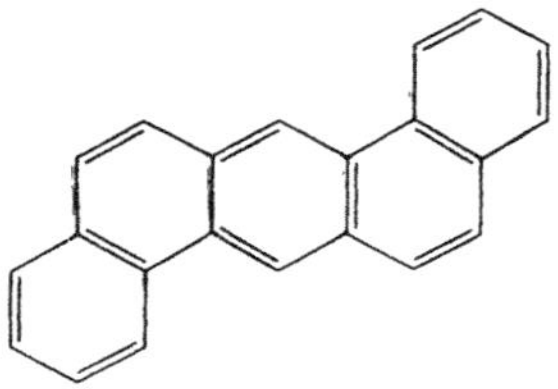	509, 512, 513, (*516*)	508, 510, 511		514, (*515*)

TABLE 6.12 (cont.)

Chemical or common name (C.A. registry number)	Structure	Point mutation: Other micro-organisms	Point mutation: Insects	Chromosome aberrations: Plant	Chromosome aberrations: Mammalian cells, mammals
20-Methylcholanthrene 3-Methylcholanthrene (56495)	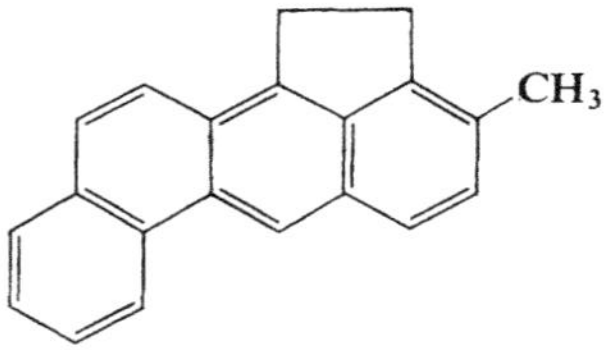	509, 512, 513, (*523*)	508, 510, 517, (*521*, *522*)		518–520
9,10-Dimethyl-1,2-benzanthracene 7,12-Dimethylbenz[*a*]anthracene (57976)	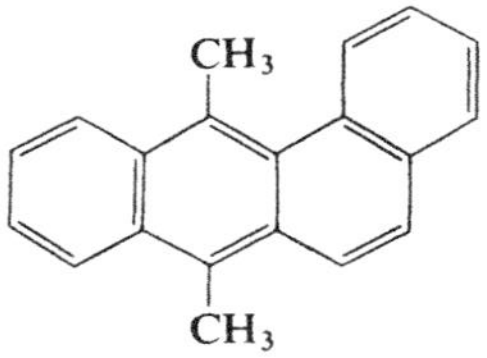	512, 513, 516			
Manganous chloride (7773015)	$MnCl_2$	68, 240, 247, 525–527, (*247*)			

Manganous salts				
Manganous acetate (638380)	$Mn(C_2H_3O_2)_2$	240		
Manganous nitrate (10377669)	$Mn(NO_3)_2$			
Manganous sulfate (7785877)	$MnSO_4$			
Ferrous chloride (7758943)	$FeCl_2$	240		
Lead acetate (646678)	$Pb(C_2H_3O_2)_2$			528
Cadmium nitrate	$Cd(NO_3)_2$		503	
Aluminum chloride (7446700)	$AlCl_3$		479, 503, 529	

REFERENCES

1. I. A. Rapoport, *Byul. Mosk. Obshchestva Ispytatelei Prirody, Otd. Biol.* **67**, 109 (1962).
2. H. Lüers and G. Röhrborn, *Mutation Res.* **2**, 29 (1965).
3. M. L. Alexander and E. Glanges, *Proc. Natl. Acad. Sci. U.S.* **53**, 282 (1965).
4. M. L. Alexander, *Genetics* **56**, 274 (1967).
5. J. K. Lim and L. A. Snyder, *Mutation Res.* **6**, 129 (1968).
6. W. E. Ratnayake, *Mutation Res.* **5**, 271 (1968).
7. M. Westergaard, *Experientia* **13**, 224 (1957).
8. F. K. Zimmermann and U. von Laer, *Mutation Res.* **4**, 377 (1967).
9. T. H. Chang and F. T. Elequin, *Mutation Res.* **4**, 83 (1967).
10. J. J. Biesele, F. S. Philips, J. B. Thiersch, J. H. Burchenal, S. M. Buckley, and C. C. Stock, *Nature* **166**, 1112 (1950).
11. G. Röhrborn, *Humangenetik* **1**, 576 (1965).
12. G. Röhrborn, *Humangenetik* **2**, 81 (1966).
13. E. Schleiermacher, *Humangenetik* **3**, 134 (1966).
14. A. J. Bateman, *Nature* **210**, 205 (1966).
15. S. S. Epstein and H. Shafner, *Nature* **219**, 385 (1968).
16. A. J. Bateman, *Genet. Res.* **1**, 381 (1960).
17. B. M. Cattanach, *Z. Vererbungslehre* **90**, 1 (1959).
18. O. G. Fahmy and M. J. Fahmy, *J. Genet.* **52**, 603 (1954).
19. O. G. Fahmy and M. J. Fahmy, *J. Genet.* **53**, 563 (1955).
20. L. A. Snyder and I. I. Oster, *Mutation Res.* **1**, 437 (1964).
21. L. A. Snyder, *Z. Vererbungslehre* **94**, 182 (1963).
22. I. H. Herskowitz, *Genetics* **41**, 605 (1956).
23. I. H. Herskowitz, *Genetics* **40**, 574 (1955).
24. D. T. North, *Mutation Res.* **4**, 225 (1967).
25. M. M. Crystal, *J. Econ. Entomol.* **56**, 468 (1963).
26. J. Moutschem, *Genetics* **46**, 291 (1961).
27. B. M. Cattanach, *Mutation Res.* **3**, 346 (1966).
28. B. M. Cattanach, *Nature* **180**, 1364 (1957).
29. B. M. Cattanach, *Mutation Res.* **4**, 73 (1967).
30. B. M. Cattanach and R. G. Edwards, *Proc. Roy. Soc. Edinburgh* **B67**, 54 (1958).
31. H. Nicoloff and K. Gecheff, *Mutation Res.* **6**, 257 (1968).
32. H. Michaelis and R. Rieger, *Nature* **199**, 1014 (1963).
33. C. H. Ockey, *J. Genet.* **55**, 525 (1957).
34. U. Read, *Intern. J. Radiation Biol.* **3**, 95 (1961).
35. R. Wakonig and T. J. Arnason, *Can. J. Botany* **37**, 403 (1959).
36. V. N. Ronchi, M. Buiatti, and P. L. Ipata, *Mutation Res.* **4**, 315 (1967).
37. G. Röhrborn and F. Vogel, *Deut. Med. Wochschr.* **92**, 2315 (1967).
38. G. Obe, *Mutation Res.* **6**, 467 (1968).
39. J. Palmquist and L. E. LaChance, *Science* **154**, 915 (1966).
40. A. R. Kaney and K. C. Atwood, *Nature* **201**, 1006 (1964).
41. W. Szybalski, *Ann. N. Y. Acad. Sci.* **76**, 475 (1958).
42. J. W. Drake, *Nature* **197**, 1028 (1963).

43. S. S. Epstein, *Toxicol. Appl. Pharmacol.* **14**, 653 (1969).
44. K. D. Wuu and W. F. Grant, *Nucleus* (*Calcutta*) **10**, 37 (1967).
45. L. E. LaChance, M. Degrugillier, and A. P. Leverich, *Mutation Res.* **7**, 63 (1969).
46. L. E. LaChance and M. M. Crystal, *Biol. Bull.* **125**, 280 (1963).
47. M. A. Arsen'eva and A. V. Golovkina, *Vliyanie Ioniz. Izluch. Nasledstvennost, Akad. Nauk SSSR* p. 122 (1966).
48. N. V. Pankova, *Genetika* p. 62 (1967).
49. G. Röhrborn, *Z. Vererbungslehre* **93**, 1 (1962).
50. G. Röhrborn, *Drosophila Inform. Serv.* **33**, 156 (1959).
51. E. L. Tatum, *Cold Spring Harbor Symp. Quant. Biol.* **11**, 278 (1946).
52. K. A. Jensen, I. Kirk, G. Kolmark, and M. Westergaard, *Cold Spring Harbor Symp. Quant. Biol.* **16**, 245 (1951).
53. G. Höhne, C. Bertram, and G. Schubert, *Acta Unio Intern. Contra Cancrum* **16**, 658 (19б0).
54. S. Kawate, *Tech. Rept. Kansai Univ.* **1**, 45 (1959); *Chem. Abstr.* **54**, 14366G (1960).
55. C. M. Stevens and A. Mylroie, *Nature* **166**, 1019 (1950).
56. D. S. Falconer, B. M. Slaynski, and C. Auerbach, *J. Genet.* **51**, 81 (1952).
57. C. Auerbach and H. Moser, *Nature* **166**, 1019 (1950).
58. O. G. Fahmy and M. J. Fahmy, *Heredity* **15**, 115 (1960).
59. A. Schalet, *Genetics* **40**, 594 (1955).
60. A. Schalet, *Am. Naturalist* **90**, 329 (1956).
61. I. I. Oster and E. Pooley, *Genetics* **45**, 1004 (1960).
62. I. H. Herskowitz and A. Schalet, *Genetics* **39**, 970 (1954).
63. W. J. Burdette, *Cancer Res.* **12**, 366 (1952).
64. E. A. Löbbecke and R. C. von Borstel, *Genetics* **47**, 853 (1962).
65. S. W. Glover, *Carnegie Inst. Wash. Publ.* **612**, 121 (1956).
66. V. N. Iyer and W. Szybalski, *Appl. Microbiol.* **6**, 23 (1958).
67. A. Loveless, "Genetic and Allied Effects of Alkylating Agents," p. 103. Penn. State Univ. Press, University Park, Pennsylvania, 1966.
68. M. Demerec, *Caryologia*, Suppl., 201 (1954).
69. A. Zampieri and J. Greenberg, *Biochem. Biophys. Res. Commun.* **14**, 172 (1964).
70. J. Nečašek, *Proc. 11th Intern. Congr. Genet., The Hague, 1963* Vol. 1, p. 59. Pergamon Press Oxford, 1965.
71. L. Silvestri, *Boll. Ist. Sieroterap. Milan.* **28**, 326 (1949).
72. E. A. Löbbecke, *Genetics* **48**, 91 (1965).
73. J. F. Stauffer and M. B. Bachus, *Ann. N. Y. Acad. Sci.* **60**, 35 (1954).
74. F. R. Roegner, M. A. Stanman, and J. F. Stauffer, *Am. J. Botany* **41**, 1 (1954).
75. E. C. Gasiorkiewitz, R. H. Larson, J. C. Walker, and M. A. Stahmann, *Phytopathology* **42**, 183 (1952).
76. M. A. Stahmann and J. F. Stauffer, *Science* **106**, 35 (1947).
77. W. D. McElroy, J. E. Cushing, and H. Miller, *J. Cellular Comp. Physiol.* **30**, 331 (1947).
78. D. Bonner, *Cold Spring Harbor Symp. Quant. Biol.* **11**, 14 (1946).
79. G. Kølmark and M. Westergaard, *Hereditas* **35**, 490 (1949).
80. E. L. Tatum, R. W. Barratt, N. Fries, and D. Bonner, *Am. J. Botany* **37**, 38 (1950).
81. N. Fries, *Physiol. Plantarum* **1**, 330 (1948).
82. N. Iguchi, *J. Agr. Chem. Soc. Japan* **27**, 229 (1953); *Chem. Abstr.* **48**, 10114F (1954).
83. A. Loveless, *Proc. Roy. Soc.* **B150**, 497 (1959).
84. A. Loveless, *Proc. Roy. Soc.* **B159**, 348 (1964).
85. C. E. Ford, *Hereditas* Suppl., p. 570 (1949).
86. S. H. Revell, *Heredity* **6**, Suppl., 107 (1953).

87. B. A. Kihlman, *Hereditas* **41**, 384 (1955).
88. R. Rieger and A. Michaelis, *Chromosoma* **11**, 573 (1961).
89. G. Cardinali and G. Morpurgo, *Ric. Sci.* Suppl. 22, 55 (1955); *Chem. Abstr.* **52**, 15756G (1958).
90. C. Auerbach and J. M. Robson, *Nature* **157**, 302 (1946).
91. C. Auerbach, *Biol. Rev.* **24**, 355 (1949).
92. C. Auerbach and E. M. Sonbati, *Z. Vererbungslehre* **91**, 237 (1960).
93. C. Auerbach, *Proc. Roy. Soc. Edinburgh* **B62**, Part 2, 211 (1945).
94. G. E. Nasrat, W. D. Kaplan, and C. Auerbach, *Z. Vererbungslehre* **86**, 249 (1954).
95. A. Gilman and F. S. Philips, *Science* **103**, 409 (1946).
96. C. Auerbach, J. M. Robson, and J. C. Carr, *Science* **105**, 243 (1947).
97. C. Auerbach and J. M. Robson, *Proc. Roy. Soc. Edinburgh* **B62**, 271 (1947).
98. N. H. Horowitz, M. B. Houlahan, M. G. Hungate, and B. Wright, *Science* **104**, 233 (1946).
99. K. A. Jensen, I. Kirk, and M. Westergaard, *Nature* **166**, 1021 (1950).
100. D. Hockenhull, *Nature* **161**, 100 (1948).
101. P. Brookes and P. D. Lawley, *Biochem. J.* **89**, 138 (1963).
102. C. D. Darlington, *Publ. Staz. Zool. Napoli* **22**, Suppl., 22 (1950); *Chem. Abstr.* **45**, 221 (1951).
103. P. B. Gibson, R. A. Brink, and M. A. Stahmann, *J. Heredity* **41**, 232 (1950).
104. P. C. Koller, *Ann. N. Y. Acad. Sci.* **68**, 783 (1958).
105. P. C. Koller, *Progr. Biophys. Biophys. Chem.* **4**, 195 (1953).
106. D. Brittinger, *Humangenetik* **3**, 156 (1966).
107. C. Bertram and G. Höhne, *Strahlentherapie* **43**, 388 (1959).
108. S. Frye, *Brit. Empire Cancer Campaign, Ann. Rept.* **38**, 670 (1960).
109. G. Röhrborn, *Mol. Gen. Genet.* **102**, 50 (1968).
110. I. W. Schmid and G. R. Staiger, *Mutation Res.* **7**, 99 (1969).
111. F. E. Arrighi, T. C. Hsu, and D. E. Bergsagel, *Texas Rept. Biol. Med.* **20**, 545 (1962).
112. K. E. Hampel, M. Fritzsche, and D. Stopik, *Humangenetik* **7**, 28 (1969).
113. M. Vrba, *Humangenetik* **4**, 362 (1967).
114. A. Michaelis and R. Rieger, *Biol. Zentr.* **80**, 301 (1961).
115. L. Pasternak, *Arzneimittel-Forsch.* **14**, 802 (1964).
116. L. Pasternak, *Naturwissenschaften* **49**, 381 (1962).
117. L. Pasternak, *Acta Biol. Med. Ger.* **10**, 436 (1963).
118. O. G. Fahmy and M. J. Fahmy, *Mutation Res.* **6**, 139 (1968).
119. O. G. Fahmy, M. J. Fahmy, J. Massasso, and M. Ondrej, *Mutation Res.* **3**, 201 (1966).
120. E. Geisler, *Naturwissenschaften* **49**, 380 (1962).
121. O. N. Pogodina, *Cytologia* **8**, 503 (1966).
122. H. Marquardt, F. K. Zimmermann, and R. Schwaier, *Z. Vererbungslehre* **95**, 82 (1964)
123. H. Marquardt, F. K. Zimmermann, and R. Schwaier, *Naturwissenschaften* **50**, 625 (1963).
124. H. V. Malling, *Mutation Res.* **3**, 537 (1966).
125. J. Rapoport, *Dokl. Akad. Nauk SSSR* **146**, 1418 (1962).
126. J. Rapoport, *Dokl. Akad. Nauk SSSR* **148**, 696 (1962).
127. H. O. Corwin, *Mutation Res.* **5**, 259 (1968).
128. F. K. Zimmermann, R. Schwaier, and U. von Laer, *Z. Vererbungslehre* **97**, 68 (1965).
129. R. Schwaier, *Z. Vererbungslehre* **97**, 55 (1965).
130. R. Schwaier, F. K. Zimmermann, and U. von Laer, *Z. Vererbungslehre* **97**, 72 (1965).
131. A. Loveless and C. C. Hampton, *Mutation Res.* **7**, 1 (1967).
132. N. N. Zoz and S. I. Makarova, *Tsitologiya* **7**, 405 (1965).

133. A. Michaelis, J. Schöneich, and R. Rieger, *Chromosoma* **16**, 101 (1965).
134. E. Glass and H. Marquardt, *Mol. Gen. Genet.* **101**, 307 (1968).
135. I. A. Rapoport, *Dokl. Akad. Nauk SSSR* **59**, 1183 (1948).
136. H. Henke, G. Höhne, and H. A. Künkel, *Biophysik* **1**, 418 (1964).
137. E. A. Adelberg, M. Mandel, and G. C. C. Chen, *Biochem. Biophys. Res. Commun.* **18**, 788 (1965).
138. A. Trams and H. A. Künkel, *Biophysik* **1**, 422 (1964).
139. F. K. Zimmermann and R. Schwaier, *Mol. Gen. Genet.* **100**, 63 (1967).
140. G. Zetterberg, *Exptl. Cell Res.* **20**, 659 (1960).
141. G. Zetterberg, *Hereditas* **47**, 295 (1961).
142. G. Zetterberg, *Hereditas* **48**, 371 (1962).
143. A. Abbondandolo and N. Loprieno, *Mutation Res.* **4**, 31 (1967).
144. R. Guglielminetti, S. Bonatti, and N. Loprieno, *Mutation Res.* **3**, 152 (1966).
145. N. Loprieno and C. H. Clarke, *Mutation Res.* **2**, 312 (1965).
146. N. Loprieno, *Mutation Res.* **1**, 469 (1964).
147. B. A. Kihlman, *Radiation Botany* **1**, 35 (1961).
148. B. A. Kihlman, *Exptl. Cell Res.* **20**, 657 (1960).
149. C. J. Grant and H. Heslot, *Ann. Genet.* **8**, 98 (1965).
150. E. Gläss and H. Marquardt, *Z. Vererbungslehre* **98**, 361 (1966).
151. A. Terawaki and J. Greenberg, *Biochim. Biophys. Acta* **95**, 170 (1965).
152. J. D. Mandell and J. Greenberg, *Biochem. Biophys. Res. Commun.* **3**, 575 (1960).
153. J. D. Mandell, *Bacteriol. Proc.* p. 124 (1965).
154. S. Zamenhof, L. H. Heldeumuth, and P. J. Zamenhof, *Proc. Natl. Acad. Sci. U.S.* **55**, 50 (1966).
155. L. Silengo, D. Schlessinger, G. Mangiarotti, and D. Apirion, *Mutation Res.* **4**, 701 (1967).
156. D. Botstein and E. W. Jones, *J. Bacteriol.* **98**, 847 (1969).
157. E. Cerda-Olmedo, P. C. Hanawalt, and N. Guerola, *J. Mol. Biol.* **33**, 705 (1968).
158. I. Takahashi and R. A. Bernard, *Mutation Res.* **4**, 111 (1967).
159. A. Eisenstark, R. Eisenstark, and R. Van Sickle, *Mutation Res.* **2**, 1 (1965).
160. F. Lingens and O. Oltmans, *Z. Naturforsch.* **21b**, 660 (1966).
161. K. Nordström, *J. Gen. Microbiol.* **48**, 277 (1967).
162. N. Loprieno, R. Guglielminetta, A. Abbondandolo, and S. Bonatti, *Microbiol. Genet. Bull.* **23**, 21 (1965).
163. R. Megnet, *Mutation Res.* **2**, 328 (1965).
164. L. S. Browning, *Genetics* **60**, 165 (1968).
165. L. S. Browning, *Mutation Res.* **8**, 157 (1969).
166. T. Gichner, A. Michaelis, and R. Rieger, *Biochem. Biophys. Res. Commun.* **11**, 120 (1963).
167. D. R. McCalla, *Science* **148**, 497 (1965).
168. D. R. McCalla, *J. Protozool.* **13**, 472 (1966).
169. N. Nashed and G. Jabbur, *Z. Vererbungslehre* **98**, 106 (1966).
170. K. A. Jensen, G. Kølmark, and M. Westergaard, *Hereditas* **35**, 521 (1949).
171. E. Cerda-Olmedo and H. Hanawalt, *Mol. Gen. Genet.* **101**, 191 (1968).
172. H. J. Teas and J. G. Dyson, *Proc. Soc. Exptl. Biol. Med.* **125**, 988 (1967).
173. D. W. E. Smith, *Science* **152**, 1273 (1966).
174. M. G. Gabridge, A. Denunzio, and M. S. Legator, *Science* **163**, 689 (1969).
175. H. J. Teas, H. J. Sax, and K. Sax, *Science* **149**, 541 (1965).
176. A. M. Clark, *Nature* **183**, 731 (1959).
177. A. M. Clark, *Z. Vererbungslehre* **91**, 74 (1960).

178. L. M. Cook and A. C. Holt, *J. Genet.* **59**, 273 (1966).
179. N. G. Brink, *Mutation Res.* **3**, 66 (1966).
180. N. G. Brink, *Mutation Res.* **8**, 139 (1969).
181. N. G. Brink, *Z. Vererbungslehre* **94**, 321 (1963).
182. T. Alderson and A. M. Clark, *Nature* **210**, 593 (1966).
183. S. Avanzi, *Caryologia* **14**, 251 (1961).
184. I. A. Evans, *Cancer Res.* **28**, 2252 (1968).
185. I. A. Rapoport, *Dokl. Akad. Nauk SSSR* **60**, 469 (1948).
186. M. J. Bird, *J. Genet.* **50**, 480 (1952).
187. H. G. Kølmark and B. J. Kilbey, *Mol. Gen. Genet.* **101**, 89 (1968).
188. H. G. Kølmark, B. J. Kilbey, and S. Kondo, *Proc. 11th Intern. Congr. Genet., The Hague, 1963* Vol. 1, p. 61. Pergamon Press, Oxford, 1965.
189. H. G. Kølmark and B. J. Kilbey, *Mol. Gen. Genet.* **101**, 185 (1968).
190. G. Kølmark and M. Westergaard, *Hereditas* **39**, 209 (1953).
191. A. C. Faberge, *Genetics* **40**, 171 (1955).
192. J. Moutschen-Dahmen, M. Moutschen-Dahmen, and L. Ehrenberg, *Hereditas* **60**, 267 (1968).
193. A. Loveless, *Heredity* **6**, Suppl., 293 (1953).
194. A. Schalet, *Drosophila Inform. Serv.* **28**, 155 (1954).
195. C. E. Voogd and P. V. D. Vet, *Experientia* **25**, 85 (1969).
196. G. Kølmark and N. H. Giles, *Genetics* **40**, 890 (1955).
197. A. Loveless and S. Howarth, *Nature* **184**, 1780 (1959).
198. B. S. Strauss and S. Okubo, *J. Bacteriol.* **79**, 464 (1960).
199. A. Loveless and J. C. Stock, *Proc. Roy. Soc.* **B150**, 497 (1959).
200. S. H. Revell, *Brit. Empire Cancer Campaign, Ann. Rept.* **30**, 42 (1953).
201. S. H. Revell, *Heredity* Suppl., p. 107 (1953).
202. C. D. Darlington and J. McLeish, *Nature* **167**, 407 (1951).
203. J. McLeish, *Heredity* **6**, Suppl., 125 (1953).
204. J. McLeish, *Heredity* **8**, 385 (1954).
205. B. A. Kihlman, *J. Biophys. Biochem. Cytol.* **2**, 543 (1956).
206. S. H. Revell, *Ann. N. Y. Acad. Sci.* **68**, 802 (1958).
207. G. R. Lane, in "Radiobiology Symposium" (Z. M. Bacq and P. Alexander, eds.), p. 265. Academic Press, New York, 1954.
208. O. G. Fahmy and M. J. Fahmy, *J. Genet.* **54**, 146 (1956).
209. O. G. Fahmy and M. J. Fahmy, *Proc. 10th Intern. Congr. Genet., Montreal, 1958* Vol. 2, p. 78. Univ. of Toronto Press, Toronto, 1959.
210. M. J. Bird and O. G. Fahmy, *Proc. Roy. Soc.* **B140**, 556 (1953).
211. Y. Nakao and C. Auerbach, *Z. Vererbungslehre* **92**, 457 (1961).
212. O. G. Fahmy and M. J. Bird, *Heredity* **6**, Suppl., 149 (1953).
213. G. Kølmark and M. Westergaard, *J. Biophys. Biochem. Cytol.* **2**, 543 (1956).
214. H. G. Kølmark and B. J. Kilbey, *Z. Vererbungslehre* **93**, 356 (1962).
215. C. Auerbach and D. Ramsey, *Japan. J. Genet.* **43**, 1 (1968).
216. G. Kølmark and N. H. Giles, *Genetics* **38**, 674 (1953).
217. G. Kølmark and N. H. Giles, *Hereditas* **39**, 209 (1953).
218. C. H. Clarke, *Mutation Res.* **8**, 35 (1969).
219. J. Kilbey, *Mutation Res.* **8**, 73 (1969).
220. J. Moutschen-Dahmen, M. Moutschen-Dahmen, and R. Loppes, *Nature* **199**, 406 (1963).
221. N. S. Cohn, *Nature* **192**, 1093 (1961).
222. N. S. Cohn, *Exptl. Cell Res.* **24**, 569 (1961).

223. R. Matagne, *Radiation Botany* **8**, 489 (1968).
224. J. Nemenzo and C. H. Hine, *Abstr. 8th Ann. Meeting Soc. Toxicol., Williamsburg, 1969*; *Toxicol. Appl. Pharmacol.* **14**, 653 (1969).
225. I. A. Rapoport, *Dokl. Akad. Nauk SSSR* **54**, 65 (1946).
226. I. A. Rapoport, *Zh. Obshch. Biol.* **8**, 359 (147).
227. I. A. Rapoport, *Dokl. Akad. Nauk SSSR* **56**, 537 (1947).
228. I. A. Rapoport, *Dokl. Akad. Nauk SSSR* **51**, 713 (1948).
229. W. D. Kaplan, *Science* **108**, 43 (1948).
230. C. Auerbach, *Science* **110**, 419 (1949).
231. C. Auerbach, *Hereditas* **37**, 1 (1951).
232. T. Alderson, *Proc. 11th Intern. Congr. Genet., The Hague, 1963* Vol. 1, p. 65. Pergamon Press, Oxford, 1965.
233. C. Auerbach, *Am. Naturalist* **86**, 330 (1952).
234. A. F. E. Khishin, *Mutation Res.* **1**, 202 (1964).
235. W. J. Burdette, *Cancer Res.* **11**, 241 (1951).
236. C. Auerbach, *Z. Vererbungslehre* **81**, 621 (1956).
237. C. Auerbach, *Nature* **210**, 104 (1966).
238. F. H. Sobels, *Am. Naturalist* **88**, 109 (1954).
239. F. H. Dickey, G. H. Cleland, and C. Lotz, *Proc. Natl. Acad. Sci. U.S.* **35**, 581 (1949).
240. M. Demerec, G. Bertani, and J. Flint, *Am. Naturalist* **75**, 119 (1951).
241. I. A. Rapoport, *Dokl. Akad. Nauk SSSR* **61**, 713 (1948).
242. A. Barthelmess, *Arzneimittel-Forsch.* **6**, 157 (1956).
243. A. Goldstein, *in* "Mutations" (W. J. Schull, ed.), p. 172. Univ. of Michigan Press, Ann Arbor, Michigan, 1960.
244. I. A. Rapoport, *Dokl. Akad Nauk. SSSR* **58**, 119 (1947).
245. F. Mukai, S. Belman, W. Troll, and I. Hawryluk, *Proc. Am. Assoc. Cancer Res.* **8**, 49 (1967).
246. H. H. Smith and A. M. Srb, *Science* **114**, 490 (1951).
247. R. W. Kaplan, *Naturwissenschaften* **49**, 457 (1962).
248. C. P. Swanson and T. Merz, *Science* **129**, 1364 (1959).
249. H. H. Smith and T. A. Lofty, *Am. J. Botany* **42**, 750 (1955).
250. I. A. Rapoport, *Byul. Mosk. Obshchestva Ispytatelei Prirody, Otd. Biol.* **67**, 109 (1962).
251. T. Alderson, *Nature* **203**, 1404 (1964).
252. I. A. Rapoport, *Dokl.—Biol. Sci. Sect.* (*English Transl.*) **141**, **1476** (1961).
253. M. Pelecanos and T. Alderson, *Mutation Res.* **1**, 173 (1964).
254. T. Alderson and M. Pelecanos, *Mutation Res.* **1**, 182 (1964).
255. M. Pelecanos, *Nature* **210**, 1294 (1965).
256. T. Alderson, *Brit. Empire Cancer Campaign, Ann. Rept.* **40**, 416 (1963).
257. S. Zamenhof, G. Leidy, E. Hahn, and H. Alexander, *J. Bacteriol.* **72**, 1 (1956).
258. M. Westergaard, *Abhandl. Deut. Akad. Wiss. Berlin, Kl. Med.* **1**, 30 (1960).
259. H. Heslot, *Abhandl. Deut. Akad. Wiss. Berlin, Kl. Med.* **1**, 98 (1960).
260. G. Kølmark, *Compt. Rend. Trav. Lab. Carlsberg., Ser. Physiol.* **26**, 205 (1956).
261. A. Loveless and W. C. J. Ross, *Nature* **166**, 1113 (1950).
262. E. Bautz and E. Freese, *Proc. Natl. Acad. Sci. U.S.* **46**, 1585 (1960).
263. N. Loprieno, *Mutation Res.* **3**, 486 (1966).
264. A. T. Natarajan and M. S. Ramanna, *Nature* **211**, 1099 (1966).
265. M. Partington and H. Jackson, *Genet. Res.* **4**, 333 (1963).
266. M. Partington and A. J. Bateman, *Heredity* **19**, 191 (1964).
267. E. Gläss and H. Marquardt, *Z. Vererbungslehre* **98**, 167 (1966).
268. O. G. Fahmy and M. J. Fahmy, *Nature* **180**, 31 (1957).

269. J. L. Epler, *Genetics* **54**, 31 (1966).
270. J. B. Jenkins, *Mutation Res.* **4**, 90 (1967).
271. I. A. Rapoport, *Dokl. Vses. Akad. Sel'skokhoz. Nauk* **12**, 12 (1947).
272. O. G. Fahmy and M. J. Fahmy, *Genetics* **46**, 1111 (1961).
273. O. G. Fahmy and M. J. Fahmy, *Genetics* **46**, 447 (1961).
274. N. M. Schwartz, *Genetics* **48**, 1357 (1963).
275. B. S. Strauss, *Nature* **191**, 730 (1961).
276. W. G. Verly, H. Barbason, J. Dusart, and A. Petispas-Dewandre, *Biochim. Biophys. Acta* **145**, 752 (1967).
277. J. Corban, *Mol. Gen. Genet.* **103**, 42 (1968).
278. J. Nečašek, P. Pikalek, and J. Drobnik, *Mutation Res.* **4**, 409 (1967).
279. M. Osborn, S. Person, S. Philips, and F. Funk, *J. Mol. Biol.* **26**, 437 (1967).
280. D. R. Krieg, *Genetics* **48**, 561 (1963).
281. A. Loveless, *Nature* **181**, 1212 (1958).
282. H. V. Malling and F. J. DeSerres, *Mutation Res.* **6**, 181 (1968).
283. A. Nasim and C. Auerbach, *Mutation Res.* **4**, 1 (1967).
284. A. Nasim, *Mutation Res.* **4**, 753 (1967).
285. H. Heslot, *Abhandl. Deut. Akad. Wiss. Berlin, Kl. Med.* **2**, 193 (1961).
286. G. Lindegren, Y. L. Hwang, Y. Oshima, and C. C. Lindegren, *Can. J. Genet. Cytol.* **7**, 491 (1965).
287. F. Lingens and O. Oltmans, *Z. Naturforsch.* **19b**, 1058 (1964).
288. E. H. Y. Chu and H. V. Malling, *Proc. Natl. Acad. Sci. U.S.* **61**, 1306 (1968).
289. U. H. Ehling, R. B. Cumming, and H. V. Malling, *Mutation Res.* **5**, 417 (1968).
290. B. M. Cattanach, C. E. Pollard, and J. H. Isaacson, *Mutation Res.* **6**, 297 (1968).
291. R. N. Rao and A. T. Natarajan, *Mutation Res.* **2**, 132 (1965).
292. M. S. Swaminathan, V. L. Chopra, and S. Bhaskaran, *Indian J. Genet.* **22**, 192 (1962).
293. A. T. Natarajan and M. D. Upadhya, *Chromosoma* **15**, 156 (1964).
294. J. Moutschen-Dahmen, A. Moes, and J. Gilot, *Experientia* **20**, 494 (1964).
295. V. V. Shevchenko, *Genetika* **4**, 24 (1968).
296. G. Röhrborn, *Z. Vererbungslehre* **90**, 116 (1959).
297. G. Röhrborn, *Z. Vererbungslehre* **90**, 457 (1959).
298. O. G. Fahmy and M. J. Fahmy, *Nature* **177**, 996 (1956).
299. E. Gebhart, *Humangenetik* **7**, (1970) (in press).
300. E. Gebhart, *Mutation Res.* **7**, 254 (1969).
301. A. Michaelis and R. Rieger, *Zuschter* **30**, 150 (1960).
302. K. Rieger and A. Michaelis, *Kulturpflanze* **8**, 23 (1960).
303. J. Moutschen and M. Moutschen-Dahmen, *Hereditas* **44**, 415 (1958).
304. J. Moutschen and M. Moutschen-Dahmen, *Experientia* **15**, 320 (1959).
305. J. Moutschen, R. Matagne, and J. Gilot, *Nature* **210**, 762 (1966).
306. J. Moutschen, R. Matagne, and J. Gilot, *Bull. Soc. Botan. Belg.* **100**, 11 (1967).
307. J. Moutschen and M. Reekmans, *Caryologia* **17**, 495 (1964).
308. J. Moutschen, *Cellule* **65**, 163 (1965).
309. W. Ostertag and W. Kersten, *Exptl. Cell Res.* **39**, 29 (1965).
310. R. I. DeMars, *Nature* **172**, 964 (1953).
311. D. A. Ritchie, *Genet. Res.* **5**, 168 (1964).
312. D. A. Ritchie, *Genet. Res.* **6**, 474 (1965).
313. C. M. Calberg-Bacq, M. Delmelle, and J. Duchesne, *Mutation Res.* **6**, 15 (1968).
314. A. Orgel and S. Brenner, *J. Mol. Biol.* **3**, 762 (1961).
315. E. M. Witkin, *Cold Spring Harbor Symp. Quant. Biol.* **12**, 256 (1947).
316. C. J. Avers and C. D. Dryfuss, *Nature* **206**, 850 (1965).

317. C. J. Avers, C. R. Pfeifer, and M. W. Rancourt, *J. Bacteriol.* **90**, 481 (1965).
318. W. Lotz, R. W. Kaplan, and H. D. Mennigmann, *Mutation Res.* **6**, 329 (1968).
319. T. Alderson and A. H. Khan, *Nature* **215**, 1080 (1967).
320. P. P. Puglesi, *Mutation Res.* **4**, 289 (1967).
321. G. E. Magni, R. C. von Borstel, and S. Sora, *Mutation Res.* **1**, 227 (1964).
322. E. A. Carlson and I. I. Oster, *Genetics* **46**, 856 (1961).
323. J. L. Southin, *Mutation Res.* **3**, 54 (1966).
324. E. A. Carlson and I. I. Oster, *Genetics* **47**, 561 (1962).
325. B. N. Ames and H. J. Whitfield, Jr., *Cold Spring Harbor Symp. Quant. Biol.* **31**, 221 (1966).
326. H. E. Brockman and W. Goben, *Science* **147**, 750 (1965).
327. H. V. Malling, *Mutation Res.* 265 (1967).
328. H. V. Malling and F. J. deSerres, *Mutation Res.* **4**, 425 (1967).
329. S. Kumar, U. Aggarwal, and M. S. Swaminathan, *Mutation Res.* **4**, 155 (1967).
330. R. Mukherjee, *Genetics* **51**, 947 (1965).
331. D. T. Suzuki, *Genetics* **51**, 635 (1965).
332. R. H. Smith, *Mutation Res.* **7**, 231 (1969).
333. M. W. Shaw and M. M. Cohen, *Genetics* **51**, 181 (1965).
334. M. M. Cohen and M. W. Shaw, *J. Cell Biol.* **23**, 386 (1964).
335. P. C. Nowell, *Exptl. Cell Res.* **33**, 445 (1964).
336. G. Zetterberg and B. A. Kihlman, *Mutation Res.* **2**, 470 (1965).
337. M. M. Cohen, M. W. Shaw, and A. P. Craig, *Proc. Natl. Acad. Sci. U.S.* **50**, 16 (1963).
338. T. T. Puck, *Science* **144**, 565 (1964).
339. B. A. Kihlman, *Mutation Res.* **1**, 54 (1964).
340. B. A. Kihlman and G. Odmark, *Mutation Res.* **2**, 494 (1965).
341. M. G. Gabridge, E. J. Oswald, and M. S. Legator, *Mutation Res.* **7**, 117 (1969).
341a. S. M. Kolbye and M. S. Legator, *Mutation Res.* **6**, 387 (1968).
342. V. W. Mayer and M. S. Legator, *J. Agr. Food Chem.* **17**, 454 (1969).
343. P. Sentein, *Compt. Rend. Soc. Biol.* **149**, 1621 (1955).
344. R. F. J. Withers, *Symp. Mutational Process, Mech. Mutation Inducing Factors, Prague, 1965* p. 359.
345. V. N. Iyer and W. Szybalski, *Proc. Natl. Acad. Sci. U.S.* **44**, 446 (1958).
346. A. Zampieri and J. Greenberg, *Genetics* **57**, 41 (1967).
347. S. Nakamura, S. Omura, M. Hamada, T. Nishimura, H. Yamaki, N. Tanaka, Y. Okami, and H. Umezawa, *J. Antibiotics* (*Tokyo*) **A20**, 217 (1967).
348. C. A. Amman and R. S. Safferman, *Antibiot. Chemotherapy* **8**, 1 (1958).
349. R. Rieger and A. Michaelis, *Kulturpflanze* **10**, 212 (1962).
350. N. Tanaka and A. Sugimura, *Proc. Intern. Genet. Symp., Tokyo Kyoto, 1956* p. 189. Sci. Council Japan, Veno Park, Tokyo, 1957.
351. N. F. Jacobs, R. L. Neu, and L. I. Gardner, *Mutation Res.* **7**, 251 (1969).
352. G. M. Jagiello, *Mutation Res.* **6**, 28 (1968).
353. K. Kajiwara, V. H. Kim, and G. C. Mueller, *Cancer Res.* **26**, 233 (1966).
354. B. Djordjevic and J. H. Kim, *Cancer Res.* **27**, 2255 (1967).
355. E. Mattingly, *Mutation Res.* **4**, 51 (1967).
356. B. A. Kihlman, G. Odmark, and B. Hartley, *Mutation Res.* **4**, 783 (1967).
357. A. DiMarco, M. Soldati, A. Fionette, and T. Dasdia, *Tumori* **49**, 235 (1963).
358. A. Theologides, J. W. Yarbro, and J. Kennedy, *Cancer* **21**, 16 (1968).
359. B. K. Vig, S. B. Kontras, and L. D. Samuels, *Experientia* **24**, 271 (1968).
360. B. K. Vig, S. B. Kontras, E. F. Paddock, and L. D. Samuels, *Mutation Res.* **5**, 279 (1968).
361. B. K. Vig, S. B. Kontras, and A. M. Aubele, *Mutation Res.* **7**, 91 (1969).

362. P. P. Slonimski, G. Perrodin, and J. H. Croft, *Biochem. Biophys. Res. Commun.* **30**, 232 (1968).
363. K. Yamagata and M. Oda, *Hakko Kogaku Zasshi* **35**, 67 (1957); *Chem. Abstr.* **51**, 16687B (1957).
364. K. Yamagata, M. Oda, and T. Ando, *J. Ferment. Technol.* **34**, 378 (1956); *Chem. Abstr.* **51**, 12231 (1957).
365. M. R. Hanna, *Can. J. Botany* **39**, 757 (1961).
366. B. O. Gillberg, G. Zetterberg, and G. Swanbeck, *Nature* **214**, 415 (1967).
367. H. Lüers, R. Gönnert, and H. Mauss, *Z. Vererbungslehre* **87**, 93 (1955).
368. G. Obe, *Mol. Gen. Genet.* **103**, 326 (1969).
369. H. R. Mahler and M. B. Baylor, *Proc. Natl. Acad. Sci. U.S.* **58**, 256 (1967).
370. S. Irwin and J. Egozcue, *Science* **157**, 313 (1967).
371. M. M. Cohen, M. J. Marinello, and M. Back, *Science* **155**, 1417 (1967).
372. M. M. Cohen, K. Hirschhorn, and W. A. Frosch, *New Engl. J. Med.* **277**, 1043 (1967).
373. L. F. Jarvik and T. Lato, *Lancet* **I**, 250 (1968).
374. J. Egozcue, S. Irwin, and C. A. Maruffo, *J. Am. Med. Assoc.* **204**, 214 (1968).
375. N. E. Skakkebaek, J. Philip, and O. J. Rafaelson, *Science* **160**, 1246 (1968).
376. M. M. Cohen and A. B. Mukherjee, *Nature* **219**, 1072 (1968).
377. W. D. Loughman, T. W. Sargeant, and D. M. Israelstam, *Science* **158**, 508 (1967).
378. R. S. Sparkes, J. Melnyk, and L. P. Bozzetti, *Science* **160**, 1343 (1968).
379. L. Bender and S. Sankar, *Science* **159**, 749 (1968).
380. S. Sturelid and B. A. Kihlman, *Hereditas* **62**, 259 (1969).
381. L. S. Browning, *Science* **161**, 1022 (1968).
382. D. Grace, E. A. Carlson, and P. Goodman, *Science* **161**, 691 (1968).
383. G. Zetterberg, *Hereditas* **62**, 262 (1969).
384. L. E. Andrew, *Am. Naturalist* **4**, 708 (1967).
385. W. Ostertag and J. Haake, *Z. Vererbungslehre* **98**, 299 (1966).
386. S. Mittler, J. E. Mittler, and S. L. Owens, *Nature* **214**, 424 (1967).
387. A. M. Clark and E. G. Clark, *Mutation Res.* **6**, 227 (1968).
388. L. E. Andrew, *Am. Naturalist* **93**, 135 (1959).
389. K. Gezelius and N. Fries, *Hereditas* **38**, 112 (1952).
390. A. Novick, *Brookhaven Symp. Biol.* **8**, 101 (1956).
391. H. E. Kubitschek and H. E. Bendigkeit, *Mutation Res.* **1**, 113 (1964).
392. N. Fries and B. Kihlman, *Nature* **162**, 573 (1948).
393. G. Zetterberg, *Hereditas* **46**, 229 (1960).
394. W. Ostertag, *Mutation Res.* **3**, 249 (1966).
395. W. Ostertag, E. Duisberg, and M. Stürmann, *Mutation Res.* **2**, 293 (1965).
396. B. A. Kihlman and A. Levan, *Hereditas* **35**, 109 (1949).
397. W. Kuhlman and W. Ostertag, *Cancer Res.* **28**, 227 (1968).
398. W. Kuhlman, H. G. Fromme, E. M. Heege, and W. Ostertag, *Cancer Res.* **28**, 2375 (1968).
399. M. F. Lyon, J. S. R. Phillips, and A. G. Searle, *Z. Verebungslehre* **93**, 7 (1962).
400. B. M. Cattanach, *Z. Vererbungslehre* **93**, 215 (1962).
401. J. F. Jackson, *J. Cell Biol.* **22**, 291 (1964).
402. G. Zetterberg, *Hereditas* **46**, 279 (1959).
403. J. McLeish, *Heredity* **6**, 385 (1953).
404. B. A. Kihlman, *Exptl. Cell Res.* **8**, 345 (1955).
405. K. Sax and H. J. Sax, *Japan. J. Genet.* **43**, 89 (1968).
406. D. Stone, E. Lamson, Y. S. Chang, and K. Pickering, *Abstr. 19th Ann. Meeting Tissue Culture Assoc., Puerto Rico, 1968* p. 60.

407. D. Stone, E. Lamson, Y. S. Chang, and K. W. Pickering, *Science* **164**, 568 (1969).
408. M. Legator, *Med. World News* **9**, 25 (1968).
409. M. Legator, K. A. Palmer, S. Green, and K. W. Petersen, *Chem. Eng. News* **48**, 37 (1968).
410. L. S. Tsarapin, *Zashch. Vosstanov. Luchevykh Povrezhdeniyakh, Akad. Nauk SSSR* p. 142 (1966); *Chem. Abstr.* **67**, 18400Y (1967).
411. N. L. Delone, *Biofizika* **3**, 717 (1958); *Chem. Abstr.* **53**, 4448 (1959).
412. N. L. Delone, *Dokl. Akad. Nauk SSSR* **119**, 800 (1958); *Chem. Abstr.* **52**, 15660 (1958).
413. E. R. McDonald and E. P. Kaufman, *Exptl. Cell Res.* **12**, 415 (1957).
414. R. Wakonig and T. J. Arnason, *Proc. Benet. Soc. Can.* **3**, 37 (1958).
415. R. Rieger, H. Nicoloff, and A. Michaelis, *Biol. Zentr.* **82**, 393 (1963).
416. B. P. Kaufman and M. R. McDonald, *Proc. Natl. Acad. Sci. U.S.* **43**, 255 and 262 (1957).
417. N. B. K. Yubova, *Dokl. Akad. Nauk SSSR* **138**, 681 (1961); *Chem. Abstr.* **55**, 21396 (1961).
418. C. Auerbach and J. M. Robson, *Nature* **154**, 81 (1944).
419. C. Auerbach and J. M. Robson, *Proc. Roy. Soc. Edinburgh* **B62**, 284 (1947).
420. C. Auerbach, M. Y. Ansari, and J. M. Robson, *Rept. Min. Supply* **Y18171** (1943).
421. K. Sharma and A. Sharma, *Nucleus (Calcutta)* **5**, 127 (1962).
422. H. B. Strack, E. B. Freese, and E. Freese, *Mutation Res.* **1**, 10 (1964).
423. E. E. Horn and R. M. Herriot, *Proc. Natl. Acad. Sci. U.S.* **48**, 1409 (1962).
424. R. Litman and H. Ephrussi-Taylor, *Compt. Rend.* **249**, 838 (1959).
425. S. E. Bresler, V. L. Kalinin, and D. A. Perumov, *Mutation Res.* **5**, 209 (1968).
426. F. Kaudewitz, *Abhandl. Deut. Akad. Wiss. Berlin, Kl. Med.* p. 86 (1960); *Chem. Abstr.* **55**, 671 (1961).
427. A. Reisenstark and J. L. Rosner, *Genetics* **49**, 343 (1964).
428. R. Rudner, *Z. Vererbungslehre* **92**, 336 (1961).
429. F. J. deSerres, H. E. Brockman, W. E. Barnett, and H. G. Kølmark, *Mutation Res.* **4**, 415 (1967).
430. H. V. Malling, *Mutation Res.* **2**, 320 (1965).
431. F. K. Zimmermann, R. Schwaier, and U. von Laer, *Z. Vererbungslehre* **98**, 230 (1966).
432. R. Schwaier, N. Nashed, and F. K. Zimmermann, *Mol. Gen. Genet.* **102**, 290 (1968).
433. A. Nasim and C. H. Clarke, *Mutation Res.* **2**, 395 (1965).
434. N. Loprieno, R. Guglielminetti, S. Bonatti, and A. Abbondandolo, *Mutation Res.* **8**, 65 (1969).
435. R. A. Steinberg and C. Thom, *Proc. Natl. Acad. Sci. U.S.* **26**, 363 (1940).
436. I. Tessman, *Virology* **9**, 375 (1959).
437. I. Tessman, R. K. Poddar, and S. Kumar, *J. Mol. Biol.* **9**, 352 (1964).
438. S. Benzer, *Proc. Natl. Acad. Sci. U.S.* **47**, 403 (1961).
439. W. Vielmetter and C. M. Wiedner, *Z. Naturforsch.* **14b**, 312 (1959).
440. C. E. Nasrat, *Nature* **207**, 439 (1965).
441. J. H. Northrup, *J. Gen. Physiol.* **46**, 971 (1963).
442. D. Scott, *Mutation Res.* **5**, 65 (1968).
443. H. G. Evans and D. Scott, *Genetics* **49**, 17 (1964).
444. M. A. McManus, *Nature* **185**, 44 (1960).
445. K. D. Wuu and W. F. Grant, *Botan. Bull. Acad. Sinica* **8**, 191 (1967).
446. G. E. Graf, *J. Heredity* **48**, 155 (1957).
447. J. M. Carlson, *Iowa State Coll. J. Sci.* **29**, 105 (1954).
448. V. V. Shevchenko, *Genetika* **1**, 86 (1965).
449. L. G. Dubinina and N. P. Dubinin, *Genetika* **4**, 5 (1968).
450. M. Legator and J. Verrett, *Conf. Biol. Effects Pesticides Mammal. Systems, New York, 1967* Abstr. No. 6, p. 17. N.Y. Acad. Sci., 1967.

451. E. Freese, E. Bautz, and E. B. Freese, *Proc. Natl. Acad. Sci. U.S.* **47**, 845 (1961).
452. H. K. Jain and R. N. Raut, *Nature* **211**, 652 (1966).
453. F. Lingens, *Z. Naturforsch.* **19b**, 151 (1964).
454. A. Rutishauser and W. Bollag, *Experientia* **19**, 131 (1963).
455. F. K. Zimmermann and R. Schwaier, *Naturwissenschaften* **54**, 251 (1967).
456. E. Freese, S. Sklarow, and E. B. Freese, *Mutation Res.* **5**, 343 (1968).
457. E. B. Freese and E. Freese, *Proc. Natl. Acad. Sci. U.S.* **52**, 1289 (1964).
458. E. Freese and H. B. Strack, *Proc. Natl. Acad. Sci. U.S.* **48**, 1796 (1962).
459. E. Freese and E. B. Freese, *Radiation Res.* Suppl. 6, 97 (1966).
460. I. Tessman, H. Ishiwa, and S. Kumar, *Science* **148**, 507 (1965).
461. E. Freese, E. B. Freese, and E. Bautz, *J. Mol. Biol.* **3**, 133 (1961).
462. H. V. Malling, *Mutation Res.* **3**, 470 (1966).
463. W. Engel, W. Krone, and U. Wold, *Mutation Res.* **4**, 353 (1967).
464. C. F. Somers and T. C. Hsu, *Proc. Natl. Acad. Sci. U.S.* **48**, 937 (1962).
465. E. Borenfreund, M. Krim, and A. Bendich, *J. Natl. Cancer Inst.* **32**, 667 (1964).
466. H. V. Malling, *Mutation Res.* **4**, 559 (1967).
467. J. J. Oppenheim and W. N. Fishbein, *Cancer Res.* **25**, 980 (1965).
468. A. Bendich, E. Borenfreund, G. C. Korngold, and M. Krim, *Federation Proc.* **22**, 582 (1963).
469. W. K. Sinclair, *Science* **24**, 1729 (1965).
470. B. A. Kihlman, T. Eriksson, and G. Odmark, *Hereditas* **55**, 386 (1966).
471. S. Belman, W. Troll, G. Teebor, R. Reinhold, B. Fishbein, and F. Mukai, *Proc. Am. Assoc. Cancer Res.* **7**, 6 (1966).
472. S. Belman, W. Troll, G. Teebor, and F. Mukai, *Cancer Res.* **28**, 535 (1968).
473. G. Perez and J. L. Radomski, *Ind. Med. Surg.* **34**, 714 (1965).
474. M. Vogt, *Experientia* **4**, 68 (1948).
475. M. Vogt, *Publ. Staz. Zool. Napoli* **22**, Suppl., 114 (1950).
476. R. Latarjet, N. P. Buu-Hoi, and C. A. Elias, *Publ. Staz. Zool. Napoli* **22**, Suppl., 76 (1950).
477. V. Bryson, *Hereditas* Suppl., p. 545 (1945).
478. A. Bateman, *Mutation Res.* **4**, 710 (1967).
479. F. Oehlkers, *Z. Induktive Abstammungs-Vererbungslehre* **81**, 313 (1943).
480. E. Boyland, R. Nery, and K. S. Peggie, *Brit. J. Cancer* **19**, 878 (1965).
481. E. B. Freese, *Genetics* **51**, 953 (1965).
482. E. B. Freese, J. Gerson, H. Taber, H. J. Rhaese, and E. Freese, *Mutation Res.* **4**, 517 (1967).
483. T. Okabayashi, *Ferment. Ind.* **33**, 513 (1955).
484. T. Okabayashi, M. Ide, A. Yoshimoto, and M. Otsubo, *Chem. & Pharm. Bull.* (*Tokyo*) **13**, 610 (1965).
485. S. Mashima and Y. Ikeda, *J. Appl. Microbiol.* **6**, 45 (1958).
486. I. H. Pan, *J. Formosan Med. Assoc.* **59**, 41 (1960).
487. I. H. Pan, *J. Formosan Med. Assoc.* **62**, 107 (1963).
488. T. H. Yoshida, Y. Kurita, and K. Moriwaki, *Gann* **56**, 513 (1956).
489. T. Mita, R. Tokuzen, F. Fukuoka, and W. Nakahara, *Gann* **56**, 293 (1965).
490. Y. Kurita, T. H. Yoshida, and K. Moriwaki, *Japan. J. Genet.* **40**, 365 (1965).
491. T. Okabayashi, *Chem. & Pharm. Bull.* (*Tokyo*) **10**, 1127 (1962).
492. O. Wyss, W. S. Stone, and J. B. Clark, *J. Bacteriol.* **54**, 767 (1947).
493. O. Wyss, J. B. Clark, F. Haas, and W. Stone, *J. Bacteriol.* **56**, 51 (1948).
494. F. L. Haas, J. B. Clark, O. Wyss, and W. S. Stone, *Am. Naturalist* **74**, 261 (1950).
495. R. P. Wagner, C. R. Haddox, R. Fuerst, and W. S. Stone, *Genetics* **35**, 237 (1950).

496. D. Luzzati, H. Schweitz, M. L. Bach, and M. R. Chevallier, *J. Chim. Phys.* **58**, 1021 (1961).
497. A. Zamenhof, H. E. Alexander, and G. Leidy, *J. Exptl. Med.* **98**, 373 (1953).
498. R. Latarjet, N. Rebeyrotte, and P. Demersen, *in* "Organic Peroxides in Radiobiology" (M. Haissinski, ed.) p. 61. Pergamon Press, Oxford, 1958.
499. C. O. Doudney, *Mutation Res.* **6**, 345 (1968).
500. L. S. Altenberg, *Proc. Natl. Acad. Sci. U.S.* **40**, 1037 (1940).
501. L. S. Altenberg, *Genetics* **43**, 662 (1958).
502. A. Loveless, *Nature* **167**, 338 (1951).
503. F. Oehlkers, *Heredity* Suppl. 6, 95 (1953).
504. R. Latarjet, *Ciba Found. Symp. Ionizing Radiations Cell Metab.* p. 275 (1957).
505. D. Luzzati and M. R. Chevallier, *Ann. Inst. Pasteur* **93**, 366 (1957).
506. F. H. Sobels, *Drosophila Inform. Serv.* **28**, 150 (1954).
507. F. H. Sobels, *Nature* **177**, 979 (1956).
508. M. Demerec, *Brit. J. Cancer* **2**, 114 (1948).
509. G. H. Scherr, M. Fishman, and R. H. Weaver, *Genetics* **39**, 141 (1954).
510. M. Demerec, *Nature* **159**, 604 (1947).
511. M. Demerec, *Genetics* **33**, 337 (1948).
512. R. W. Barratt and E. L. Tatum, *Cancer Res.* **11**, 234 (1951).
513. R. W. Barratt and E. L. Tatum, *Ann. N.Y. Acad. Sci.* **71**, 1072 (1958).
514. J. C. Carr, *Brit. J. Cancer* **1**, 152 (1947).
515. C. Auerbach, *Proc. Roy. Soc. Edinburgh* **60**, 559 (1940).
516. F. K. Zimmermann, *Z. Krebsforsch.* **72**, 65 (1969).
517. L. C. Strong, *Proc. Natl. Acad. Sci. U.S.* **31**, 290 (1945).
518. L. C. Strong, *Genetics* **32**, 108 (1946).
519. L. C. Strong, *Am. Naturalist* **81**, 50 (1947).
520. L. C. Strong, *A.M.A. Arch. Pathol.* **39**, 232 (1945).
521. C. A. Auerbach, *Proc. Roy. Soc. Edinburgh* **60**, 164 (1939).
522. S. Bhattacharaya, *Nature* **162**, 573 (1948).
523. R. Latarajet, *Compt. Rend. Soc. Biol.* **142**, 453 (1948).
524. M. R. Chevallier and D. Luzatti, *Compt. Rend.* **250**, 1572 (1960).
525. M. Demerec and J. Hanson, *Cold Spring Harbor Symp. Quant. Biol.* **16**, 215 (1951).
526. I. D. Steinman, V. N. Iyer, and W. Szybalski, *Arch. Biochem. Biophys.* **76**, 78 (1958).
527. N. N. Durham and O. Wyss, *J. Bacteriol.* **74**, 548 (1957).
528. L. A. Muro and R. A. Goyer, *Arch. Pathol.* **87**, 660 (1969).
529. J. Deufel, *Chromosoma* **4**, 239 (1951).

CHAPTER 7

Alkylating Agents I (Aziridines, Mustards, Nitrosamines, Nitrosamides, and Related Derivatives)

The main objective of the following three chapters on chemical mutagens is to focus on agents primarily from an environmental health consideration, and hence to stress their preparation and/or occurrence, areas of utility, salient biological and physical properties, and their identification and analysis. In the latter regard, *selected* references concerned with the detection and identification of a mutagen (metabolite or its degradation product) in biological tissues or an environmental parameter, e.g., air, water, food residues, were considered germane.

It is convenient to divide the chemical mutagens into discrete entities: (a) the alkylating agents which constitute the largest category are subdivided on the basis of functional groups (aziridines, nitrogen, sulfur, and oxygen mustards, nitrosamines, nitrosamides, epoxides, lactones, aldehydes, dialkyl sulfates, alkane sulfonic esters, and related derivatives) and (b) other chemical mutagens classified as drugs, food additives, pesticides, and, finally, miscellaneous agents. (There are overlaps in each category, e.g., alkylating agents which are drugs and pesticides, etc.)

The general literature on alkylating agents is extensive. Their chemistry and mechanism of action have been reviewed by Ross (1, 2), Johnson and Bergel (3), Alexander (4), Stacey *et al.* (5), and Warwick (6); their distribution and fate were reviewed by Smith *et al.* (7); degradation and residue products by Fishbein (8); chromatographic analysis by Fishbein and Falk (9, 10); pharmacology and therapeutic utility by Goodman and Gilman (11) and Karnofsky and Clarkson (12); carcinogenic activity by Clayson (13), Brookes and Lawley (14), and Walpole (15); and their mutagenic action by Auerbach (16), Orgel (17), Krieg (18), and Loveless (19).

1. *Ethylenimine*

Ethylenimine (aziridine) is produced via the reaction of β-chloroethylamine with sodium hydroxide and is available in commercial quantities. It is an extremely reactive compound undergoing two major types of reactions; (1) ring-opening reactions similar to those undergone by ethylene oxide and (2) ring-preserving reactions in which ethylenimine acts as a secondary amine. The ease with which the ring opening occurs arises from the strained nature of the three-membered ring. Traces of acid cause exothermic polymerization to a water-soluble polyamine (polyethylenimine).

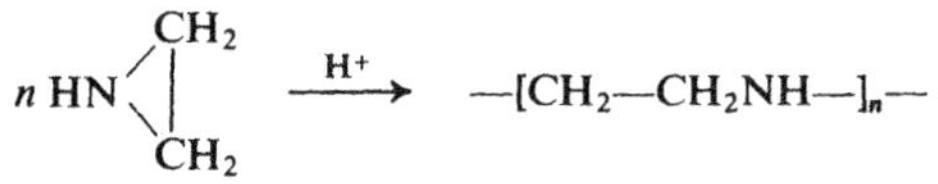

A great variety of aminoethylated derivatives can be formed by nucleophilic attack at one of the methylene groups:

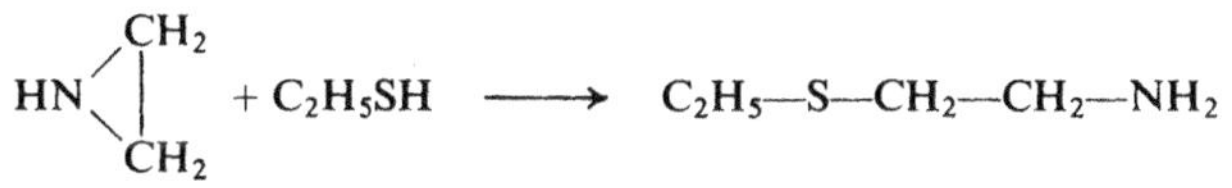

Under proper conditions ethylenimine reacts with many organic functional groups containing an active hydrogen to yield an aminoethyl derivative or products derived therefrom.

In ring-preserving reactions, ethylenimine can undergo replacement of the hydrogen atom on the nitrogen in various ways or it can form salts and metallic complexes. This type of reaction is most readily performed under basic conditions or in the presence of acid acceptors.

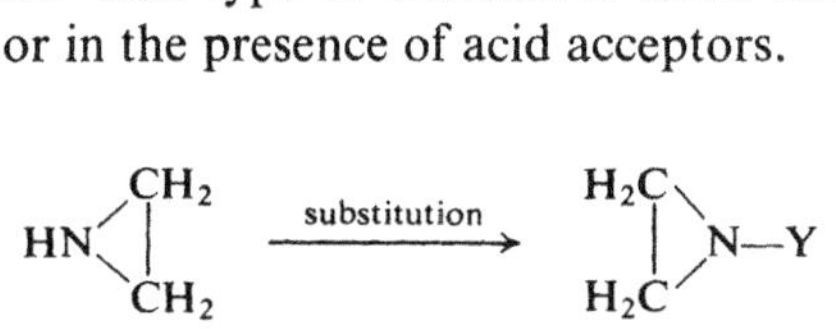

Y = alkyl or acyl group

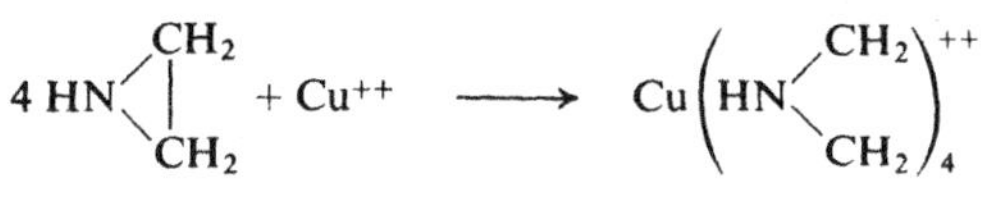

Ethylenimine, because of its dual functionality and high degree of reactivity, exhibits actual or potential utility in a broad and expanding range of applications as depicted in Table 7.1.

TABLE 7.1

Applications of Ethylenimine and Its Derivatives

Application	Component and references
1. Textiles	
a. Crease proofing	Aziridinyl ureas and amides (20)
b. Dyeing and printing	Polyethylenimines (21)
c. Flame proofing	Aziridinyl phosphorus derivatives (22, 23)
d. Shrink proofing, form stabilization, stiffening	Polyalkylenimines (24, 25)
e. Waterproofing	Carbamoyl aziridines (26)
2. Adhesives and binders	Ethylenimine and polyethylenimines (27)
3. Petroleum products and synthetic fuels	
a. Lubricant additives	Ethylenimine condensates (28)
b. Rocket and jet fuels	Ethylenimine and substituted ethylenimines (29)
4. Coatings	Polyethylenimines and ethylenimine derived polyurethans (30, 31)
5. Agricultural chemicals	
a. Insecticides	Aziridinyl phosphonates and ethylenimine-chlorinated phenol conjugates (32)
b. Chemosterilants	Tris-aziridinyl derivatives (33, 34)
c. Soil conditioners	Polyethylenimines (35)
6. Antimicrobials	Aziridinyl benzoquinones (36, 37)
7. Flocculants	Polyethylenimines (38)
8. Ion-exchange resins	Polyethylenimines (39)
9. Photographic chemicals	Polyethylenimines and phosphoryl aziridines (40)
10. Curing and vulcanizing polymers	*N*-Acylated aziridines (41, 42)
11. Surfactants	Aziridinyl alkarylureas (43)
12. Paper and printing	Polyethylenimines (44)
13. Chemotherapeutics	Tris(aziridinyl)phosphine oxides and sulfides (45, 46)

The physiological effects (47), toxicology (48–50), carcinogenicity (48), and metabolism (51, 52) of ethylenimine have been described.

Ethylenimine has been shown to cause mutation in *Drosophila* (53–58), *Neurospora* (59), wheat (60, 61), barley (62–64), and *Saccharomyces cerevisiae* (65), and chromosome aberrations in cultured human cells (66), mouse embryonic skin cultures (67), Crocker mouse Sarcoma 188 (67), and root tips of *Allium cepa* (67).

The analysis of ethylenimine has been achieved by titrimetry (50, 68), colorimetry (69–71), and gas chromatography (52).

2. *Triethylenemelamine*

Triethylenemelamine [2,4,6-tris(1-aziridinyl)-*s*-triazine; tretamine; TEM] is prepared from ethylenimine and cyanuric chloride:

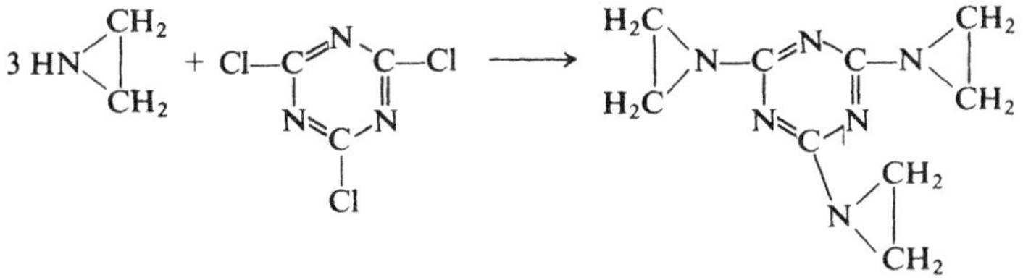

TEM is used in the manufacture of resinous products, as a cross-linking agent in textile technology, the finishing of rayon fabrics, and in the waterproofing of cellophane. Major interest, however, has been related to its medical utility as an antineoplastic agent (72, 73) and more recently its use as a chemosterilant for the housefly (*Musca domestica*) (74), screwworm (75), and oriental, melon, and Mediterranean fruit flies (76).

Similar to all aziridine chemosterilants, TEM is extremely susceptible to moisture and acidic conditions (77) (at pH 3.0, degradation is complete and occurs almost immediately, whereas minor degradation occurs in buffered or unbuffered solutions at pH 7.5). Crystalline TEM stored at room temperatures polymerizes to an inactive material.

The clinical and biological effects (78, 79) and toxicity (80–83) of TEM have been described. The pharmacological properties of TEM are very similar to those of the nitrogen mustards. Presumably, the reaction within the cell is appropriate for conversion of TEM to the quaternary ethylenimonium form.

The metabolism of ^{14}C-TEM has been studied in normal and tumor-bearing mice and in man (84–86). The radiosynthesis was carried out via the sequence:

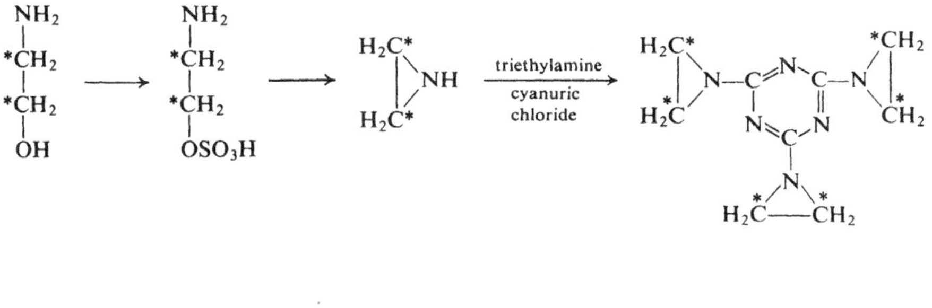

The urinary excretion pattern in the human indicated that the alkylating portion of the molecule was separated from the carrier (84) [confirming the suggestion of Mandel (79)]. *In vivo* conversions of ^{14}C-TEM to cyanuric acid occurred rapidly in the mouse (84).

The mutagenic activity of TEM has been demonstrated in mice (87–93), *Drosophila* (58, 94–99), *Musca domestica* (100), *Neurospora* (59), screwworm fly, *Cochliomyia hominivorax* (75). TEM has been shown to induce chromosome aberrations in mice (92, 101–105), *Drosophila* (97, 98, 106, 107), cultured human leukocytes (66, 108), barley (109), *Vicia faba* (110–114), *Allium cepa* (113), *S. typhimurium* (115), and *E. coli* (116).

The analysis of TEM has been carried out by paper (86), thin-layer chromatography (77), nmr spectroscopy (77), and colorimetry (69).

3. *Trenimon*

Trenimon (2,3,5-trisethylenimino-1,4-benzoquinone) was prepared by Domagk (117); it possesses clearly pronounced cytostatic activity (72, 73). The mutagenicity of Trenimon has been demonstrated in mice (87–89, 118), *Drosophila* (54), and human leukocyte chromosomes *in vitro* (119). Its causation of chromosome aberrations in barley (109), as well as its radiomimetic effects in *Vicia faba* (109), have been reported.

4. *TEPA*

TEPA [tris(1-aziridinyl)phosphine oxide; triethylene phosphoramide; aphoxide; APO] is prepared by the reaction of ethylenimine with phosphorus oxychloride in base (120), viz.:

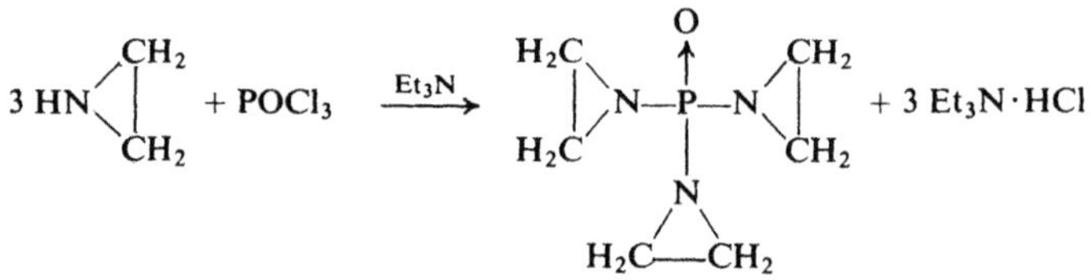

TEPA has been the most extensively employed of all the aziridine phosphine oxides. Its industrial applications include the flameproofing of textiles (121, 122), water-repellent, wash-and-wear, and crease-resistant fabrics (123), dyeing and printing (124), adhesives and binders (125), treatment of paper (126, 127) and wood (128). It is also a component in rocket fuels (129).

The above utility of TEPA (as well as other aziridines) is illustrative of the ability of these compounds to act as cross-linking agents for polymers containing active hydrogen groups such as carboxyl, phenol, sulfhydryl amide, and hydroxyl. This can be represented schematically as follows with ZH representing the active hydrogen group in the polymer chain:

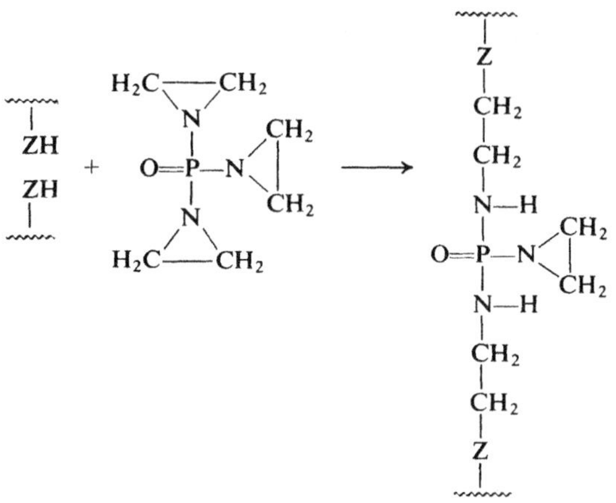

The first report of the tumor-inhibitory properties of *N*-ethylenimine-substituted phosphoramides was made by Buckley *et al.* (130), who demonstrated the effect of TEPA against Sarcoma 180. The therapeutic applications of TEPA (131) and related aziridinyl derivatives are the same as those of the nitrogen mustards. The biological, physical (132), and pharmacological properties (133) of these derivatives have been described.

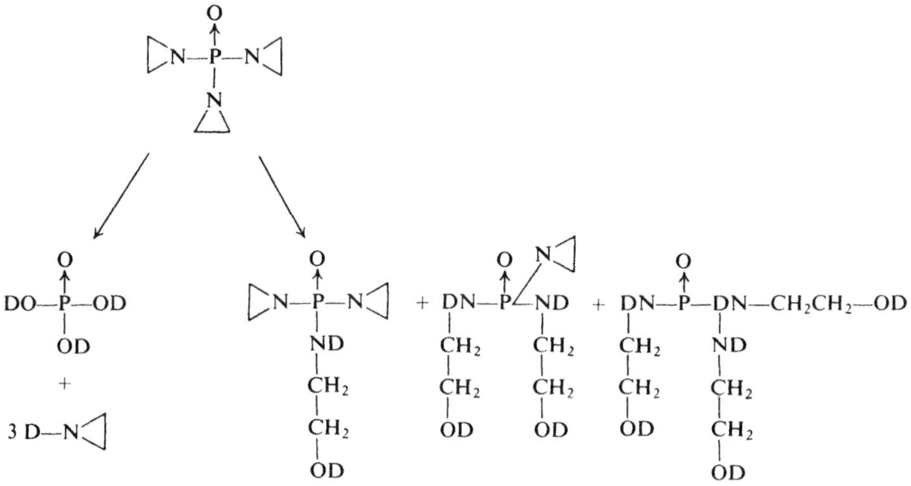

FIG. 7.1. Decomposition of TEPA in D_2O. Beroza, M. and Borkovec, A. B., *J. Med. Chem.* 7, 44 (1964).

TEPA is unstable in aqueous media [with the accompanying liberation of inorganic phosphate (77, 134)] as well as in acidic solutions. TEPA, however, has been stabilized in liquid anhydrous polyethylene glycols (135).

Beroza and Borkovec (77) investigated the effects of temperature and pH on the stability of TEPA and other related aziridine derivatives (apholate, metepa, and tretamine).

Based on nmr data, it was suggested that the decomposition of TEPA in D_2O can be represented schematically, as shown in Fig. 7.1.

The metabolic fate of ^{32}P-TEPA in tumor-bearing and control mice (136), rats (134, 137), dogs (138), and cancer patients (7, 139) has been described. Craig *et al.* (134, 137) suggested that although TEPA is a potent tumor inhibitor, it did not appear to be highly reactive *in vivo*. They also suggested that its cytotoxic activity may be due to the interference by very small proportions of the dose administered with some highly susceptible cellular mechanism which might be produced by simple hydrolysis (a), or subsequent to reaction of the compound with tissue components (b).

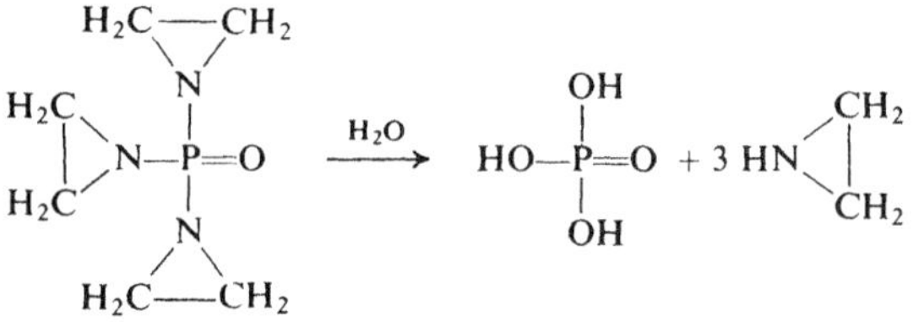
(a)

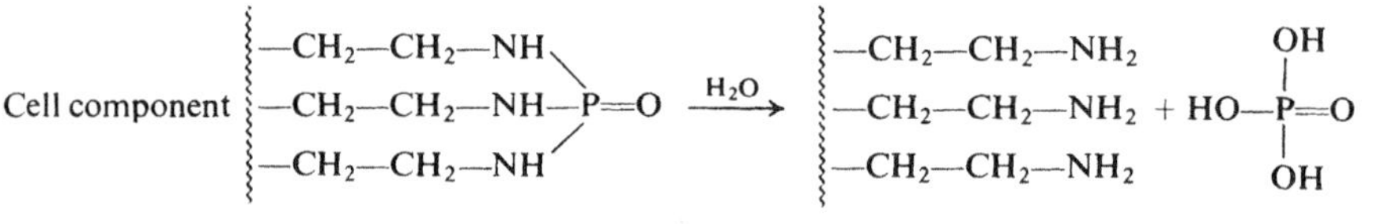
(b)

A study of the urinary metabolites of ^{32}P-TEPA (as well as ^{14}C-TEM and ^{35}S-Myleran) following IV administration to cancer patients suggested the detachment of the alkylating moiety from the "carrier" portion of the drug molecule (139).

Interest in TEPA and related compounds has been intensified dramatically in recent years by the discovery of their activity as insect chemosterilants (antifertility agents) and their potential utility as new and powerful tools for insect control and eradication. The chemistry (33, 34, 140–143), chemosterilant application (33, 34, 140–143), and toxicity (143–146) of these agents have been extensively described. Historically there is a connection between the development of cancer chemotherapeutic agents and insect chemosterilants.

The cancerous cell is in certain respects similar to embryonic cells (sperm, ovum), and the ultimate aim, i.e., the interference with cellular reproduction, may be similar in cancer chemotherapy and insect sterilization. There is inherent in such programs of eradication the dispersion of large numbers of treated insects into the environment. The determination of how much active chemosterilant remains on and in the insect that is to be released and how long it will persist is of paramount interest.

The degradation and residues of the chemosterilant aziridines have been reviewed by Fishbein (8), and the fate of ^{14}C-TEPA in houseflies (147), weevils (148), gypsy moths (149), Japanese beetles (150), Mexican fruit flies (151), armyworm moths (152), and codling moths (153) has been described.

TEPA has been shown to inhibit nucleic acid biosynthesis (154), increase nuclease activity in tumor-bearing rats (154), and reduce concentration of RNA and DNA in malignant tissues (155, 156).

The mutagenicity of TEPA has been demonstrated in the parasitic wasp, *Habrobracon* (157), in the dominant lethal test in mice (91), in *Neurospora* (158), *E. coli* (159), and bacteriophage T4 (160).

The analysis of the aziridinyl chemosterilants (e.g., TEPA, thio-TEPA, metepa, and apholate) is primarily performed using (a) colorimetric techniques (69, 161–164) based on the reaction of the alkylating agent with 4-(*p*-nitrobenzyl)pyridine under acidic conditions and subsequent formation of the colored product with alkali, and (b) thiosulfate titration (138, 165), gas chromatography (166), thin-layer chromatography (166), and nmr spectroscopy (166). Photofluorometry (138) and paper chromatography (134, 136, 139) have also been employed for the analysis of TEPA.

5. *Metepa*

Metepa [tris(2-methyl-1-aziridinyl)phosphine oxide; mapo; methaphoxide] is prepared by the reaction of 1-methyl ethylenimine and phosphoryl chloride. Metepa is as sensitive as the other aziridine chemosterilants to acidic conditions (degradation in acidic media is complete within 2 hours) (167). The effect of temperature and pH on the stability of metepa was studied by Beroza and Borkovec (77).

The utility of metepa in the chemosterilization of *Musca domestica* (168), *Aedes aegypti* (168), *Anopheles quadrimaculatus* (168), *Cochliomyia hominivorax* (169), *Stomoxys calcitrans* (169), and *Popilla japonica* (170) has been described.

The rates of absorption, degradation, and excretion of ^{32}P-mapo in mosquitoes, houseflies, and mice (171), screwworm fly (172), stable fly (172), two species of mosquitoes (168), and houseflies (168, 173) have been described.

The teratogenesis of metepa in the rat (174), its mutagenicity in mice (dominant lethal test) (91, 175), *Neurospora* (158), and bacteriophage T4

(160), as well as its induction of chromosome aberrations in *Vicia faba* (176), have been reported.

Metepa has been analyzed by paper (171, 172), thin-layer (77), and gas chromatography (77, 166), as well as by nmr spectroscopy (77).

6. *Apholate*

Apholate [2,2,4,4,6,6-hexakis(1-aziridinyl)-2,2,4,4,6,6-hexahydro-1,3,5,2,4,6-triazatriphosphorine] is prepared as illustrated:

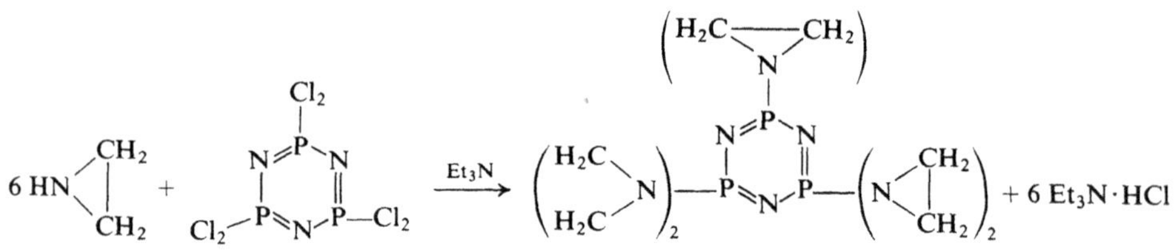

The hydrolysis and acid-catalyzed polymerization of apholate is suggested to proceed as follows (177):

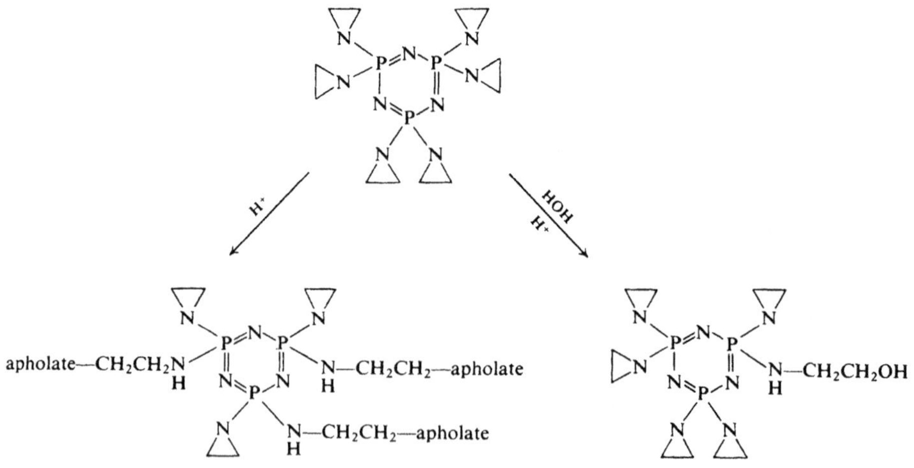

The utility of apholate as a chemosterilant for *Drosophila* (178), *Musca domestica* (179, 180), *Aedes aegypti* (181), *Cochliomyia hominivorax* (182, 183), and red bollworm (*Diparopsis castanea*) (184) has been described.

Apholate has been reported to inhibit DNA and lactic dehydrogenase

(LDH) synthesis in housefly eggs (185) as well as to interfere with the incorporation of tritiated thymidine into ovarian DNA of some of the nuclei of nurse cells and follicular cells of stable flies (186).

Apholate has been shown to induce mutations in *Neurospora* (158), to induce dominant lethal mutations in mature sperm and gonial cell death in *Musca domestica* (187), and to induce chromosome breaks (110, 188). Apholate has been analyzed by colorimetry (77) and thin-layer (77) and gas chromatography (166).

7. *Thio-TEPA*

Thio-TEPA [tris(1-aziridinyl)phosphine sulfide; thiophosphoramide] is prepared via the reaction of $PSCl_3$ and ethylenimine in the presence of base (189).

Thio-TEPA has patented applicability in dyeing (190, 191), flameproofing (192, 193), and waterproofing of textiles (194), as well as in stabilization of polymers and photographic emulsion hardening. Thio-TEPA has demonstrated clinical utility in the temporary palliation of certain cancers (195–197) (malignant breast cancer, Wilm's tumor, chronic lymphatic leukemia, ovarian carcinoma, and Hodgkin's disease).

The distribution and metabolism of ^{14}C-thio-TEPA in normal and tumor-bearing rats (198–200), dogs (138, 201), and humans (138, 201, 202) and ^{32}P-thio-TEPA in the rat, mouse, dog, and rabbit (199) have been described. The above studies revealed that thio-TEPA is metabolized rapidly *in vivo*. The primary stage is the rapid replacement of sulfur by oxygen with the formation of TEPA.

The salient differences in species metabolism are as follows: (a) in the rat TEPA is the main metabolite; (b) in the dog and rabbit three other metabolites are excreted although the major product is TEPA; and (c) in the mouse thio-TEPA is degraded so that the principal metabolite excreted is inorganic phosphate.

Thio-TEPA, similar to most of the aziridine chemosterilants, tends to polymerize in aqueous solutions, or in the presence of moisture, especially at acidic pH.

Benckhuijsen (203) studied the behavior of thio-TEPA (I) in acid medium and presented evidence of an intramolecular alkylation of the sulfur on protonation of the molecule. The five-membered ring (III) formed in this manner was hydrolyzed slowly in neutral solution at room temperature liberating an SH group. In the SH derivative (IV), two ethylenimine rings were retained, which were demonstrated to be more reactive toward water and sulfhydryl than were those of thio-TEPA. The sequence of above reactions can be depicted:

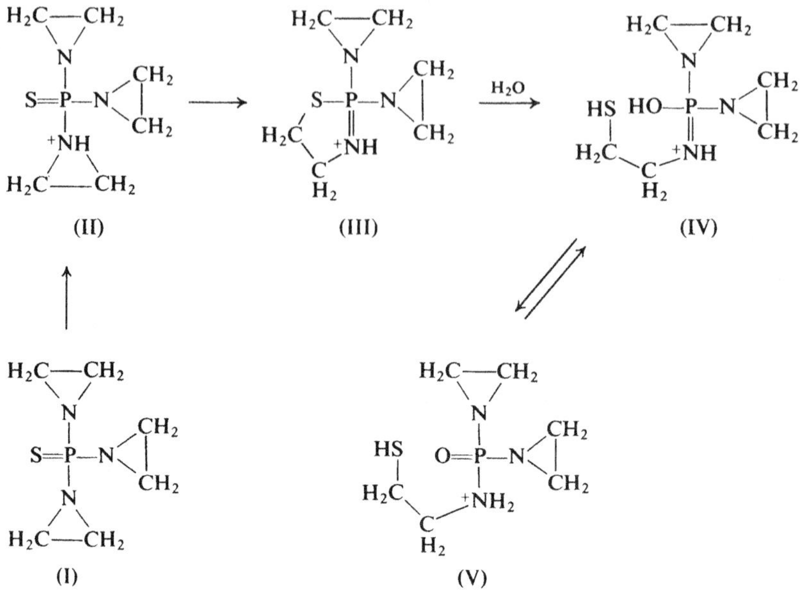

The isomerization of thio-TEPA via II to III (above) was postulated to occur by the following mechanism:

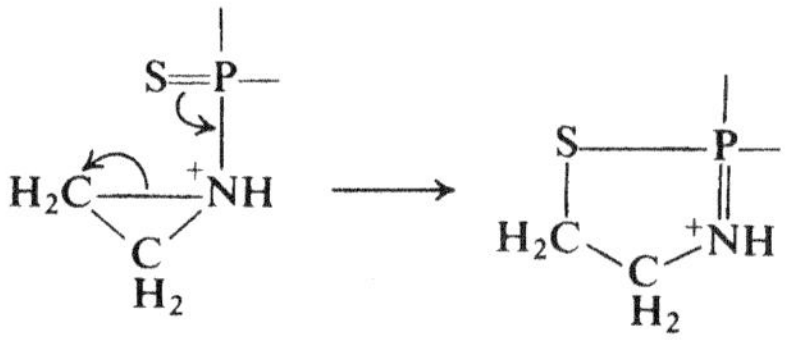

The utility of thio-TEPA in the chemosterilization of *Musca domestica* (74, 179), *Aedes aegypti* (204), *Anopholes gambiae* (204), and *Cochliomyia hominivorax* (75) has been described.

The metabolism, distribution, and residues of ^{32}P-thio-TEPA following topical application to the German cockroach (*Blatella germanica*), housefly (*Musca domestica*), stable fly (*Stomoxys calcitrans*), and boll weevil (*Anthonomus grandis*) were studied by Parrish and Arthur (205).

Thio-TEPA has been shown to reduce incorporation of ^{3}H-thymidine into DNA (206), inhibit incorporation of adenine-8-^{14}C into DNA (207), alter depolymerase activity of normal and tumor cells (154), prolong metaphase of dividing cells in tissue culture (208), and cause sticky and condensed chromosomes (209).

The mutagenic activity of thio-TEPA in mice (dominant lethal) (91), screwworm fly (*Cochliomyia hominivorax*) (210), bone marrow cells of mice (211), as well as its induction of chromosome aberrations in human chromosomes (212) and bone marrow cells of mice (211) have been described.

Thio-TEPA has been analyzed by colorimetric (69, 161–163), spectrophotofluorimetric (138), paper (138, 199, 202, 213), thin-layer (203), and gas chromatographic procedures (166).

8. *Trimethylolmelamine*

Trimethylolmelamine (2,4,6-trioxymethylamino-1,3,5-triazine; TMM) is prepared by the condensation of melamine with 3 moles formaldehyde. TMM is used as a curing agent for cellulose acetate-polyacrylate compounds (214), in crease-resistant cottons (215), water-resistant fabrics (216), in polymers with thio-TEPA for flameproofing textiles (217), as a tanning agent (218), in leather treatment (219), in the improvement of mechanical properties of elastomers by formation of resinous condensation products with dry rubber (220), and in cosmetic powder bases (221).

TMM has been shown to be effective against Walker carcinoma 256 (222, 223) and to possess significant cytostatic activity analogous to urethan or the nitrogen mustards (224, 225). The utility of TMM in human tumor therapy has also been reported (226, 227).

The kinetics of formation of TMM (228), its ionization (229), ultraviolet spectra (230), pharmacology and toxicology (231, 232), and mutagenicity in *Drosophila* (233, 234) have been described.

9. *2,2′-Dichloroethylamine*

2,2′-Dichloroethylamine (nor-HN2) is prepared by the reaction of diethanolamine and thionyl chloride (235) and has exhibited neoplasm inhibition (236, 237). The toxicology and pharmacology (238, 239) and *in vivo* and *in vitro* reactions of nor-HN2 have been described (240, 241).

Nor-HN2 has been reported to be mutagenic in *E. coli* (159, 242) but nonmutagenic in *Neurospora* (243), and is most conveniently analyzed colorimetrically using the 4-(*p*-nitrobenzyl)pyridine reagent (244).

10. *Tris(2-chloroethyl)amine*

Tris(2-chloroethyl)amine (HN3) is prepared by the reaction of tris(hydroxyethyl)amine with thionyl chloride (245). The therapeutic action of HN3 (246, 247), toxicity (248), pharmacology (249), and inhibition of growth of *Saccharomyces* (250) and HeLa cells (251) have been described.

The mutagenesis of HN3 in *Drosophila* (252), *Aspergillus niger* (253), and *Neurospora* (254) have been reported.

11. *Methyl Di(2-chloroethyl)amine*

Methyl di(2-chloroethyl)amine, (nitrogen mustard; HN2), is prepared by the action of thionyl chloride on 2,2′-(methylimino)diethanol in trichloroethylene, the free base being obtained by subsequent treatment with caustic.

Nitrogen mustards are chemically very reactive electrophilic compounds that react with and alkylate nucleophilic substances and such biologically important moieties as phosphate, amino, sulfhydryl, hydroxyl, imidazole, and carboxyl groups. The nitrogen mustards, in neutral or alkaline aqueous solution, rapidly undergo intramolecular transformation with release of chloride ion to form a cyclic ethylenimonium derivative (a quaternary ammonium compound) that is highly reactive chemically. The reaction for mechlorethamine (HN2) can be depicted as follows:

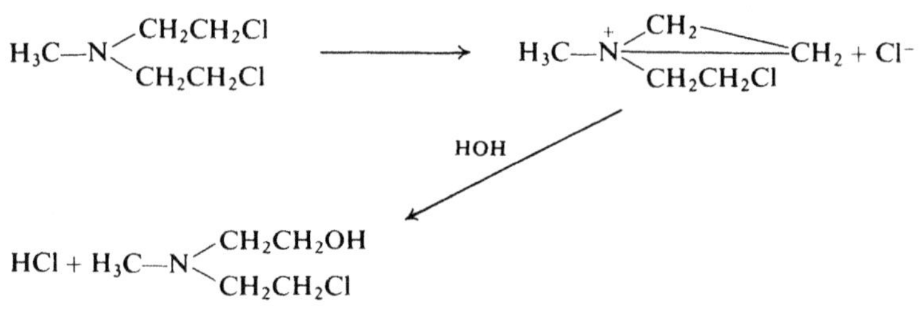

HN2, the first of the nitrogen mustards to be introduced into clinical medicine, has been more widely studied than any of its congeners. The beneficial results of HN2 in Hodgkin's disease and less predictably in other lymphomas has been well established.

The comparative clinical and biological effects (255, 256), pharmacology (257, 258), metabolism (259, 260), induction of tumors (261, 262), teratogenicity in rat (263–265), mouse (266, 267), and chick embryo (265) of HN2 have been described.

The mutagenicity of nitrogen mustard in mice (268), *Drosophila* (106, 269–274), *Habrobracon* (275), *E. coli* (159, 242, 276–280), corynebacterium (281), bacteriophage (282, 283), *P. chrysogenum* (284), *Pyrenochaeta terrestris* (285), *Penicillium notatum* (286), *Neurospora* (254, 287–291), *Ophiostoma* (292), *Aspergillus soya* (293), *A. niger* (252), maize (294), barley and wheat (295), the nonmutagenicity in phage (296, 297), and the induction of chromosome aberrations in *Vicia* (298–301) and *Allium cepa* (67) have been described.

The nitrogen mustards have been analyzed by a variety of techniques, including colorimetry (69, 161, 244, 302), fluorimetry (303), polarography (304), and paper chromatography (305).

12. *Methyl Bis(2-chloroethyl)amine N-Oxide*

Methyl bis(2-chloroethyl)amine *N*-oxide (nitromin; *N*-oxide mustard) is prepared by treating nitrogen mustard with hydrogen peroxide and acetic anhydride (306). Its use as an antineoplastic agent in cancer therapy (307), as well as its pharmacology (308, 309), mode of action (310, 311), and colorimetric determination in body fluids (312) have been described.

The degradation of *N*-oxide mustard in aqueous solution (at 37°C, for 2 hours) was shown by Arnold *et al.* (313) to proceed as follows:

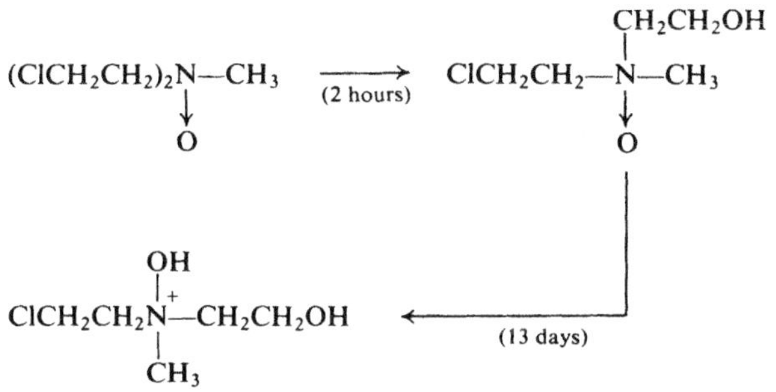

N-Oxide mustard and *aqueous* solutions of *N*-oxide mustard in the presence of copper or other heavy metals are degraded to formaldehyde and bis(2-chloroethyl)amine, and to dimethylamine, formaldehyde, and an unknown amine, respectively (314).

The mutagenicity of *N*-oxide mustard in *Drosophila* has been reported (252, 315).

13. *Sulfur Mustard*

Bis(2-chloroethyl)sulfide (mustard gas; sulfur mustard), first synthesized in 1822, represents the oldest of the alkylating agents. It received usage as a vesicant during World War I and, although it possesses tumor-inhibiting activity similar to that of the various other alkylating agents, its clinical employment has been minimal due to the inconvenience of handling the volatile, water-insoluble drug.

It has, however, patented applications in a variety of polymer areas, e.g., the preparation of polycarboxy polyethers (316), polyfunctional vinyl ethers

(317), rubberlike resins of the polysulfide type (318), and in the preparation of vat and sulfur dyes (319).

The kinetics of hydrolysis (320–322) and displacement reactions of sulfur mustard and half-sulfur mustard (320), the thermal decomposition of sulfur

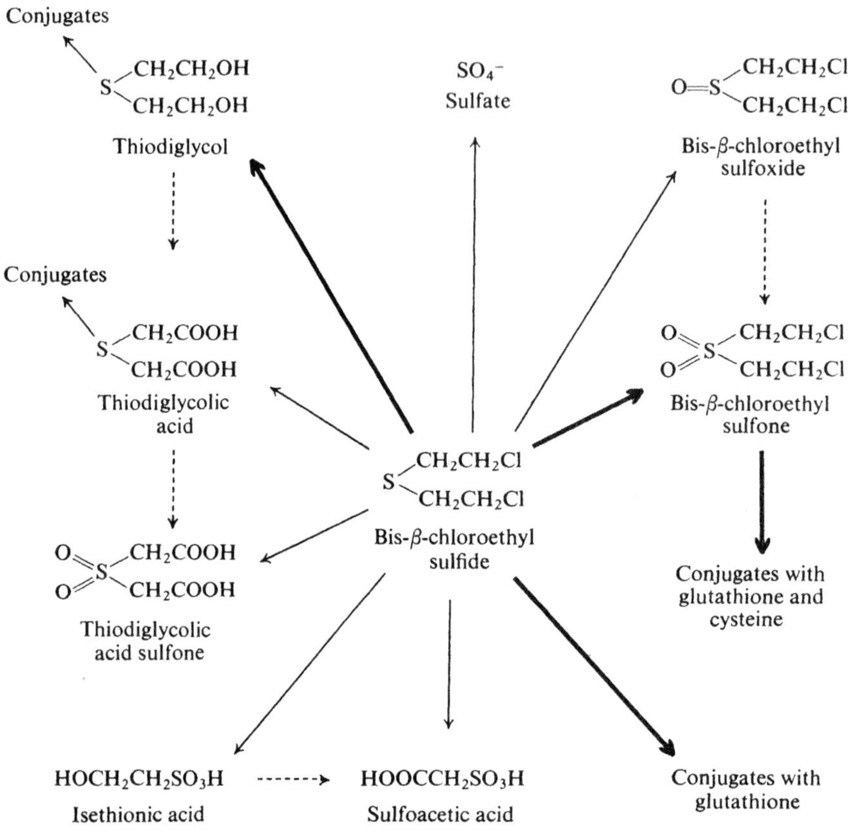

FIG. 7.2. Metabolic pathways of sulfur mustard. Heavy arrows indicate major pathways. Dashed arrows indicate possible pathways. Davison, C., Rozman, R. S., and Smith, P. K., *Biochem. Pharmacol.* 7, 65 (1961).

mustard (323) and its biological effects (324), reaction with nucleic acids *in vitro* and *in vivo* (325, 326), utility in cancer chemotherapy (327), toxicity (328, 329), and pharmacology (257) have been described.

The metabolism of sulfur mustard has been studied in the rat (7, 330), mouse (7, 331), rabbit (332), and man (330, 331). Davison *et al.* (331) described the synthesis of ^{35}S-sulfur mustard as outlined below and suggested metabolic pathways (333) (Fig. 7.2).

In the rat large amounts of thiodiglycol and of bis(2-chloroethyl)sulfone were detected chiefly as conjugates. Much of the sulfur mustard reacts immediately with glutathione and this complex is excreted. The major portions of radioactivity excreted in the urine represent compounds formed from alkylation by the drug rather than metabolites formed by enzymic action. In rodents, the major portion of radioactivity was excreted in the urine within 24 hours, whereas in man the isotope was retained for long periods.

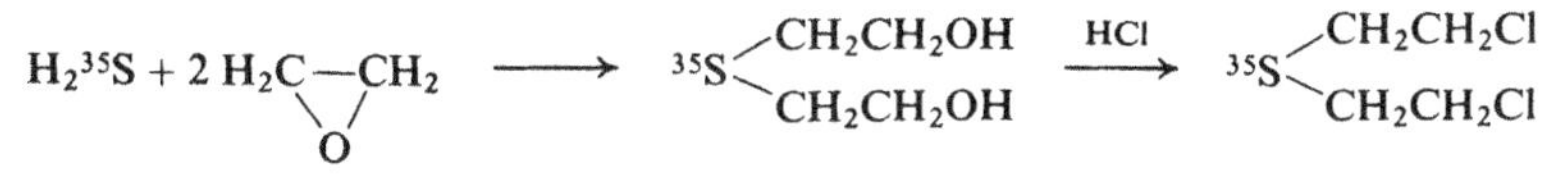

Carcinogenicity of sulfur mustard following intravenous or subcutaneous injection or inhalation, the production of sarcomas at the site, and of pulmonary adenoma, mammary and hepatic tumors, and leukemia have been observed in mice (334–336), rats (337), and respiratory neoplasia in man (338).

The teratogenicity (15, 265) of sulfur mustard, its mutagenicity in *Drosophila* (339–346), *Neurospora* (254, 347, 348), *Aspergillus nidulans* (349), *E. coli* (242), maize pollen (294), inactivation effects on T2 and T4 bacteriophages of *E. coli* (350), chromosome aberrations in *Drosophila* (343), maize pollen (294, 351), and the plant *Tradescantia* (352, 353) have been described.

Analytic procedures for sulfur mustard include colorimetry (302, 354), potentiometric titration (355), ultraviolet (356) and chlorine nuclear quadrupole resonance (357) spectroscopy, and paper (325, 333, 358) and thin-layer (359) (for the separation of both sulfur and half-sulfur mustard) chromatography.

14. *Half-Sulfur Mustard (HSM)*

(2-Chloroethyl 2-hydroxyethyl) sulfide has been prepared by (a) the partial chlorination of thioglycol (360), (b) the photochemically catalyzed addition of mercaptoethanol to vinyl chloride (361), and (c) via the reaction of sodium β-mercaptoethanol and ethylene dichloride (362, 363). (The latter synthesis yields a product free of sulfur mustard contaminant.)

The toxicity of HSM is about one-sixth that of mustard gas (364) and one-thirteenth that of nitrogen mustard (363). Clinical trials with HSM have been reported (363).

The extent of reaction of half-sulfur mustard with protein, RNA and DNA of liver, transplanted hepatoma, and leukemic spleen of mice *in vivo* was determined by Brookes and Lawley (326, 365, 366). In all cases the three cellular constituents had reacted with the formation of 7-(2-hydroxyethylthioethyl)guanine.

A comparison of the reactivity of ^{35}S-sulfur mustard and ^{35}S-half-sulfur mustard indicated that the compounds reacted very similarly with DNA both *in vitro* and *in vivo* and the only detectable difference was the formation of the di(7-guaninyl) derivative by the mustard gas (325, 326, 365, 366).

HSM has been reported to be mutagenic in *Drosophila melanogaster* (269) and inhibits the growth of yeast cells (367).

15. *2,2'-Dichloroethyl Ether*

2,2'-Dichloroethyl ether ($ClCH_2CH_2OCH_2CH_2Cl$) is obtained by (1) treating β-chloroethyl alcohol with sulfuric acid and (2) as a by-product in the production of ethylene glycol from ethylene chlorohydrin. It is extensively used in the paint and varnish industry as a solvent for many resins, including glyceryl phthalate resins, ester gums, paraffin, gum camphor, castor, linseed and other fatty oils, turpentine, polyvinyl acetate, and ethyl cellulose. It is also used as a solvent for rubber or cellulose esters but only in the presence of 10% to 30% alcohol. In the textile industry it is used for grease-spotting and the removal of paint and tar brand marks from raw wool and is incorporated in scouring and fueling soaps. Its other uses include its utility as an extractive for lubricating oils in the petroleum industry and its application as a soil insecticide (fumigant).

The acute (368–370) and chronic (371) toxicity of 2,2'-dichloroethyl ether have been described. Although no specific investigations of the metabolic fate of dichloroethyl ether have been reported, it might be assumed that it is rapidly distributed throughout the body tissues—the kidney and lung taking up the greatest amounts (372). This assumption is based on the similarity of structure to that of mustard gas and also because dichloroethyl ether has some of the vesicant properties of mustard gas with regard to the respiratory system.

2,2'-Dichloroethyl ether has been found to be mutagenic in *Drosophila* (345), and has been analyzed by titrimetry (371, 373).

It is of interest to note that the related α-halo ether, bis(chloromethyl)ether (BCME; $ClCH_2OCH_2Cl$) is representative of a group of compounds which have found wide laboratory and industrial utility as intermediates in organic synthesis (374) in the treatment of textiles, for the manufacture of polymers, insecticides, as solvents for polymerization reactions, and in the preparation of ion-exchange resins. The α-halo ethers are much more reactive than their β- or α-isomers in which the halogen atom is situated one or two carbons distant from the oxygen bearing carbon atom.

Van Duuren *et al.* (375) have found BCME to be a potent alkylating carcinogen for mouse skin and, most recently, Gargus *et al.* (376) have reported its potency in the induction of lung adenomas in newborn mice.

16. *Cyclophosphamide*

Cyclophosphamide [*N*,*N*-bis(2-chloroethyl)-*N'*,*O*-propylenephosphoric acid ester diamide; endoxan] was synthesized in 1957 by Arnold and Bourseaux (377) as an antitumor agent with an inert transport moiety and activation only

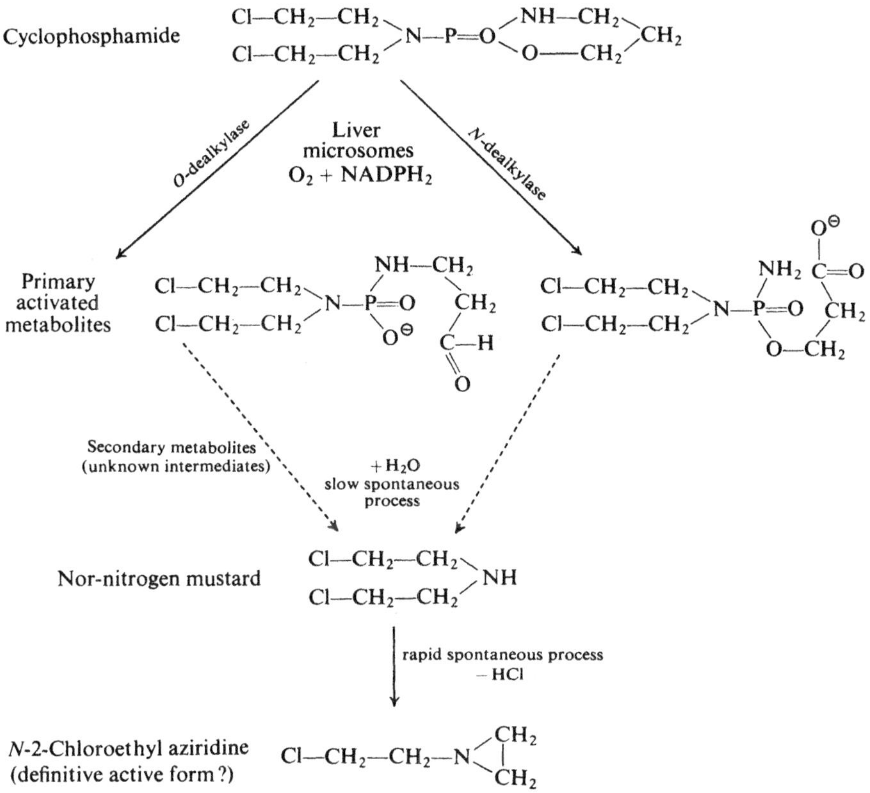

FIG. 7.3. Activation and breakdown of cyclophosphamide *in vivo*. Brock, N. and Hohorst, H. J., *Cancer* **20**, 900 (1967).

in vivo (to nor-HN2) in an attempt to circumvent some of the toxic effects of the other alkylating agents. Cyclophosphamide shows a significantly greater selectivity against many kinds of tumor cells than other cytostatic agents of the nitrogen mustard series. Its antitumor spectrum and chemotherapy in neoplastic diseases have been well documented (378–383).

The *in vivo* metabolism of cyclophosphamide in rats has been studied by Brock, Hohorst, and co-workers (384–386), Rauen and Norpoth (387, 388), and Friedman (389, 390); the distribution and biological activity of tritiated cyclophosphamide in patients was studied by Bolt *et al.* (391, 393), and the fate of cyclophosphamide labeled with ^{14}C or tritium acting on animal cells *in vitro* was described by Graul *et al.* (393).

Figure 7.3 illustrates the activation and breakdown of cyclophosphamide *in vivo* according to Brock and Hohorst (384).

The degradation of cyclophosphamide in aqueous media has been studied by Hirata *et al.* (394) and Friedman and co-workers (389, 395). Figure 7.4 illustrates the overall reaction sequence of the hydrolysis of cyclophosphamide

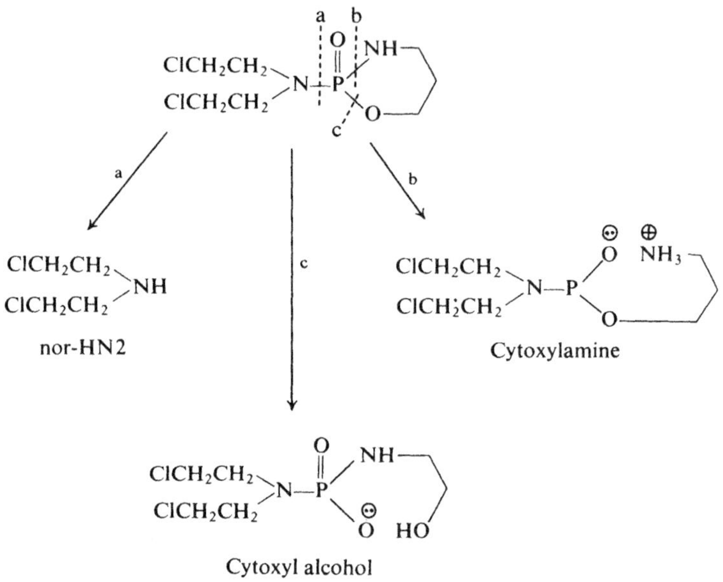

FIG. 7.4. Hydrolysis of cyclophosphamide *in vitro*. Friedman, O. M., *Cancer Chemotherapy Rept.* **51**, 327 (1967).

(heated to reflux in distilled water) according to Friedman (389). The main pathway for the spontaneous hydrolysis of cyclophosphamide in boiling water apparently involves an initial intramolecular alkylation followed by a sequence of simple hydrolytic cleavages of P–N and P–O bonds.

Friedman (389) concluded that the results on the spontaneous hydrolysis of cyclophosphamide do not support the notion of various biological mechanisms which involve release of discrete entities, nor-HN2 (377) or its hydrolytic products (396), as the therapeutically effective metabolite. The results of Friedman (389) make it appear unlikely that the metabolism of cyclophosphamide *in vivo* parallels the path of spontaneous hydrolysis *in vitro*, at least in the primary pathway.

Antitumor agents such as cyclophosphamide exhibit marked alopecic properties (397). Dolnick *et al.* (398) have recently described the utility of cyclophosphamide as a chemical defleecing agent for sheep.

The teratogenicity of cyclophosphamide in the rat (399), and some of its degradation products in mice (267, 400), as well as its mutagenicity in mice (401) and *Drosophila* (402–404), and its induction of chromosome aberrations in Chinese hamster bone marrow *in vivo* (405), murine and human cells *in vivo* (406), human leukocytes *in vitro* (407), mouse cells (89), HeLa cells (408), and in *Vicia faba* (409) have been reported.

Cyclophosphamide has been analyzed by colorimetry (394) and paper chromatography (384, 387, 395, 410, 411).

17. *Dimethylnitrosamine*

Nitroso derivatives represent a category of compounds of great activity and considerable diversity of action (especially as carcinogens). Their occurrence, whether as synthetic derivatives, natural products, or accidental products in food processing or tobacco smoke has sparked intensive investigation.

N-Nitroso compounds are generally prepared from the respective alkylamino compound by the action of nitrous acid. A number of nitrosamines have been patented for use as gasoline and lubricant additives, antioxidants, and pesticides. Dimethylnitrosamine (DMN) is used primarily in the electrolytic production of the hypergolic rocket fuel 1,1-dimethylhydrazine (412, 413). Other areas of utility include the control of nematodes (414), the inhibition of nitrification in soil (415), use as plasticizer for acrylonitrile polymers (416), use in active metal anode–electrolyte systems (high-energy batteries) (417), in the preparation of thiocarbonyl fluoride polymers (418), in the plasticization of rubber (419), and in rocket fuels (420). The preparation (421, 422) and chemistry of nitrosamines (423), kinetics and mechanism of decomposition in acid (424), and photolysis (425, 426) have been described.

The nitrosamines have been shown to induce a great variety of tumors at different sites in many species. In man, the exposure to high doses of DMN can lead to cirrhosis of the liver (47).

The general area of nitrosamine toxicity, carcinogenicity, and metabolism has been reviewed by Magee (427, 428) and Druckrey *et al.* (429). As a consequence of the extensive work of Magee and Druckrey and their respective co-workers, it seems established that the dialkylnitrosamines are metabolized *in vivo* to yield an active alkylating agent which can be diazoalkane itself or an active alkene or an alkyl carbonium ion derived from it by further decomposition (430, 431).

Figure 7.5 illustrates the mechanism for metabolism as suggested by

Druckrey *et al.* (431). Studies of Magee *et al.* (432, 433) have shown that the hepatotoxin DMN is capable of methylating both RNA and DNA *in vivo* to produce 7-methylguanine. Figure 7.6 illustrates metabolic pathways of DMN as proposed by Magee and Farber (432).

The possible significance of nitrosamines as etiological factors in human carcinogenesis is based on the widespread occurrence of nitrosamine precursors (secondary amines and nitrites) in biological systems, and the remark-

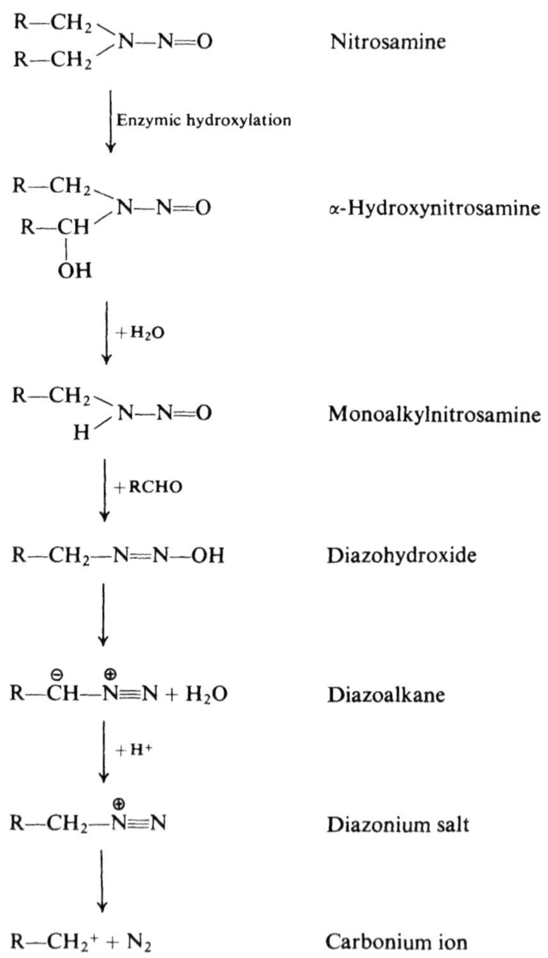

FIG. 7.5. Possible mechanism for the *in vivo* metabolism and reaction of dialkylnitrosamines. Druckrey, H., Schildbach, A., Schmähl, D., Preussmann, R., and Ivankovic, S., *Arzneimittel-Forsch.* **83**, 841 (1963).

able manner in which the various nitrosamines can affect different organs in the same species of experimental animal.

It is possible that nitrosamines occur in foods, since many foods contain large amounts of amines and small quantities of nitrite as preservative. The

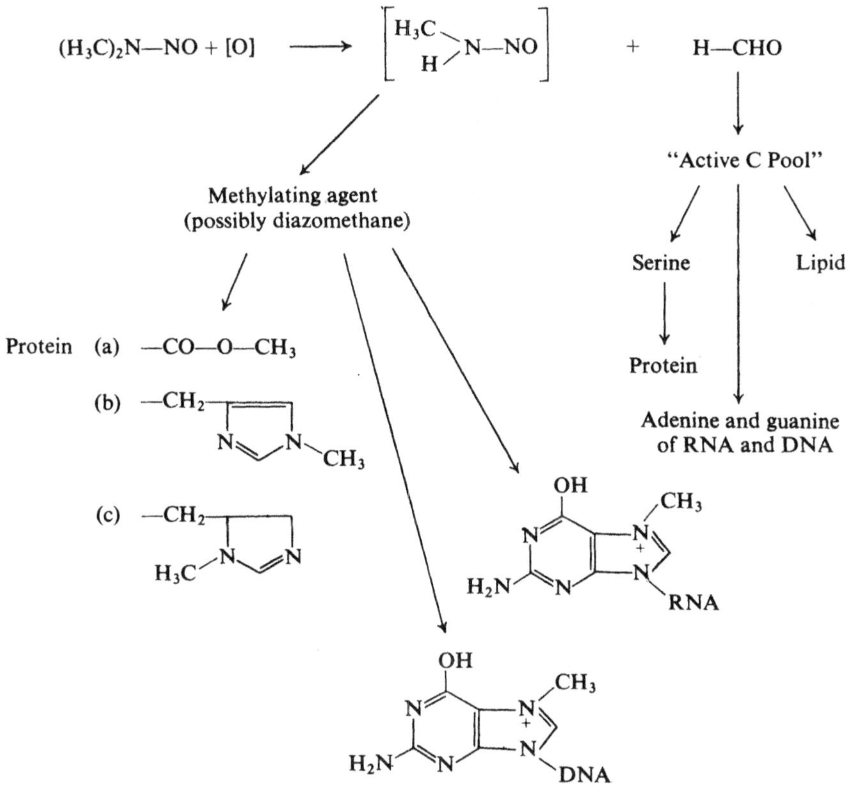

FIG. 7.6. Some postulated metabolic pathways of dimethylnitrosamine. Magee, P. N. and Farber, E., *Biochem. J.* **83**, 114 (1962).

presence of nitrosamines has been suggested in tobacco (434, 435), tobacco smoke (436), fish meals (437–439), wheat kernels and flour (435, 440), dairy products (milk and cheese) (441, 442), and in various smoked fish, meat, and mushrooms (443). Table 7.2 illustrates the levels of nitrosamine in a variety of foodstuffs.

The formation of dimethylnitrosamine in herring meal is caused by various methylamines which occur normally in fish. Dimethylamine, the most potent producer of dimethylnitrosamine, reacts with nitrite even at temperatures below 0°C (443).

It has been shown that amines and nitrite in food could produce nitrosamines in the stomach under suitable conditions, e.g., *in vitro* for stomach

TABLE 7.2

Levels of Nitrosamine in Foodstuffs[a]

Type of food	Nitrosamine (μg/kg)[b]	No. of samples analyzed
Fish		
Smoked herring	0.5–9.5	5
Kippers	0.5–2.4	4
Kippers	40	1
Smoked haddock (from Iceland)	15	1
Smoked mackerel	0.6	1
Meat		
Smoked sausage	0.8, 1.1, 2.4	3
Bacon	0.6, 1.2, 6.5	3
Smoked ham (from Iceland)	5.7	1
Mushrooms		
Polyporus ovinus	11.6	1
Boletus scaber	1.4	1
Amanita muscaria	30	1
Champignon	0.4, 5	2
Hydnum imbricatum	3, 15	2
Armillaria mellea	12	1
Lactarius trivalis	9.2	1
Russula emetica	10.2	1
A mixture of various edible mushrooms	14	1

[a] From F. Ender and L. Ceh, *Food Cosmetic Toxicol.* **6**, 569 (1968).

[b] Determined by the hydrazine method.

contents of rabbits and dogs (444) and rats (444–446), and *in vivo* in rabbits and cats (444).

Most recent investigations by DuPlessis *et al.* (447) have revealed the presence of dimethylnitrosamine in the fruit of a solanaceous bush (*Solanum incanum*), the juice of which is used to curdle milk. The resulting curds are the chief source of sustenance of the Bantu people in localized areas of the Transkei where there is a high incidence of esophageal cancer.

Although primary attention has been focused on dimethylnitrosamine because of its significance in a broad spectrum of environmental health considerations, it is of interest to note that a number of other nitrosamines, e.g., dialkyl-, *N*-alkyl-*N*-aryl-, *N*-nitroso cyclic amines, and dinitroso derivatives have been patented for utility as rubber additives, insecticides, stabilizers, antioxidants, and gasoline and fuel additives. Their toxicity and carcinogenic activity has been reviewed by Magee and Barnes (427). The structures of a number of nitrosamines as well as their target species and organs are illustrated in Table 7.3.

With regard to the mutagenicity of nitroso compounds, it is of note that the nitrosamines which are believed to require enzymic decomposition before becoming active carcinogens (e.g., dimethyl- and diethylnitrosamines) are mutagenic in *Drosophila* (448–452) and *Arabidopsis thaliana* (453) and inactive in microorganisms such as *E. coli* (454, 455), *Serratia marscesens* (454), *Saccharomyces cerivisiae* (456, 457) and *Neurospora* (458) [but active in *Neurospora* (459) in the hydroxylating model system of Udenfriend (460), or in the presence of oxygen].

Methylvinyl-, methylbenzyl-, and *N*-methylpiperazine nitrosamine have all been found mutagenic in *Drosophila* (448). However, ethyl *tert*-butylnitrosamine (which has no carcinogenic action) was found also to be nonmutagenic in *Drosophila* (448). Presumably, this substance is not degraded *in vivo*.

The analysis of nitrosamines has been achieved by a variety of techniques including polarography (430, 461), colorimetry (462, 463), thin-layer (444, 464, 465) and gas chromatography (440, 447, 460, 466), and infrared (467) and nmr spectroscopy (468).

18. *N-Methyl-N-nitrosourea*

N-Methyl-*N*-nitrosourea is prepared by the nitrosation of methylurea (469). The inactivation of biologically active *N*-methyl-*N*-nitroso compounds in aqueous solution as well as the effect of various conditions of pH was studied by McCalla *et al.* (470). Exposure of these nitroso derivatives to high concentrations of hydrogen ion results in the rapid destruction of the compound, e.g., (a) *N*-methyl-*N*-nitrosourea yields methylamine among the degradation products, suggesting that methylnitrosourea may first lose its nitroso group to yield methylurea which is then slowly hydrolyzed to ammonia, methylamine, and carbon dioxide; (b) MNNG is quantitatively converted to *N*-methyl-*N*-nitroguanidine.

It is well known that methylnitroso compounds yield diazomethane under alkaline conditions (a property which has led to their extensive use for synthetic

TABLE 7.3

Carcinogenic Activity of Some Nitrosamines

Compound	Structure	Species	Effected organ
Dimethylnitrosamine (DMN)	H_3C ⟍ N—NO H_3C ⟋	Rat Mouse Hamster Trout	Liver, kidney, lung, nasal sinus Liver, kidney, lung Liver Liver
Diethylnitrosamine (DEN)	C_2H_5 ⟍ N—NO C_2H_5 ⟋	Rat Mouse Guinea pig Rabbit Dog Monkey Hamster Fish	Liver, kidney, esophagus Liver, stomach, esophagus, nose Liver, lung Liver Liver Liver Liver, lung and bronchi, nose Liver
Di-*n*-propylnitrosamine	C_3H_7 ⟍ N—NO C_3H_7 ⟋	Rat	Liver
Methyl vinylnitrosamine	$CH_2{=}CH$ ⟍ N—NO H_3C ⟋	Rat	Nose
Methyl benzylnitrosamine	$C_6H_5CH_2N$—NO | CH_3	Rat	Esophagus
Methyl phenylnitrosamine	C_6H_5—N—NO | CH_3	Rat	Esophagus

N-Nitrosomorpholine	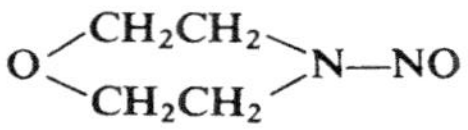	Rat Mouse Hamster	Liver, kidney, nose Liver Lung
N-Nitrosopiperidine	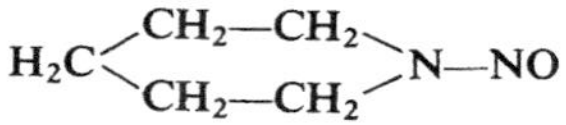	Rat Hamster	Nose, esophagus, liver Lung
N-Nitrosohexamethyleneimine (NHMI)	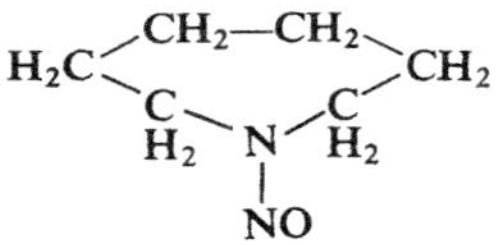	Rat	Liver, esophagus
N-Nitrosoazetine	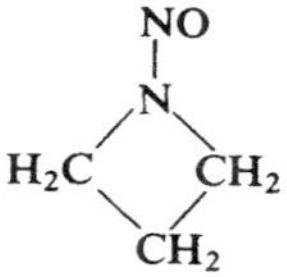	Rat	Liver, lung
N,*N*′-Dinitrosopiperazine	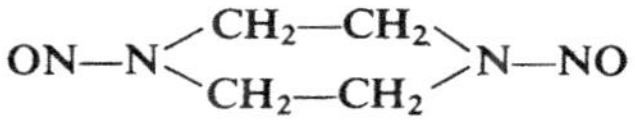	Rat Mouse	Nose, esophagus Liver, lung
N,*N*′-Dinitroso-*N*,*N*′-dimethylethylenediamine	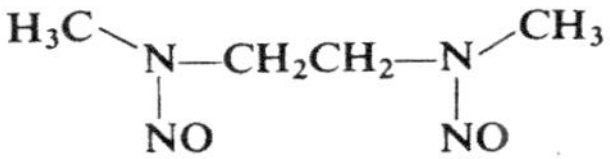	Rat	Esophagus

purposes). However, none of these compounds is particularly stable even in neutral aqueous solution and the rates at which the different compounds degrade differ.

The kinetics of solvolysis of various *N*-alkyl-*N*-nitrosoureas in neutral and alkaline solution has also been studied by Garrett *et al.* (471) and the rate of decomposition of *N*-methyl-*N*-nitrosourea in the rat by Swann (472).

N-Methyl-*N*-nitrosourea is one of the most effective carcinogenic substances known, causing primary gastric cancer in rats after a single oral administration (473). Its carcinogenicity in newborn mice and rats (474), teratogenicity in rat (475, 476), as well as its ability to induce morphological conversion, hyperconversion, and reversion of mammalian cells *in vitro* (477, 478) has also been reported. In this latter study it was found that the mode of growth of cells from a lung tissue cell line of a Chinese hamster was altered, and the effect accompanied its ability to cause tumors when such cells were later injected into foreign host animals.

The mutagenicity of *N*-methyl-*N*-nitrosourea has been demonstrated in *Drosophila* (448, 450, 479–482), *Saccharomyces cerevisiae* (456, 457, 483–485), *Arabidopsis* (486–488), and its inactivation and mutation of coliphage T2 (489), as well as its induction of chromosome aberrations in wheat seeds (490), *Vicia faba* (491), ascites tumors of the mouse (491), and *Bellevalia romana* (492) also have been described.

The analysis of *N*-methyl-*N*-nitrosourea has been achieved by polarographic (472, 493) and thin-layer chromatographic techniques (471, 494).

19. *N-Methyl-N-nitrosourethan*

N-Methyl-*N*-nitrosourethan (MNU) is prepared by the nitrosation of methylurethan (495) and is used primarily as a convenient source of diazomethane by its treatment with alcoholic potassium hydroxide.

MNU and its ethyl homolog (ENU) have been found to be effective carcinogens which can induce stomach, esophageal, intestinal, and lung tumors with one or a few doses (473, 496–498). The interaction of MNU with thiols (499), its effects on cells in tissue culture (500) and a comparison of its biological and metabolic effects with methylnitrosourea in *E. coli* (494) as well as a comparison of the methylation of DNA by MNU and MNNG (501) have been described.

In this last study (501) the similarity in nature and yields of the principal products from MNU, MNNG, and MMS (methyl methanesulfonate) indicated at least a partial similarity in their mode of action with DNA, e.g., nucleophilic substitutions with the proximate reactive species being a methyl carbonium ion. A possible mechanism by which hydrolysis of MNU could yield the methyl carbonium ion in aqueous solution was suggested by Gutsche and Johnson (502):

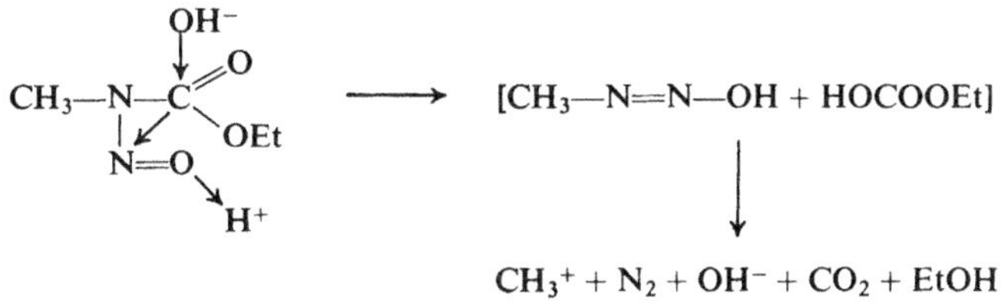

The mutagenicity of MNU has been demonstrated in *Drosophila* (448, 503, 504), *E. coli* (505, 506), *Saccharomyces cerevisiae* (457, 483, 484, 507), *Ophiostoma multiannulatum* (508–510), *Schizosaccharomyces pombe* (511–514), *Colleotrichum* (515), barley seeds (516), and ameba (517, 518), and the induction of chromosome aberrations has been demonstrated in *Vicia faba* (519–521) and *Bellevalia romana* (522).

N-Methyl-*N*-nitrosourethan has been analyzed by thin-layer chromatography (494).

20. *N-Methyl-N'-nitro-N-nitrosoguanidine*

N-Methyl-*N'*-nitro-*N*-nitrosoguanidine (MNNG) is synthesized via the nitrosation of methylnitroguanidine (523). Its overall preparation is illustrated as follows:

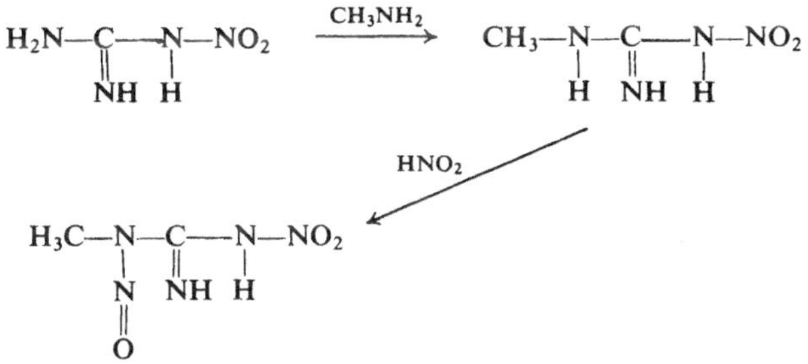

MNNG decomposes slowly to diazomethane under alkaline conditions (524) and to nitrous acid under more acidic conditions (it is chemically stable at pH 5). MNNG reacts with nucleophiles such as amino moieties to form products containing the nitroguanidine group (as well as methylated derivatives) (525). The reaction of MNNG with a lysine residue of a protein molecule is illustrated in Fig. 7.7. MNNG also has been shown to alkylate DNA to form 7-methylguanine as the principal product (526–529).

The antitumor (530–533), antileukemic (534, 535), antimicrobial (536), antimalerial (537), and carcinogenic properties (538–541) of MNNG have been described.

MNNG has been shown to inactivate transforming DNA *in vitro* (542), to induce mutants in *E. coli* (505, 543–549), *Salmonella typhimurium* (550), *Saccharomyces cerevisiae* (456, 483–485, 507, 551, 552), *Schizosaccharomyces pombe* (513, 553, 554), *Arabidopsis thaliana* (455, 488, 555), barley (556),

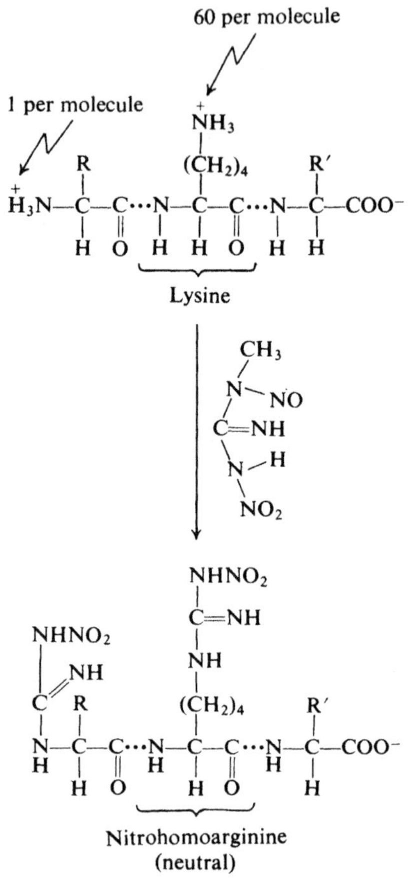

FIG. 7.7. Reaction of *N*-methyl-*N'*-nitro-*N*-nitrosoguanidine with a lysine residue of a protein molecule. McCalla, D. R., and Reuvers, A., *Can. J. Biochem.* **46**, 1411 (1968).

Drosophila (557, 558), the heterothallic unicellular green alga *Chlamydomonas reinhardi* (559), and *Euglena gracilis* (560), and to induce chromosome aberrations in *Vicia faba* (561).

The analysis of MNNG has been achieved primarily by paper chromatography (531, 562).

21. *N-Methyl-N-nitroso-p-toluenesulfonamide*

N-Methyl-*N*-nitroso-*p*-toluenesulfonamide (diazald; MNTS) is prepared by the action of nitrous acid on *p*-tolylsulfonylmethylamide (562) and is the most useful precursor for the laboratory preparation of diazomethane (563).

MNTS is quantitatively converted to *N*-methyl-*p*-toluenesulfonamide when exposed to high concentrations of hydrogen ions (470). MNTS has also been found to be photosensitive, yielding *N*-methyl-*p*-toluenesulfonamide (470). It is suggested that the nitroso group is lost from nitrosoamides on photolysis in the form of nitroso radicals which are rapidly converted into nitrous acid (470).

MNTS has been shown to methylate DNA *in vitro* (528, 564) and *in vivo* (528), yielding 7-methylguanine as the main product. The fact that increasing amounts of labeled MNTS becomes incorporated into DNA as the pH of the reaction mixture is increased supports the suggestion that methyl-*N*-nitroso compounds yield diazomethane as an active methylating intermediate (the rate of formation of diazomethane is accelerated by increasing concentrations of OH^-).

In contrast to the potent activity of *N*-methyl-*N'*-nitro-*N*-nitrosoguanidine (MNNG), MNTS has been reported to be inactive as a mutagen (456) and carcinogen (473). However, McCalla (565) has shown MNTS to be as active as MNNG in causing mass mutation of the *Euglena* chloroplast system. The apparent contradiction of the activity of MNTS is suggested to be a consequence of its low solubility and/or rapid destruction in some biological systems.

22. *1-Nitroso-2-imidazolidone-2* (*NIL*)

1-Nitroso-2-imidazolidone-2 (NIL) is prepared by the nitrosation of 1-amino-2-imidazolidone. NIL has been reported to be mutagenic in *Saccharomyces cerevisiae* (456, 457, 484, 507, 566, 567) and to induce chromosome aberrations in *Bellevalia romana* (492).

23. *Diazomethane*

Diazomethane is a very toxic gas and, as noted earlier, is usually prepared in ethereal solution by the decomposition with alkali of a variety of *N*-methyl-*N*-nitroso compounds, e.g., respective derivatives of urea, urethan, nitroguanidine and *p*-toluenesulfonamide.

It is a powerful methylating agent for acidic compounds such as carboxylic acids, phenols, and enols, and, as a consequence, is both an important laboratory reagent and has industrial utility.

Its synthesis and properties (568–570), mechanism of action (571), toxicity (572, 573), carcinogenicity in rats and mice (497), and its role as the active agent responsible for the carcinogenic action of many compounds (574–576) has been described. It is of interest to note, however, that amino acid derivatives of diazomethane such a *O*-diazoacetyl-2-serine and 6-diazo-5-oxonorleucine have been used as tumor inhibitors.

Diazomethane can react with many biological molecules, especially nucleic acids and their constituents. For example, its action on DNA includes methylation at several positions on the bases and the deoxyribose moiety as well as structural alterations that result in lower resistance to alkaline hydrolysis and altered hyperchromicity (577–580). Methylation by diazomethane involves loss of nitrogen and liberation of highly reactive $:CH_2$ radicals. [The formation of methylene radicals is greatly accelerated by ultraviolet radiation (581–583). However, reactions are known where it undergoes addition without loss of nitrogen.]

The mutagenicity of diazomethane in *Drosophila* (503), *Neurospora* (243, 503, 584), and *Saccharomyces cerevisiae* (457, 458) has been described. Recently, Cerda-Olmedo and Hanawalt (585) have established that diazomethane is the principal agent of nitrosoguanidine mutagenesis in *E. coli* at pH *above* 5, and they considered it likely that nitrosoguanidine is also a mutagen by itself. The implication of diazomethane in the mutagenesis of nitroso compounds has been alluded to previously (456, 483–485).

24. *Cycasin*

The high incidence of human and animal neurological diseases in areas of the world where cycads are utilized as food and medicines (586) (principally in the tropics and subtropics) has resulted in an intensive investigation of the constituents of cycads. The cycad nuts are a source of starch and are often used as food after appropriate preparation which includes soaking in water (586, 587). Both the wide scope of the application of cycads as medicines and their geographical area of usage are illustrated in Table 7.4.

Cycasin [methylazoxymethane-β-D-glucoside] and its aglycone methylazoxymethanol (MAM) are extractable from nuts, seeds, and roots of cycad plants primarily from the very widespread species *Cycas circinalis* and *Cycas revoluta*.

$$\underset{\substack{\downarrow \\ O}}{CH_3{-}N{:}N{-}CH_2O{-}\beta\text{-D-glucopyranosyl}} \qquad \underset{\substack{\downarrow \\ O}}{CH_3N{:}N{-}CH_2OH}$$

Methylazoxymethane-β-D-glucoside MAM

The aglycone has recently been synthesized as the stable acetate ester by Matsumoto *et al.* (588) and has been found to be as active as the natural aglycone itself (589).

Cycasin has been shown to be hepatoxic and carcinogenic in rats (590). The compound is nontoxic when administered parenterally to mice and is toxic when administered orally, but only after a latent period. It is suggested that cycasin must be deglucosylated before it is carcinogenic and that the intestinal flora provide the β-glucosidase activity necessary for this hydrolysis (591, 592).

Recent studies (593) have shown that the aglycone of cycasin (MAM) is the proximate carcinogen since MAM causes neoplasms after subcutaneous and intraperitoneal injection as well as feeding while cycasin causes tumors in mature animals only if fed.

In addition to its hepatoxic and carcinogen effects, MAM has been reported to be teratogenic to the golden hamster (591, 593–595).

A direct action by the aglycone seems likely following proof by thin-layer chromatography of its presence in the fetuses of the rat and hamster (595).

Alkylation of liver RNA and DNA with cycasin and MAM has been described both *in vitro* (596) and *in vivo* (597). In both studies increased amounts of 7-methylguanine were observed in both hepatic RNA and DNA. The similarity of biological effects between cycasin and dimethylnitrosamine have been documented (597, 598). For example, dimethylnitrosamine also inhibits hepatic protein synthesis and methylates liver and kidney nucleic acids *in vivo*, and MAM shows a pattern of hepatoxic and carcinogenic activity in the rat similar to that of dimethylnitrosamine (this includes both the acute toxic manifestations in the liver, and in the chronic feeding experiments and the histological types of neoplasms in liver and kidney). The chemical structure of MAM suggests that it may have a mechanism of action similar to that of dimethylnitrosamine at the molecular level (598, 599).

Figure 7.8 (p. 176) depicts the proposed degradation and metabolic pathways for dimethylnitrosamine and cycasin as proposed by Miller (598).

The potent mutagenicity of cycasin aglycone has been demonstrated in *Drosophila melanogaster* (600) and *Salmonella typhimurium* (601). Both cycasin and its aglycone have also been found to increase the mutant frequency of *Salmonella typhimurium* histidine auxotrophs when tested in the host-mediated assay (602). (The degree of cycasin-related mutagenic activity depended on the facility with which it could be enzymically deglucosylated by the normal intestinal flora.) Cycasin aglycone also induced chromosome aberrations in *Allium* (onion) seedlings (603) (the equivalent of 200 r of γ-rays).

The chemistry (598) and toxicity (587) of cycads have been reviewed and the analysis of cycasin and its aglycone by colorimetric (604, 605), polarographic (604), nmr spectrographic (606), paper (605, 607–609), thin-layer

TABLE 7.4

Cycads as Medicine[a]

Genus, species	Plant part	Preparation	Use	Geographic area
1. *Zamia*	Fruit	Mash	As a therapeutic shampoo	Dominican Republic
2. *C. pectinata*	Stem	Pound	As a wash for diseased hair	India
3. *Z. multi.*	Roots	Extract starch	In a bath to give strength	Cuba
4. *Cycas*	Gum	—	For insect bites	—
5. *C. circinalis*	Gum	—	For snake bites	India
6. *Zamia*	—	—	For snake bites	Mexico
7. *C. rumphii*	Seeds	Grate	Removes framboesia scars, cures ulcerations	Indonesia
8. *C. rumphii*	Resin	—	Cures malignant ulcers	India
9. *C. circinalis*	Seeds	Squeeze and grate	For tropical ulcers	Guam and Manus
10. *Zamia*	Stem	Extract gum	To treat ulcers	Dominican Republic
11. *C. circinalis*	Seeds	Roast, powder, put in coconut oil	For wounds, boils, itch	Philippines
12. *C. circinalis*	Buds	Crush	For wounds, swollen glands, boils	Indochina
13. *C. rumphii*	Stem	Chew with betel	As poultice to relieve swelling	Indonesia
14. *C. circinalis*	Cones	Crush	As poultice to relieve nephritic pains	India
15. *C. circinalis*	Bark, seeds	Make tincture, or grind to paste and mix with coconut oil	As poultice for sores and swelling	India

16. *C. revoluta*	Megasporaphyll	Grind	To stop bleeding	W. China
17. *C. revoluta*	Fruit	Boil in water	As an expectorant	China
18. *Zamia*	Roots	Chew	To relieve cough and improve singing voice	Guatemala
19. *C. revoluta*	Seeds	Make a tincture	For headache, giddiness, sore throat	India
20. *C. rumphii*	Seeds	—	For asthma	Malaya, Indonesia
21. *C. rumphii*	Leaves	Squeeze	To check hematemesis; to relieve colic	Indonesia
22. *C. revoluta*	Seeds	—	As an emmenagogue and astringent, dose 10–20 g per day	Ryukyus, S. Japan
23. *C. revoluta*	Seeds	Use raw	To check diarrhea	Oshima, Ryukyus
24. *C. circinalis*	Fruit	Mix with sugar	As a laxative	India
25. *C. circinalis*	Seeds	Boil starch	For hemorrhoids and dysentery	Ceylon
26. *C. rumphii*	Fruit	Boil	As an emetic	Indonesia
27. *Z. muri.*	Sap	Squeeze fruit	As a drastic purge	Venezuela
28. *C. circinalis*	Stem	Extract starch	As a strengthener and restorative	India, Europe
29. *C. revoluta*	Fruit	—	As a tonic; produces plumpness	China
30. *D. edule*	Seeds	Make decoction	To relieve neuralgia	Mexico
31. *C. circinalis*	Male	Mix with sugar	As an aphrodisiac and stimulant	India
32. *C. circinalis*	Scales	—	As an anodyne, dose 30–60 grains or more	India
33. *C. circinalis*	Male bracts	—	As a narcotic	India

[a] From M. J. Whiting, *Econ. Botan.* **17**, 270 (1963).

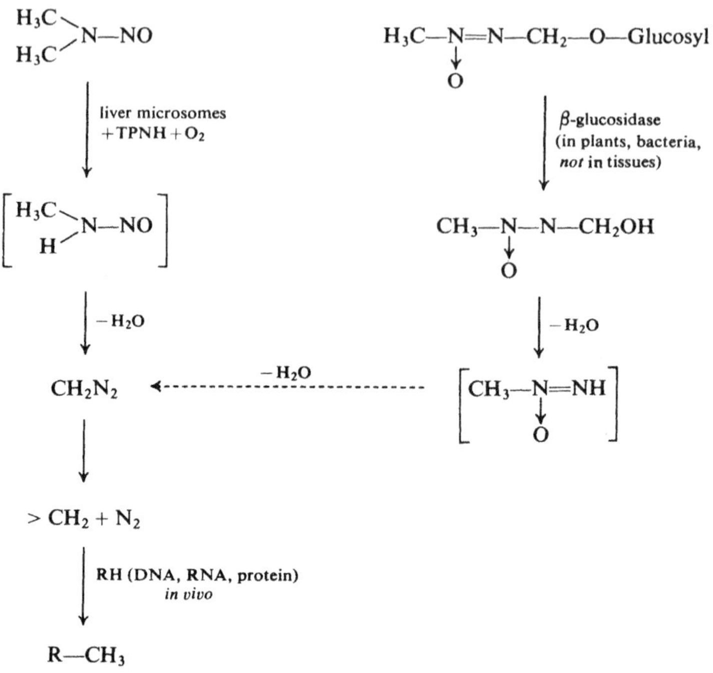

FIG. 7.8. Proposed degradation and metabolic pathways for dimethylnitrosamine and cycasin.

(595) and gas (610, 611) chromatographic and bioassay (605, 609, 612) techniques described.

25. *Pyrrolizidine Alkaloids*

Alkaloids of the pyrrolizidine class are found in members of the *Senecio*, *Crotolaria*, *Amsinkia*, and other genera. These common plants are widely distributed throughout range areas and are hepatotoxic to livestock. The alkaloids they contain have been established as the responsible compounds (613–615). Many of the plants containing such alkaloids have been and are being traditionally used as herbal folk medicines for various disorders, chronic or recurrent, as emmenagogues and abortifacients (616).

A common structural feature in these alkaloids (e.g., in lasiocarpine and heliotrine) is the presence in the molecule of the pyrrolizidine ring system.

Pyrrolizidine alkaloids, of which the carcinogenic retrosine is an example, were among the first to be suggested as possible natural etiological factors in

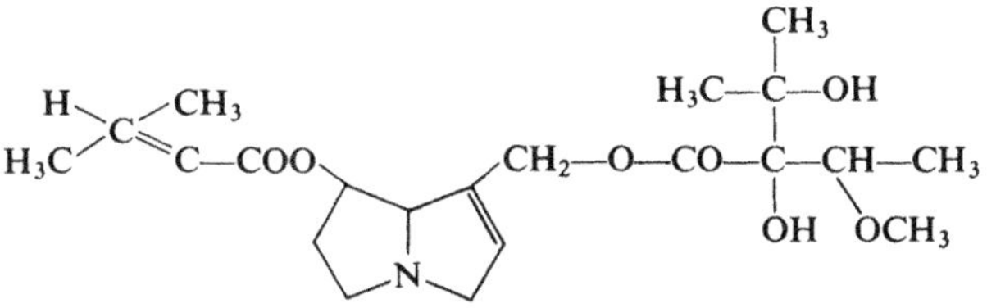

(Angelic acid) (Heliotridine) (Lasiocarpic acid)
Lasiocarpine

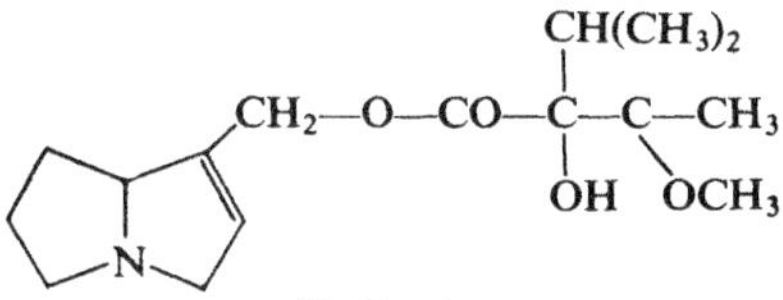

Heliotrine

the high incidence of liver diseases (kwashiorkor, cirrhosis, and primary liver tumors) in the tropics and subtropics (617).

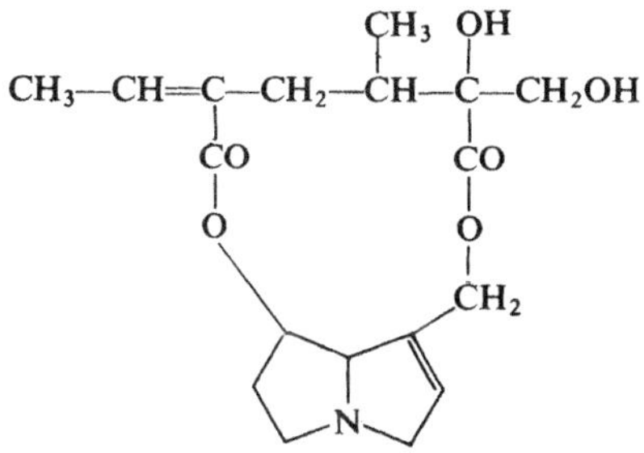

Retrosine

The first observable effect of chronic pyrrolizidine alkaloid poisoning is the production of nuclear abnormalities in the cells of the liver (618). These abnormalities resemble those which follow the actions on living cells of compounds which are known to be alkylating agents. Culvenor, Dann, and Dick (619) found that these compounds possess alkylating activity *in vitro* against nucleophilic reagents such as benzyl mercaptan, viz:.

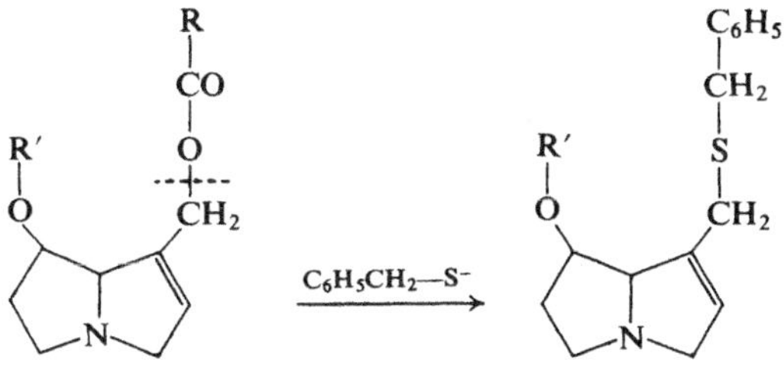

Protein sulfhydryl groups might be expected to react in a similar manner.

Only the pyrrolizidine alkaloids with a sterically hindered allylic ester group (> C=C—CH_2OCOR) have alkylating activity and hepatotoxic properties. For example, the alkaloids represented by retrosine may function by a mechanism involving alkyl–oxygen fission of the ester linkage. This reaction which is known to occur with carboxylic esters of allylic alcohols but not with those of saturated alcohols would result in displacement of the anion R—CO_2^- by a nucleophilic agent X^-. Since combinations may occur at either end of the allylic system, the product may be of type (B) or (C). The alkylation may be depicted as follows:

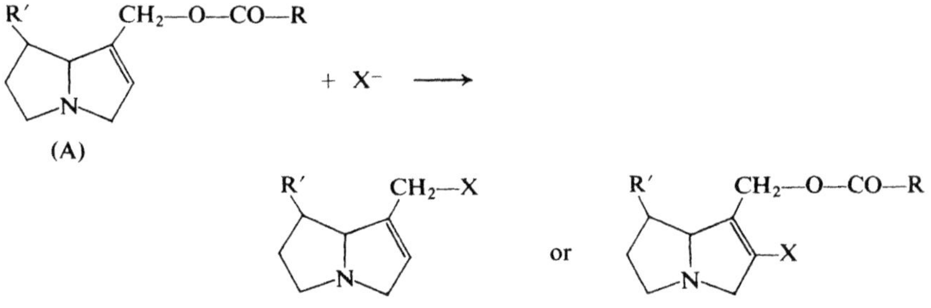

(A) (B) (C)

R = $CH_3(CH_2)_2^-$ or $CH_3(CH_2)_3^-$
R′ = H, OH, or acyl

Christie (620) has shown that a sudden loss of activity of liver enzymes whose activity depends on the presence of nicotinamide adenine dinucleotide (NAD) occurs in acute heliotrine poisoning. Possibly this phenomenon was due to a reductive cleavage by NAD of the alkaloid which might be expected to involve fission at the alkyl–oxygen bond. (The metabolite which might be expected to arise from such a reaction was present in the content of sheep rumen following ingestion of heliotrine.)

In most recent studies with ^{14}C-pyrrolizidine alkaloids (621), it was found that the alkaloids were readily converted *in vitro* and to some extent *in vivo* into pyrrole derivatives which were highly reactive and exerted cytotoxic and other biological actions suggestive of attack on nucleic acids. The main soluble pyrrolic metabolite from lasiocarpine and heliotrine (see above) is 7-hydroxy-1-hydroxymethyldihydro-5*H*-pyrrolizidine. The present evidence suggests that the alkaloids or a metabolite react with DNA in such a way as to inhibit synthesis of messenger RNA.

Mattocks (622) suggested that pyrrolizidine alkaloids were not per se hepatotoxic but are transformed in the liver by ring dehydrogenation to pyrrole-like derivatives which react with tissue constituents to form soluble "bound pyrroles" (subsequently excreted in the urine) and insoluble "bound pyrroles" taken up in tissues. Such pyrrole products would be expected to react with sulfhydryls more readily than do pyrrolizidine alkaloids (623).

Barnes (624) has reported that many pyrrolizidines not only exert their effects as liver poisons but may also produce lung damage via the formation of a pyrrole derivative (II) such as that formed from monocrotaline (I).

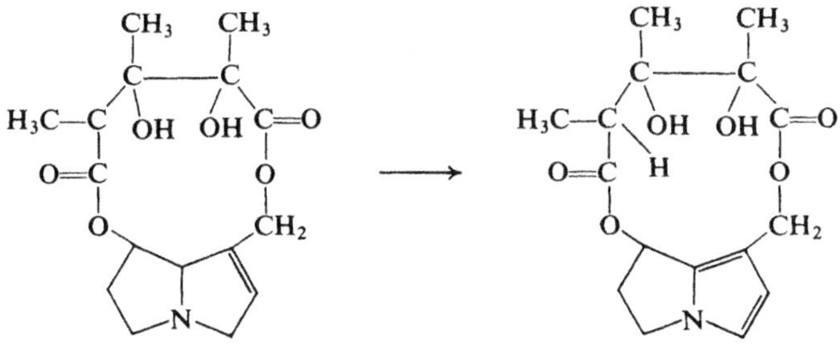

Monocrotaline (I) (II)

The chemistry and distribution (625), pharmacology (626), hepatotoxicity (617, 618, 623, 627, 628), teratogenicity (629), carcinostatic properties (630), mutagenicity in *Drosophila melanogaster* (631–637), in *Aspergillus nidulans* (638), and chromosome aberrations in *Drosophila* (635, 637, 639) and in *Allium cepa* (639) of the pyrrolizidine alkaloids have been reported. The analyses of these alkaloids have been achieved by spectrophotometric (640, 641), and paper (642), thin-layer (643–646) and gas chromatographic techniques (647).

26. *Bracken Fern*

Bracken fern, *Pteridium aquilinum*, has been often associated with the fatal poisoning of cattle observed in various parts of the world such as Turkey, Yugoslavia, Brazil, Panama (648), and Wales (649). The poisoning of cattle with bracken fern has been reproduced by feeding fresh foods, sun-dried bracken, or powdered rhizomes mixed with an otherwise adequate diet and an alcoholic fraction of the bracken plant (648, 650, 651). The bracken plant is known to contain radiomimetic activity (652, 653). Ingestion by cattle of the whole plant or of extracts produces a syndrome in which panmyeloid bone marrow damage, pyrexin, and, often, intestinal musocal damage and ulceration are found. It also induces polyps in the bladder mucosa; also typical are the widespread petechial hemorrhages (649).

Inclusion of bracken in the diet of young rats produced malignant adenocarcinoma of the intestinal mucosa (652).

It is of interest that in Japan a plant similar to bracken fern, zen mai (*Osmunda japonica*), is used as a vegetable and seasoning. The suggestion has been made that bracken may be confused with zen mai and thus might contribute to the relatively high incidence of stomach cancer found in Japan (654).

A variety of chemicals have been identified in bracken such as astragalin, isoquercitrin, rutin (655), catechol tannins, pteraquilin, sugar, starch, aliphatic nondrying oil, and pectose micrin (656); however, there is no indication that these chemicals or their metabolites may be bladder carcinogens.

Van Duuren (657) has suggested that the observed carcinogenicity and alkylating potential of bracken fern when fed to cattle may be due to dimethylsulfoniumpropionic acid hydrochloride.

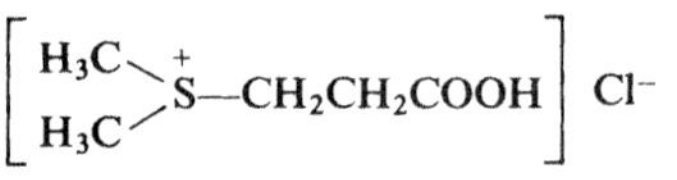

The isolation of two major molting hormones of insects, α-ecdysone (I) and 20-hydroxyecdysone (II) in crystalline form from dry pinnae of bracken fern has been recently reported (658). Three unidentified substances with molting hormone activity were also detected. Bracken is the first plant found to contain both of the major insect ecdysones as well as the first known plant source of α-ecdysone.

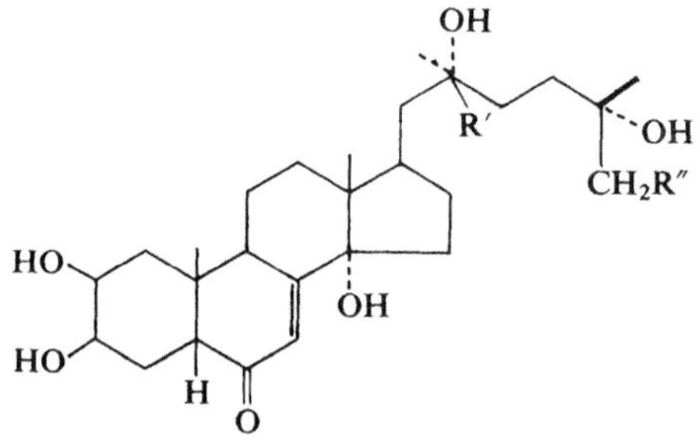

(I) $R' = R'' = H$ = α-Ecdysone
(II) $R' = OH$; $R'' = H$ = 20-Hydroxyecdysone

The mechanism of antianeurin activity of bracken has been reported by Jones and Evans (659). The inclusion of fresh, green, or air-dried bracken in the diet of simple-stomached animals, e.g., rat or horse, gives rise to an avitaminosis B_1 (660), which arises as a result of the presence of thiaminase in bracken (661).

In contrast to monogastric animals, ruminants poisoned by bracken fern show primarily hematological alterations. The predominant feature is a depression of bone-marrow activity, giving rise to a severe leukopenia and thrombocytopenia, with no avitaminosis B_1.

The toxicity in sheep (662) and cattle (663–665), carcinogenicity (652, 666, 667) and mutagenicity in *Drosophila* and mice (653), and sterility effect in quail (due to direct action on spermatogenesis and/or the expression of dominant lethal mutations (653) have been described for bracken fern.

REFERENCES

1. W. C. J. Ross, *Advan. Cancer Res.* **1**, 397 (1953).
2. W. C. J. Ross, "Biological Alkylating Agents." Butterworth, London and Washington, D.C., 1962.
3. J. M. Johnson and F. Bergel, *in* "Metabolic Inhibitors" (R. M. Hochster and J. H. Quastel, eds.), Vol 2, p. 161. Academic Press, New York, 1963.
4. P. Alexander, *Advan. Cancer Res.* **2**, 1 (1954).
5. K. A. Stacey, M. Cobb, S. F. Cousens, and P. Alexander, *Ann. N.Y. Acad. Sci.* **68**, 682 (1958).
6. G. P. Warwick, *Cancer Res.* **23**, 1315 (1963).
7. P. K. Smith, M. V. Nadkarni, E. G. Trams, and C. Davison, *Ann. N.Y. Acad. Sci.* **68**, 834 (1958).
8. L. Fishbein, *Ann. N. Y. Acad. Sci.* **163**, 869 (1970).
9. L. Fishbein and H. L. Falk, *Chromatog. Rev.* **11**, 101 (1969).
10. L. Fishbein and H. L. Falk, *Chromatog. Rev.* **11**, 365 (1969).
11. L. S. Goodman and A. Gilman, "The Pharmacological Basis of Therapeutics," 3rd ed. Macmillan, New York, 1965.
12. D. A. Karnofsky and B. D. Clarkson, *Ann. Rev. Pharmacol.* **3**, 357 (1963).
13. D. B. Clayson, "Chemical Carcinogenesis." Little, Brown, Boston, Massachusetts, 1962.
14. P. Brookes and P. D. Lawley, *Brit. Med. Bull.* **20**, 91 (1964).
15. A. L. Walpole, *Ann. N. Y. Acad. Sci.* **68**, 750 (1958).
16. C. Auerbach, *Ann. N. Y. Acad. Sci.* **68**, 731 (1958).
17. L. E. Orgel, *Advan. Enzymol.* **27**, 289 (1965).
18. D. R. Krieg, *Progr. Nucleic Acid Res.* **2**, 125 (1963).
19. A. Loveless, "Genetic and Allied Effects of Alkylating Agents." Pennsylvania State Univ. Press, University Park, Pennsylvania, 1966.
20. F. B. Jones, H. G. Hammon, R. I. Leniger, and R. G. Heiligman, *Textile Res. J.* **31**, 57 (1961).
21. A. Bonvicini and C. Caldo, U.S. Patent 3,037,835 (1962); *Chem. Abstr.* **58**, W38 (1963).
22. L. H. Chance, U.S. Patent 2,901,444 (1959); *Chem. Abstr.* **53**, 23079 (1959).
23. W. A. Reeves, U.S. Patent 2,933,367 (1960); *Chem. Abstr.* **54**, 14717 (1960).
24. G. C. Tesoro and S. B. Sello, *J. Textile Res.* **34**, 523 (1964).
25. J. E. Moore and C. E. Pardo, Jr., U.S. Patent 2,925,317 (1960); *Chem. Abstr.* **54**, 12606 (1960).
26. Farbwerke Hoechst, British Patent 795,380 (1958); *Chem. Abstr.* **53**, 221 (1959).

27. D. K. Pattilloch, U.S. Patent 3,016,325 (1962); *Chem. Abstr.* **56**, 10428 (1962).
28. C. H. Adams and B. H. Shoemaker, U.S. Patent 2,372,244 (1945); *Chem. Abstr.* **40**, 1029 (1946).
29. A. F. Graefe, U.S. Patent 2,944,051 (1960); *Chem. Abstr.* **54**, 22681 (1960).
30. J. L. Justice, U.S. Patent 2,940,889 (1960); *Chem. Abstr.* **54**, 18976 (1960).
31. E. Windemuth, German Patent 1,112,286 (appl. 1959); *Chem. Abstr.* **56**, 4958 (1962).
32. E. H. Birum, U.S. Patent 2,813,819 (1957); *Chem. Abstr.* **52**, 2326 (1958).
33. A. B. Borkovec, *Science* **137**, 1034 (1962).
34. A. B. Borkovec, *Residue Rev.* **6**, 87 (1964).
35. D. R. Mussell, U.S. Patent 2,745,815 (1956); *Chem. Abstr.* **50**, 12385 (1956).
36. W. Gauss and G. Domagk, British Patent 833,067 (1961); *Chem. Abstr.* **55**, 18771 (1961).
37. A. Marxer, U.S. Patent 2,841,581 (1958); *Chem. Abstr.* **53**, 295 (1959).
38. L. E. Nagan, U.S. Patent 3,131,144 (1964); *Chem. Abstr.* **61**, 1613 (1964).
39. L. Longmaack, German Patent 1,020,225 (1957); *Chem. Abstr.* **54**, 20007 (1960).
40. Gevaert Photo-Producten, British Patent 918,950 (1963); *Chem. Abstr.* **60**, 1701 (1964).
41. J. W. Britain, U.S. Patent 3,054,757 (1962); *Chem. Abstr.* **58**, 3568 (1963).
42. P. Fram and R. R. Charbonneau, U.S. Patent 3,079,367 (1963); *Chem. Abstr.* **68**, 14224 (1963).
43. H. S. Bloch and D. R. Strehlau, U.S. Patent 2,839,568 (1958); *Chem. Abstr.* **52**, 17763 (1958).
44. D. K. Pattilloch, U.S. Patent 3,027,295 (1962); *Chem. Abstr.* **57**, 1132 (1962).
45. H. Druckrey, *Arzneimittel.-Forsch.* **6**, 539 (1956).
46. S. Petersen, *Angew. Chem.* **67**, 217 (1955).
47. H. C. Hodge and S. H. Sterner, *Am. Ind. Hyg. Assoc. Quart.* **10**, 93 (1949).
48. A. L. Walpole, D. C. Roberts, F. L. Rose, J. A. Hendry, and R. F. Homer, *Brit. J. Pharmacol.* **9**, 306 (1954).
49. C. P. Carpenter, H. J. Smyth, and C. B. Shaffer, *J. Ind. Hyg. Toxicol.* **30**, 2 (1948).
50. S. D. Silver and F. P. McGrath, *J. Ind. Hyg. Toxicol.* **30**, 7 (1948).
51. J. P. Danehy and D. J. Pflaum, *Ind. Eng. Chem.* **30**, 778 (1938).
52. G. F. Wright and V. K. Rowe, *Toxicol. Appl. Pharmacol.* **11**, 575 (1967).
53. I. L. Rapoport, *Byul. Mosk. Obshchestva Ispytatelei Prirody, Otd. Biol.* **67**, 109 (1962).
54. H. Lüers and G. Röhrborn, *Mutation Res.* **2**, 29 (1965).
55. M. L. Alexander and E. Glanges, *Proc. Natl. Acad. Sci. U.S.* **53**, 282 (1965).
56. M. L. Alexander, *Genetics* **56**, 274 (1967).
57. J. K. Lim and L. A. Snyder, *Mutation Res.* **6**, 129 (1968).
58. W. E. Ratnayake, *Mutation Res.* **5**, 271 (1968).
59. M. Westergaard, *Experientia* **13**, 224 (1957).
60. V. V. Khvostova, U. S. Mozhaeva, N. S. Aigaes, and S. A. Valeva, *Mutation Res.* **2**, 339 (1965).
61. N. S. Aigaes, *Radiobiologyia* **4**, 170 (1964).
62. L. Ehrenberg, A. Gustafsson, and U. Lundquist, *Hereditas* **47**, 243 (1961).
63. L. Ehrenberg, A. Gustafsson, and J. Lundquist, *Hereditas* **45**, 35 (1959).
64. L. Ehrenberg, U. Lundquist, and G. Ström, *Hereditas* **45**, 351 (1959).
65. F. K. Zimmermann and U. von Laer, *Mutation Res.* **4**, 377 (1967).
66. T. H. Chang and F. T. Elequin, *Mutation Res.* **4**, 83 (1967).
67. J. J. Biesele, F. S. Philips, J. B. Thiersch, J. H. Burchenal, S. M. Buckley, and C. C. Stock, *Nature* **166**, 1112 (1950).
68. E. Allen and W. Seaman, *Anal. Chem.* **27**, 540 (1955).
69. J. Epstein, R. W. Rosenthal, and P. J. Ess, *Anal. Chem.* **27**, 1435 (1955).
70. D. H. Rosenblatt, P. Hlinka, and J. Epstein, *Anal. Chem.* **27**, 1290 (1955).

71. T. R. Crompton, *Analyst* **90**, 107 (1965).
72. D. Schmähl and W. Sattler, *Arzneimittel-Forsch.* **14**, 746 (1964).
73. F. Hölzel, H. Maass, and H. Bock, *Arzneimittel-Forsch.* **14**, 792 (1964).
74. G. C. LaBrecque, P. H. Adcock, and C. N. Smith, *J. Econ. Entomol.* **53**, 802 (1960).
75. M. M. Crystal, *J. Econ. Entomol.* **56**, 468 (1963).
76. I. Keiser, L. F. Steiner, and H. Kamasaki, *J. Econ. Entomol.* **58**, 682 (1965).
77. M. Beroza and A. B. Borkovec, *J. Med. Chem.* **7**, 44 (1964).
78. L. H. Schmidt, *Ann. N. Y. Acad. Sci.* **68**, 652 (1958).
79. H. B. Mandel, *Pharmacol. Rev.* **11**, 743 (1959).
80. F. S. Philips and J. B. Thiersch, *J. Pharmacol. Exptl. Therap.* **100**, 398 (1950).
81. H. Jackson, B. W. Fox, and A. W. Craig, *Brit. J. Pharmacol.* **14**, 149 (1959).
82. J. A. Hendry, R. F. Homer, and F. L. Rose, *Brit. J. Pharmacol.* **6**, 357 (1951).
83. L. H. Schmidt, *Ann. N. Y. Acad. Sci.* **68**, 657 (1958).
84. M. V. Nadkarni, E. I. Goldenthal, and P. K. Smith, *Cancer Res.* **14**, 559 (1954).
85. M. V. Nadkarni, E. G. Trams, and P. K. Smith, *Proc. Am. Assoc. Cancer Res.* **2**, 136 (1956).
86. E. I. Goldenthal, M. V. Nadkarni, and P. K. Smith, *J. Pharmacol. Exptl. Therap.* **122**, 431 (1958).
87. G. Röhrborn, *Humangenetik* **1**, 576 (1965).
88. G. Röhrborn, *Humangenetik* **2**, 81 (1966).
89. E. Schleirmacher, *Humangenetik* **3**, 134 (1966).
90. A. J. Bateman, *Nature* **210**, 205 (1966).
91. S. S. Epstein and H. Shafner, *Nature* **219**, 385 (1968).
92. A. J. Bateman, *Genet. Res.* **1**, 381 (1960).
93. B. M. Cattanach, *Z. Vererbungslehre* **90**, 1 (1959).
94. O. G. Fahmy and M. J. Fahmy, *J. Genet.* **52**, 603 (1954).
95. O. G. Fahmy and M. J. Fahmy, *J. Genet.* **53**, 563 (1955).
96. L. A. Snyder and I. I. Oster, *Mutation Res.* **1**, 437 (1964).
97. L. A. Snyder, *Z. Vererbungslehre* **94**, 182 (1963).
98. I. H. Herskowitz, *Genetics* **41**, 605 (1956).
99. I. H. Herskowitz, *Genetics* **40**, 574 (1955).
100. D. T. North, *Mutation Res.* **4**, 225 (1967).
101. J. Moutschen, *Genetics* **46**, 291 (1961).
102. B. M. Cattanach, *Mutation Res.* **3**, 346 (1966).
103. B. M. Cattanach, *Nature* **180, 1364** (1957).
104. B. M. Cattanach, *Mutation Res.* **4**, 73 (1967).
105. B. M. Cattanach and R. G. Edwards, *Proc. Roy. Soc. Edinburgh* **B67**, 54 (1958).
106. A. Schalet, *Genetics* **40**, 594 (1955).
107. W. Ratnayake, C. Strachen, and C. Auerbach, *Mutation Res.* **4**, 380 (1967).
108. K. E. Hampel and H. Gerhartz, *Exptl. Cell Res.* **37**, 251 (1965).
109. H. Nicoloff and K. Gecheff, *Mutation Res.* **6**, 257 (1968).
110. A. Michaelis and R. Rieger, *Nature* **199**, 1014 (1963).
111. C. H. Ockey, *J. Genet.* **55**, 525 (1957).
112. D. Read, *Intern. J. Radiation Biol.* **3**, 95 (1961).
113. R. Wakonig and T. J. Arnason, *Can. J. Botany* **37**, 403 (1959).
114. V. N. Ronchi, M. Buiatti, and P. L. Ipata, *Mutation Res.* **4**, 315 (1967).
115. Z. Hartman, *Carnegie Inst. Wash. Publ.* **612**, 107 (1956).
116. V. N. Iyer and W. Szybalski, *Proc. Natl. Sci. U.S.* **44**, 446 (1958).
117. W. Gauss and G. Domagk, U.S. Patent 2,976,279 (1951).
118. G. Röhrborn and F. Vogel, *Deut. Ned. Wochschr.* **92**, 2315 (1967).

119. G. Obe, *Mutation Res.* **6**, 467 (1968).
120. H. Bestian, *Ann. Chem.* **566**, 210 (1950).
121. G. L. Drake, Jr. *et al.*, *Am. Dysetuff Reptr.* **50**, 27 (1961); *Chem. Abstr.* **55**, 8870 (1961).
122. L. E. Friese, *Melliand Textilber.* **39**, 795 (1958); *Chem. Abstr.* **52**, 17148 (1958).
123. G. L. Drake, Jr. and J. D. Guthrie, *Textile Res. J.* **29**, 155 (1959).
124. J. Heyna *et al.*, German Patent 1,044,761 (1958); *Chem. Abstr.* **54**, 23350 (1960).
125. O. Herrmann and H. Müller, U.S. Patent 2,911,321 (1959); *Chem. Abstr.* **54**, 5159 (1960).
126. W. A. Reeves, G. L. Drake, Jr., L. H. Chance, and J. D. Guthrie, *Textile Res. J.* **27**, 260 (1957).
127. H. Osborg, J. W. Brook, and A. Goldstein, U.S. Patent 3,298,902 (1967); *Chem. Abstr.* **66**, 77163a (1967).
128. D. L. Renaga, U.S. Patent 3,312,520 (1967); *Chem. Abstr.* **67**, 230712a (1967).
129. P. S. Hudson and C. C. Bice, U.S. Patent 3,087,844 (1963); *Chem. Abstr.* **59**, 1433 (1963).
130. S. M. Buckley, C. C. Stock, R. P. Parker, M. L. Crossley, E. Kuch, and D. R. Seeger, *Proc. Soc. Exptl. Biol. Med.* **78**, 295 (1951).
131. M. P. Sykes, D. A. Karnofsky, F. S. Phillips, and J. H. Burchenal, *Cancer* **6**, 142 (1953).
132. I. R. Duvall, *Cancer Chemotherapy Rept.* **8**, 1950 (1960).
133. L. S. Goodman and A. Gilman, "The Pharmacological Basis of Therapeutics," 3rd ed. Macmillan, New York, 1967.
134. A. W. Craig and H. Jackson, *Brit. J. Pharmacol.* **10**, 321 (1955).
135. Sumimoto Chem. Co. Ltd., British Patent 845,823 (1960); *Chem. Abstr.* **55**, 74336 (1961).
136. M. V. Nadkarni, E. I. Goldenthal, and P. K. Smith, *Cancer Res.* **17**, 97 (1957).
137. A. W. Craig, B. W. Fox, and H. Jackson, *Biochem. J.* **69**, 16P (1958).
138. L. B. Mellett and L. A. Woods, *Cancer Res.* **20**, 524 (1960).
139. M. V. Nadkarni, E. G. Trams, and P. K. Smith, *Cancer Res.* **19**, 713 (1959).
140. A. B. Borkovec, "Insect Chemosterilants." Wiley (Interscience), New York, 1966.
141. A. B. Borkovec, *Proc. 12th Intern. Congr. Entomol., London, 1964* p. 314. 1965.
142. W. W. Kilgore, *in* "Pest Control: Biological, Physical and Selected Chemical Methods" (W. W. Kilgore and R. L. Doutt, eds.), p. 197. Academic Press, New York, 1967.
143. G. C. LaBrecque and C. N. Smith, "Principles of Insect Chemosterilization." Appleton, New York, 1968.
144. W. J. Hayes, *Bull. World Health Organ.* **31**, 721 (1964).
145. K. R. S. Ascher, *World Rev. Pest Control* **3**, Part I, 7 (1964).
146. J. M. Barnes, *Trans. Roy. Soc. Trop. Med. Hyg.* **58**, 327 (1964).
147. S. C. Chang, A. B. Borkovec, and C. W. Woods, *J. Econ. Entomol.* **59**, 937 (1966).
148. P. A. Hedin, G. Wiygul, D. A. Vickers, A. C. Bartlett, and N. Mitlin, *J. Econ. Entomol.* **60**, 209 (1965).
149. C. W. Collier and R. Tardif, *J. Econ. Entomol.* **60**, 28 (1967).
150. T. L. Ladd, Jr., C. W. Collier, and E. L. Plasket, *J. Econ. Entomol.* **61**, 942 (1968).
151. S. C. Chang and A. B. Borkovec, *J. Econ. Entomol.* **59**, 102 (1966).
152. H. C. Cox, J. R. Young, and M. C. Bowman, *J. Econ. Entomol.* **60**, 1111 (1967).
153. J. Maitlen and L. M. McDonough, *J. Econ. Entomol.* **60**, 1391 (1967).
154. V. A. Chernov and Z. F. Presnova, *Vopr. Onkol.* **9**, 70 (1963).
155. V. A. Chernov and Z. F. Zakharova, *Vopr. Onkol.* **3**, 289 (1957).
156. A. A. Grushina, *Vopr. Onkol.* **1**, 51 (1955).
157. J. Palmquist and L. E. LaChance, *Science* **154**, 915 (1966).
158. A. R. Kaney and K. C. Atwood, *Nature* **201**, 1006 (1964).
159. W. Szybalski, *Ann. N. Y. Acad. Sci.* **76**, 475 (1958).
160. J. W. Drake, *Nature* **197**, 1028 (1963).

161. O. Klatt, A. C. Griffin, and J. S. Stehlin, Jr., *Proc. Soc. Exptl. Biol. Med.* **104**, 129 (1960).
162. Y. L. Tan and D. R. Cole, *Clin. Chem.* **11**, 50 (1965).
163. R. Truhaut, E. Selacoux, G. Bruk, and C. Bohoun, *Clin. Chim. Acta* **8**, 235 (1963).
164. R. K. Ausman, E. E. Crevar, H. Hagedorn, T. J. Bardos, and J. L. Ambrus, *J. Am. Med. Assoc.* **178**, 735 (1961).
165. E. Allen and W. Seamen, *Anal. Chem.* **27**, 540 (1955).
166. M. C. Bowman and M. Beroza, *J. Assoc. Offic. Anal. Chemists* **49**, 1046 (1966).
167. A. B. Borkovec, S. C. Chang, and A. M. Limburg, *J. Econ. Entomol.* **57**, 815 (1964).
168. D. A. Dame and C. A. Schmidt, *J. Econ. Entomol.* **57**, 77 (1964).
169. W. F. Chamberlain and C. C. Barrett, *J. Econ. Entomol.* **57**, 267 (1964).
170. T. L. Ladd, Jr., *J. Econ. Entomol.* **59**, 422 (1966).
171. F. W. Plapp, Jr., W. S. Bigley, G. A. Chapman, and G. W. Eddy, *J. Econ. Entomol.* **55**, 607 (1962).
172. W. F. Chamberlain and E. W. Hamilton, *J. Econ. Entomol.* **57**, 800 (1964).
173. P. B. Morgan, M. C. Bowman, and G. C. LaBrecque, *J. Econ. Entomol.* **61**, 805 (1968).
174. T. B. Gaines and R. D. Kimbrough, *Bull. World Health Organ.* **34**, 317 (1966).
175. S. S. Epstein, *Toxicol. Appl. Pharmacol.* **14**, 653 (1969).
176. K. D. Wuu and W. F. Grant, *Nucleus* (*Calcutta*) **10**, 37 (1967).
177. C. E. Mendoza, Ph.D. Dissertation, Iowa State University (1964).
178. H. C. Mason and F. F. Smith, *J. Econ. Entomol.* **60**, 1127 (1967).
179. R. R. Painter and W. W. Kilgore, *J. Econ. Entomol.* **57**, 154 (1964).
180. H. K. Gouck, *J. Econ. Entomol.* **57**, 235 (1964).
181. D. A. Dame, D. B. Woodward, and H. R. Ford, *Mosquito News* **24**, 1 (1964).
182. W. F. Chamberlain, *J. Econ. Entomol.* **55**, 240 (1962).
183. M. M. Crystal and L. E. LaChance, *J. Cellular Comp. Physiol.* **125**, 270 (1963).
184. D. G. Campion, *Nature* **214**, 1031 (1967).
185. W. W. Kilgore and R. R. Painter, *Biochem. J.* **92**, 353 (1964).
186. W. F. Chamberlain and C. C. Barrett, *Nature* **218**, 471 (1968).
187. L. E. LaChance, M. Degrugillier, and A. P. Leverich, *Mutation Res.* **7**, 63 (1969).
188. W. S. Murray and W. E. Bickley, *Maryland, Univ., Agr. Expt. Sta., Bull.* **A-134**, 1 (1964).
189. E. Kuh and D. R. Seeger, U.S. Patent 2,670,347 (1954); *Chem. Abstr.* **49**, 2481a (1955).
190. J. Blinov and O. M. Cherntsov, U.S.S.R. Patent 132,188 (1960); *Chem. Abstr.* **55**, 5974 (1961).
191. Y. Jo and R. Murase, *Sen-i-Kogyo Shinkensho Hokoku* **45**, 99 (1958); *Chem. Abstr.* **55**, 4968 (1961).
192. W. A. Reeves, *Textile Res. J.* **27**, 260 (1957); *Chem. Abstr.* **51**, 7725 (1957).
193. W. A. Reeves *et al.*, U.S. Patent 2,889,289 (1959); *Chem. Abstr.* **53**, 15593 (1959).
194. A. Sumida, Japanese Patent 10,399 (1957); *Chem. Abstr.* **52**, 21150 (1958).
195. K. B. Olson, *Ann. N. Y. Acad. Sci.* **6**, 1018 (1958).
196. J. E. Ultmann, G. A. Hyman, and A. Gellhorn, *Ann. N. Y. Acad. Sci.* **68**, 1007 (1958).
197. J. C. Wright, F. M. Golumb, and S. L. Gumport, *Ann. N. Y. Acad. Sci.* **68**, 937 (1958).
198. R. W. Ruddon and L. B. Mellett, *Cancer Chemotherapy Rept.* **39**, 7 (1964).
199. A. W. Craig, B. W. Fox, and H. Jackson, *Biochem. Pharmacol.* **3**, 42 (1959).
200. I. V. Boone, B. S. Rogers, and D. L. Williams, *Toxicol. Appl. Pharmacol.* **4**, 344 (1962).
201. L. B. Mellett and L. A. Woods, *Federation Proc.* **20**, 157 (1961).
202. J. C. Bateman, H. N. Carlton, R. C. Calvert, and G. E. Lindenblad, *Intern. J. Appl. Radiation Isotopes* **7**, 287 (1960).
203. C. Benckhuijsen, *Biochem. Pharmacol.* **17**, 55 (1968).
204. D. S. Bertram, *Trans. Roy. Soc. Trop. Med. Hyg.* **57**, 322 (1963).

205. J. C. Parish and B. W. Arthur, *J. Econ. Entomol.* **58**, 976 (1965).
206. G. O. MacDonald, A. N. Stroud, A. M. Brues, and W. H. Cole, *Ann. Surg.* **157**, 785 (1963).
207. G. P. Wheeler and J. A. Alexander, *Cancer Res.* **24**, 1328 (1964).
208. H. Oishi, *J. Fac. Sci., Hokkaido Univ., Ser. VI* **14**, 629 (1961).
209. T. Umemo, *Nippon Acta Radiol.* **19**, 1597 (1959).
210. L. E. LaChance and M. M. Crystal, *Biol. Bull.* **125**, 280 (1963).
211. M. A. Arsen'eva and A. V. Golovkina, *Vliyanie Ioniz. Izluch. Nasledstvennost, Akad. Nauk SSR* p. 122 (1966); *Chem. Abstr.* **67**, 8466k (1967).
212. N. V. Pankova, *Genetika* p. 62 (1967); *Chem. Abstr.* **67**, 1071645 (1967).
213. L. B. Mellett, P. E. Hodgson, and L. A. Woods, *J. Lab. Clin. Med.* **60**, 88 (1962).
214. J. R. Caldwell and R. Gilkey, U.S. Patent 2,839,479 (1958); *Chem. Abstr.* **52**, 15966 (1958).
215. W. N. Berard, E. K. Leonard, and W. A. Reeves, *Am. Dyestuff Reptr.* **50**, 29 (1961).
216. H. H. St. Mard, C. Hamalainen, and A. S. Cooper, Jr., *Am. Dyestuff Reptr.* **55**, 1046 (1966).
217. E. Klein and J. W. Weaver, U.S. Patent 2,911,322 (1959); *Chem. Abstr.* **54**, 8106g (1960).
218. T. Pfirrmann, German Patent 1,004,336 (1957); *Chem. Abstr.* **54**, 16889g (1960).
219. P. Strakov and D. A. Kutsidi, *Kozarstvi* **14**, 232 (1964); *Chem. Abstr.* **64**, 5310c (1966).
220. C. Pinazzi and J. LeBras, French Patent 1,187,648 (1959); *Chem. Abstr.* **54**, 23400g (1960).
221. W. Mende, East German Patent 19,788 (1960); *Chem. Abstr.* **54**, 21498i.
222. A. L. Walpole, *Brit. J. Pharmacol.* **6**, 135 (1951).
223. F. L. Rose, J. A. Hendry, and A. L. Walpole, *Nature* **165**, 993 (1950).
224. J. A. Hendry, F. L. Rose, and A. L. Walpole, *Brit. J. Pharmacol.* **6**, 201 (1951).
225. J. A. Hendry, R. F. Homer, F. L. Rose, and A. L. Walpole, *Brit. J. Pharmacol.* **6**, 235 (1951).
226. H. Schmidt-Elmendorff, W. Schmidt, and K. H. Schreyer, *Med. Klin.* (*Munich*) **50**, 2189 (1955).
227. H. Gehrhartz, *Wiss. Praxix* **7**, 12 (1959).
228. R. Kveton and F. Hanouset, *Chem. Listy* **49**, 63 (1955).
229. J. R. Dudley, *J. Am. Chem. Soc.* **73**, 3007 (1951).
230. R. C. Hirt and D. J. Salley, *J. Chem. Phys.* **21**, 1181 (1953).
231. I. Staib and H. Wilz, *Arzneimittel-Forsch.* **6**, 395 (1956).
232. O. Eichler and I. Staib, *Arzneimittel-Forsch.* **6**, 119 (1956).
233. G. Röhrborn, *Z. Vererbungslehre* **93**, 1 (1962).
234. G. Röhrborn, *Drosophila Inform. Serv.* **33**, 156 (1959).
235. H. Ulrich, E. Ploetz, and M. Bögemann, U.S. Patent 2,163,181 (1939); *Chem. Abstr.* **33**, 7818 (1939).
236. M. Ishidate, Y. Sakurai, H. Imamura, and A. Moriwaki, *Chem. & Pharm. Bull.* (*Tokyo*) **8**, 444 (1960).
237. H. Druckrey, *Giorn. Ital. Chemioterap.* **3**, 21 (1956); *Chem. Abstr.* **51**, 4549 (1957).
238. W. P. Anslow, Jr., D. A. Karnovsky, B. V. Jager, and H. W. Smith, *J. Pharmacol. Exptl. Therap.* **91**, 224 (1947).
239. M. Yamamoto, Y. Takebayashi, and H. Osaki, *Japan. J. Pharmacol.* **10**, 47 (1960); *Chem. Abstr.* **55**, 15741b (1961).
240. N. Brock and H. J. Hohorst, *Arzneimittel-Forsch.* **11**, 164 (1961),
241. H. J. Hohorst, A. Zieman, and N. Brock, *Arzneimittel-Forsch.* **16**, 1529 (1966).

242. E. L. Tatum, *Cold Spring Harbor Symp. Quant. Biol.* **11**, 278 (1946).
243. K. A. Jensen, I. Kirk, G. Kølmark, and M. Westergaard, *Cold Spring Harbor Symp. Quant. Biol.* **16**, 245 (1951).
244. O. M. Friedman and E. Boger, *Anal. Chem.* **33**, 906 (1961).
245. K. Ward, Jr., U.S. Patent 2,092,348 (1937); *Chem. Abstr.* **31**, 2614 (1937).
246. H. Druckrey, *Giorn. Ital. Chemioterap.* **3**, 31 (1956).
247. E. Hannig, *Pharm. Zentralhalle* **100**, 103 (1961).
248. R. Truhaut, C. Paoletti, J. Schlumberger, J. Nevoret, R. Nogues, and M. Marlot, *J. Med. Bordeaux Sud-Ouest* **134**, 735 (1957).
249. G. Kamiya, *Yakugaku Zasshi* **53**, 190 (1957).
250. H. Kröger, *Arzneimittel-Forsch*, **7**, 147 (1957).
251. J. Kovarik and F. Suec, *Neoplasma* **13**, 57 (1966).
252. G. Höhne, C. Bertram, and G. Schubert, *Acta, Unio Intern. Contra Cancrum* **16**, 658 (1960).
253. S. Kawate, *Technol. Rept. Kansai Univ.* **1**, 45 (1959); *Chem. Abstr.* **54**, 14366 (1960).
254. C. M. Stevens and A. Mylroie, *Nature* **164**, 1019 (1950).
255. D. A. Karnofsky, *Ann. N. Y. Acad. Sci.* **68**, 657 (1958).
256. C. T. Klopp and J. C. Bateman, *Advan. Cancer Res.* **11**, 255 (1954).
257. F. S. Philips, *Pharmacol. Rev.* **2**, 281 (1950).
258. S. S. Brown, *Advan. Pharmacol.* **2**, 243 (1963).
259. H. E. Skipper, L. J. Bennett, Jr., and G. P. Wheeler, *Proc. 2nd Natl. Cancer Conf.* p. 1571 Am. Cancer Soc., New York, 1954; *Chem. Abstr.* **48**, 11632b (1954).
260. H. E. Skipper, L. L. Bennett, Jr., and W. A. Langham, *Cancer* **4**, 1025 (1951).
261. A. C. Griffin, E. L. Brant, and E. L. Tatum, *Cancer Res.* **11**, 253 (1951).
262. W. E. Heston and M. A. Schneiderman, *Science* **117**, 109 (1953).
263. D. Haskin, *Anat. Record* **493**, 511 (1948).
264. M. L. Murphy and D. A. Karnofsky, *Cancer* **9**, 955 (1956).
265. M. L. Murphy, A. D. Moro, and C. Lacon, *Ann. N. Y. Acad. Sci.* **68**, 762 (1958).
266. C. H. Danforth and E. Center, *Proc. Soc. Exptl. Biol. Med.* **86**, 705 (1954).
267. J. E. Gibson and B. A. Becker, *Toxicol. Appl. Pharmacol.* **14**, 639 (1969).
268. D. S. Falconer, B. M. Slaynski, and C. Auerbach, *J. Genet.* **51**, 81 (1952).
269. C. Auerbach and H. Moser, *Nature* **166**, 1019 (1950).
270. O. G. Fahmy and M. J. Fahmy, *Heredity* **15**, 115 (1960).
271. A. Schalet, *Am. Naturalist* **90**, 329 (1956).
272. I. I. Oster and E. Pooley, *Genetics* **45**, 1004 (1960).
273. I. H. Herskowitz and A. Schalet, *Genetics* **39**, 970 (1954).
274. W. J. Burdette, *Cancer Res.* **12**, 366 (1952).
275. E. A. Löbbecke and R. von Borstel, *Genetics* **47**, 853 (1962).
276. S. W. Glover, *Carnegie Inst. Wash. Publ.* **612**, 121 (1956).
277. V. N. Iyer and W. Szybalski, *Appl. Microbiol.* **6**, 23 (1958).
278. A. Loveless, "Genetics and Allied Effects of Alkylating Agents," p. 103. Pennsylvania State Univ. Press, University Park, Pennsylvania, 1966.
279. M. Demerec, *Caryologia* Suppl., p. 201 (1954).
280. A. Zampieri and J. Greenberg, *Genetics* **57**, 41 (1967).
281. J. Nečašek, *Proc. 11th Intern. Congr. Genet. The Hague, 1963* Vol. 1, p. 59. Pergamon Press, Oxford, 1965.
282. L. Silvestri, *Boll. Ist. Sieroterap. Milan* **28**, 326 (1949).
283. E. A. Löbbecke, *Genetics* **48**, 91 (1965).
284. J. F. Stauffer and M. B. Bachus, *Ann. N. Y. Acad Sci.* **60**, 35 (1954).
285. F. R. Roegner, M. A. Stahmann, and J. F. Stauffer, *Am. J. Botany* **41**, 1 (1954).

286. E. C. Gasiorkiewitz, R. H. Larson, J. C. Walker, and M. A. Stahmann, *Phytopathology* **42,** 183 (1952).
287. M. A. Stahmann and J. F. Stauffer, *Science* **106,** 35 (1947).
288. W. D. McElroy, J. E. Cushing, and H. Miller, *J. Cellular Comp. Physiol.* **30,** 331 (1947).
289. D. Bonner, *Cold Spring Harbor Symp. Quant. Biol.* **11,** 14 (1946).
290. G. Kølmark and M. Westergaard, *Hereditas* **35,** 490 (1949).
291. E. L. Tatum, R. W. Barratt, N. Fries, and D. Bonner, *Am. J. Botany* **37,** 38 (1950).
292. N. Fries, *Physiol. Plantarum.* **1,** 330 (1948).
293. N. Iguchi, *J. Agr. Chem. Soc. Japan* **27,** 229 (1953).
294. P. B. Gibson, R. A. Brink, and M. A. Stahmann, *J. Heredity* **41,** 232 (1950).
295. J. Mackey, *Acta Agr. Scand.* **4,** 419 (1954).
296. A. Loveless, *Proc. Roy. Soc.* **B150,** 497 (1959).
297. A. Loveless, *Proc. Roy. Soc.* **B159,** 348 (1964).
298. C. E. Ford, *Hereditas* Suppl., p. 570 (1949).
299. S. H. Revell, *Heredity* **6,** Suppl., 107 (1953).
300. B. A. Kihlman, *Hereditas* **41,** 384 (1955).
301. R. Rieger and A. Michaelis, *Chromosoma* **11,** 573 (1961).
302. E. G. Trams, *Anal. Chem.* **30,** 256 (1958).
303. L. B. Mellett and L. A. Woods, *Cancer Res.* **20,** 518 (1960).
304. R. Mantsavinos and J. E. Christian, *Anal. Chem.* **30,** 1071 (1958).
305. Y. Sakurai and K. Ito, *Chem. & Pharm. Bull.* (*Tokyo*) **8,** 655 (1960); *Chem. Abstr.* **55,** 12769a (1961).
306. I. Aiko, *J. Pharm. Soc. Japan* **72,** 1297 (1952); *Chem. Abstr.* **47,** 1289 (1953).
307. N. Brock, *Arzneimittel-Forsch.* **7,** 727 (1957).
308. Y. Sugiya, Y. Inazu, M. Arakawa, S. Shimada, R. Takeda, and Y. Aoiks, *Takamine Kenkyusho Nempo* **12,** 218 (1960); *Chem. Abstr.* **55,** 7660 (1961).
309. G. Kamiya, *Nippon Yakurigaku Zasshi* **53,** 190 (1967).
310. M. Ishidate, *Acta, Unio Intern. Contra Cancrum* **15,** 139 (1959); *Chem. Abstr.* **53,** 22496h (1959).
311. H. Satoh, *Acta, Unio Intern. Contra Cancrum* **15,** 258 (1959); *Chem. Abstr.* **53,** 22496 (1959).
312. M. Minami, *Osaka Shiritsu Daigaku Igaku Zasshi* **9,** 1483 (1960).
313. H. Arnold, N. Brock, and H. J. Hohorst, *Arzneimittel-Forsch.* **7,** 735 (1957).
314. H. Arnold and J. Venjakob, *Arzneimittel-Forsch.* **5,** 722 (1955).
315. G. Cardinali and G. Morpurgo, *Ric. Sci.* Suppl. 22, 55 (1955); *Chem. Abstr.* **52,** 15756g (1958).
316. S. O. Greenlee, U.S. 3,300,444 (1967); *Chem. Abstr.* **66,** 66208t (1967).
317. V. M. Vlasov, G. C. Balezina, and E. I. Kostsyna, *Zh. Organ. Khim.* **2,** 2137 (1966); *Chem. Abstr.* **66,** 85410o (1967).
318. J. A. Jordan and J. K. Aiken, British Patent 576,721 (1946); *Chem. Abstr.* **42,** 21323 (1948).
319. R. C. Keller, British Patent 1,035,923 (1966); *Chem. Abstr.* **66,** 11802b (1967).
320. P. D. Bartlett and C. G. Swain, *J. Am. Chem. Soc.* **71,** 1406 (1949).
321. G. Holst, *Acta Chem. Scand.* **12,** 1042 (1958).
322. A. G. Ogston, E. R. Holiday, S. L. Philpot, and C. A. Stocken, *Trans. Faraday Soc.* **44,** 45 (1948); *Chem. Abstr.* **42,** 6632c (1948).
323. A. H. Williams, *J. Chem. Soc.* p. 318 (1947).
324. E. Boyland, *Biochem. Soc. Symp.* (*Cambridge, Engl.*) **2,** 61 (1948).
325. P. Brookes and P. D. Lawley, *Biochem. J.* **77,** 478 (1960).

326. P. Brookes and P. D. Lawley, *Exptl. Cell Res.* Suppl. 9, 521 (1963).
327. Z. M. Bacq, *Chim. Ind.* (*Paris*) **59**, 468 (1948).
328. I. Graef, D. A. Karnofsky, B. V. Jager, B. Krichesky, and W. H. Smith, *Am. J. Pathol.* **24**, 1 (1948).
329. W. P. Anslow, Jr., D. A. Karnofsky, B. V. Jager, and H. W. Smith, *J. Pharmacol. Exptl. Therap.* **93**, 1 (1948).
330. S. Black and J. F. Thomson, *J. Biol. Chem.* **167**, 283 (1947).
331. C. Davison, R. S. Rozman, L. Bliss, and P. K. Smith, *Proc. Am. Assoc. Cancer Res.* **2**, 195 (1957).
332. J. C. Boursnell, J. A. Cohen, M. Dixon, G. E. Francis, G. D. Greville, D. M. Needham, and A. Wormall, *Biochem. J.* **40**, 756 (1946).
333. C. Davison, R. S. Rozman, and P. K. Smith, *Biochem. Pharmacol.* **7**, 65 (1961).
334. W. E. Heston, *J. Natl. Cancer Inst.* **11**, 415 (1950).
335. W. E. Heston, *J. Natl. Cancer Inst.* **14**, 131 (1953).
336. W. E. Heston and W. D. Levillain, *Proc. Soc. Exptl. Biol. Med.* **82**, 457 (1953).
337. A. Haddow, *in* "Physiopathology of Cancer" (F. Homburger, ed.), p. 602. Harper (Hoeber), New York, 1959.
338. S. Wada, M. Miyanishi, Y. Nishimoto, S. Kambe, and R. W. Miller, *Lancet* **1**, 1161 (1968).
339. C. Auerbach and J. M. Robson, *Nature* **157**, 302 (1946).
340. C. Auerbach, *Biol. Rev. Cambridge Phil. Soc.* **24**, 335–391 (1949).
341. C. Auerbach and E. M. Sonbati, *Z. Vererbungslehre* **91**, 237 (1960).
342. C. Auerbach, *Proc. Roy. Soc. Edinburgh* **B62**, Part 2, 211 (1945).
343. C. E. Nasrat, W. D. Kaplan, and C. Auerbach, *Z. Induktive Abstammungs- Vererbungslehre* **86**, 249 (1954).
344. A. Gilman and F. S. Philips, *Science* **103**, 409 (1946).
345. C. Auerbach, J. M. Robson, and J. G. Carr, *Science* **105**, 243 (1947).
346. C. Auerbach and J. M. Robson, *Proc. Roy. Soc. Edinburgh* **B62**, 271 (1947).
347. N. H. Horowitz, M. B. Houlahan, M. G. Hungate, and B. Wright, *Science* **104**, 2333 (1946).
348. K. A. Jensen, I. Kirk, and M. Westergaard, *Nature* **166**, 1021 (1950).
349. D. Hockenhail, *Nature* **161**, 100 (1948).
350. P. Brookes and P. D. Lawley, *Biochem. J.* **89**, 13B (1963).
351. C. D. Darlington, *Pubbl. Staz. Zool. Napoli* **22**, Suppl., 22–31 (1950); *Chem. Abstr.* **45**, 221 (1951).
352. P. C. Koller, *Progr. Biophys. Biophys. Chem.* **4**, 195 (1953).
353. P. C. Koller, *Ann. N. Y. Acad. Sci.* **68**, 783 (1958).
354. A. Koblin, *Anal. Chem.* **30**, 430 (1958).
355. P. Malatesta and A. Lorenzini, *Ric. Sci.* Suppl. 28, 1874 (1958); *Chem. Abstr.* **53**, 7864 (1959).
356. S. Imanishi and Y. Kanda, *J. Sci. Res. Inst.* (*Tokyo*) **43**, 1 (1949).
357. E. A. C. Lucken, *J. Chem. Soc.* p. 2954 (1959).
358. G. P. Wheeler, J. S. Morrow, and H. E. Skipper, *Arch. Biochem. Biophys.* **57**, 133 (1955).
359. F. G. Stanford, *Analyst* **92**, 64 (1967).
360. W. M. Grant and V. E. Kinsey, *J. Am. Chem. Soc.* **68**, 2075 (1946).
361. W. H. C. Rueggeberg, W. A. Cook, and E. E. Reid, *J. Org. Chem.* **13**, 110 (1948).
362. R. C. Fuson and J. B. Ziegler, *J. Org. Chem.* **11**, 510 (1946).
363. A. M. Seligman, A. M. Rutenburg, L. Persky, and O. M. Friedman, *Cancer* **5**, 354 (1952).

364. W. P. Anslow, Jr., D. A. Karnofsky, B. V. Jager, and H. W. Smith, *J. Pharmacol. Exptl. Therap.* **93**, 1 (1948).
365. P. Brookes and P. D. Lawley, *Biochem. J.* **80**, 496 (1961).
366. P. Brookes and P. D. Lawley, "Isotopes in Experimental Pharmacology," p. 403, Univ. of Chicago Press, Chicago, Illinois, 1965.
367. V. E. Kinsey and W. M. Grant, *J. Cellular Comp. Physiol.* **30**, 31 (1947).
368. H. F. Smyth, Jr. and C. F. Carpenter, *J. Ind. Hyg. Toxicol.* **28**, 262 (1946).
369. H. H. Schrenk, F. A. Petty, and W. P. Yant, *Public Health Rept.* (U.S.) **48**, 1389 (1933).
370. R. S. McLaughlin, *J. Pharmacol. Exptl. Therap.* **53**, 274 (1936).
371. C. L. Hake and V. K. Rowe, *in* "Industrial Hygiene and Toxicology" (F. A. Patty, ed.), 2nd ed., Vol. II, p. 1673. Wiley (Interscience), New York, 1963.
372. W. T. Williams, "Detoxification Mechanisms." Chapman & Hall, London, 1932.
373. H. Allen, *Chem. Prod.* **19**, 482 (1956).
374. L. Summers, *Chem. Rev.* **55**, 301 (1955).
375. B. L. Van Duuren, B. M. Goldschmidt, C. Katz, L. Langseth, G. Mercado, and A. Sivak, *Arch. Environ. Health* **16**, 472 (1968).
376. J. L. Gargus, W. H. Reese, Jr., and H. A. Rutter, *Toxicol. Appl. Pharmacol.* **15**, 92 (1969).
377. H. Arnold and F. Bourseaux, *Angew. Chem.* **70**, 539 (1958).
378. H. Arnold, F. Bourseaux, and N. Brock, *Naturwissenschaften* **45**, 64 (1958).
379. N. Brock, *Arzneimittel-Forsch.* **8**, 1 (1958).
380. K. Sugiura, F. Schmid, and M. Schmid, *Proc. Am. Assoc. Cancer Res.* **3**, 271 (1961).
381. D. G. Decker, E. Mussey, G. D. Malkasiam, and C. E. Johnson, *Clin. Obstet. Gynecol.* **11**, 382 (1968).
382. R. W. Rundles, J. Laszlo, F. E. Garrison, Jr., and J. B. Hobson, *Cancer Chemotherapy Rept.* **16**, 407 (1962).
383. H. Haar, G. Marshall, G. June, H. R. Bierman, and J. L. Steinfeld, *Cancer Chemotherapy Rept.* **6**, 41 (1960).
384. H. Brock and H. J. Hohorst, *Cancer* **20**, 900 (1967).
385. H. J. Hohorst, A. Zieman, and N. Brock, *Arzneimittel-Forsch.* **15**, 432 (1965).
386. N. Brock and H. J. Hohorst, *Arzneimittel-Forsch.* **13**, 1021 (1963).
387. H. M. Rauen and K. Norpoth, *Arzneimittel-Forsch.* **17**, 599 (1967).
388. H. M. Rauen and K. Norpoth, *Naturwissenschaften* **52**, 47 (1965).
389. O. M. Friedman, *Cancer Chemotherapy Rept.* **51**, 327 (1967).
390. O. M. Friedman, *Cancer Chemotherapy Rept.* **51**, 347 (1967).
391. W. Bolt, F. Ritzl, and H. Nahrmann, *Nucl. Med.* **2**, 251 (1962); *Chem. Abstr.* **63**, 8898 (1965).
392. W. Bolt, F. Ritzl, R. Toussaint, and H. Nahrmann, *Arzneimittel-Forsch.* **11**, 170 (1961).
393. E. H. Graul, H. Hundeshagen, and H. Williams, *Proc. 3rd Intern. Congr. Chemotherapy, Stuttgart, 1963* No. 2, p. 1103 (1964); *Chem. Abstr.* **65**, 1428 (1966).
394. M. Hirata, H. Kagawa, and M. Baba, *Shionogi Kenkyusho Nempo* **17**, 107 (1967); *Chem. Abstr.* **69**, 21955e (1968).
395. O. M. Friedman, S. Bien, and J. K. Chatrabarti, *J. Am. Chem. Soc.* **87**, 4978 (1965).
396. H. Arnold and H. Klose, *Arzneimittel-Forsch.* **11**, 159 (1961).
397. E. R. Homan, R. P. Zendzian, W. M. Busey, and D. P. Rall, *Nature* **221**, 1059 (1969).
398. E. H. Dolnick, I. L. Lundaehl, C. E. Terrill, and P. J. Reynolds, *Nature* **221**, 467 (1969).
399. T. Von Kreybig, *Arch. Exptl. Pathol. Pharmakol.* **252**, 173 (1965).
400. J. E. Gibson and B. A. Becker, *Cancer Res.* **28**, 475 (1968).
401. D. Brittinger, *Humangenetik* **3**, 156 (1966).
402. C. Bertram and G. Höhne, *Strahlentherapie* **43**, 388 (1959).
403. S. Frye, *Brit. Empire Cancer Campaign, Ann. Rept.* **38**, 670 (1960).

404. G. Röhrborn, *Mol. Gen. Genet.* **102**, 50 (1968).
405. I. W. Schmid and G. R. Staiger, *Mutation Res.* **7**, 99 (1969).
406. F. E. Arrighi, T. C. Hsu, and D. E. Bergsagel, *Texas Rept. Biol. Med.* **20**, 545 (1962).
407. K. E. Hampel, M. Fritzsche, and D. Stopik, *Humangenetik* **7**, 28 (1969).
408. M. Vrba, *Humangenetik* **4**, 362 (1967).
409. A. Michaelis and R. Rieger, *Biol. Zentr.* **80**, 301 (1961).
410. N. Brock, *Cancer Chemotherapy Rept.* **51**, 315 (1967).
411. A. Dede and F. Farabollini, *Boll. Soc. Ital. Biol. Sper.* **43**, 1489 (1967); *Chem. Abstr.* **68**, 48061 (1968).
412. D. Horvitz and E. Cerwonka, U.S. Patent 2,916,426 (1959); *Chem. Abstr.* **54**, 6370c (1960).
413. National Distillers and Chem. Corp., British Patent 817,523 (1959); *Chem. Abstr.* **54**, 10601c (1960).
414. E. G. Maitlen, U.S. Patent 2,970,939 (1961); *Chem. Abstr.* **56**, 11752f (1961).
415. C. A. I. Goring, U.S. Patent 3,256,083 (1966); *Chem. Abstr.* **65**, 6253d (1966).
416. M. R. Lytton, E. A. Wielicki, and E. Lewis, U.S. Patent 2,776,946 (1957); *Chem. Abstr.* **51**, 5466e (1957).
417. W. E. Elliot, J. R. Huff, R. W. Adler, and W. L. Towle, *Proc. Ann. Power Sources Conf.* **20**, 67–70 (1966); *Chem. Abstr.* **66**, 100955w (1967).
418. W. J. Middleton, U.S. Patent 3,240,765 (1966): *Chem. Abstr.* **64**, 19826f (1966).
419. Sh. L. Lel'Chuk and V. I. Sedlis, *Zh. Prikl. Khim.* **31** 128 (1958); *Chem. Abstr.* **52**, 17787g (1958).
420. K. Klager, U.S. Patent 3,192,707 (1965).
421. W. W. Hartmann and R. Philips, *Org. Syn.* **2**, 464 (1943).
422. A. H. Dutton and D. F. Heath, *J. Chem. Soc.* p. 1892 (1956).
423. L. F. Fieser and M. Fieser, "Organic Chemistry." Heath, Boston, Massachusetts, 1960.
424. R. Zahradnik, *Chem. Listy* **51**, 937 (1957).
425. C. M. Bamford, *J. Chem. Soc.* p. 124 (1939).
426. H. Ballweg and D. Schmähl, *Naturwissenschaften* **54**, 116 (1967).
427. P. N. Magee and J. M. Barnes, *Advan. Cancer Res.* **10**, 163 (1967).
428. P. N. Magee and R. Schoental, *Brit. Med. Bull.* **20**, 102 (1964).
429. H. Druckrey, R. Preussmann, and D. Schmähl, *Acta, Unio Intern. Contra Cancrum* **19**, 510 (1963).
430. D. F. Heath, *Biochem. J.* **85**, 72 (1962).
431. H. Druckrey, A. Schildbach, D. Schmähl, R. Preussmann, and S. Ivankovic, *Arzneimittel-Forsch.* **13**, 841 (1963).
432. P. N. Magee and E. Farber, *Biochem. J.* **83**, 114 (1962).
433. P. N. Magee and T. Hultin, *Biochem. J.* **83**, 106 (1962).
434. W. J. Serfontein and J. H. Smit, *Nature* **214**, 169 (1967).
435. E. Kröller, *Deut. Lebensm.-Rundschau* **10**, 303 (1967).
436. G. B. Neurath, B. Pirmann, W. Lüttich, and H. Wichern, *Bietr. Tabakforsch.* **3**, 251 (1965).
437. J. Sakshaug, E. Sögnen, M. A. Hansen, and N. Koppang, *Nature* **206**, 1261 (1965).
438. F. Ender, G. Havre, A. Helgebostad, N. Koppang, R. Madsden, and L. Ceh, *Naturwissenschaften* **51**, 637 (1964).
439. O. G. Devik, *Acta Chem. Scand.* **21**, 2302 (1967).
440. P. Marquardt and L. Hedler, *Arzneimittel-Forsch.* **16**, 778 (1966).
441. L. Hedler and P. Marquardt, *Food Cosmet. Toxicol.* **6**, 341 (1968).
442. K. Möhler and O. L. Mayrhofer, *Z. Lebensm.-Untersuch. Forsch.* **135**, 313 (1968).
443. F. Ender, *Z. Tierphysiol., Tierernalhr. Futtermittelk.* **22**, 189 (1967).

444. N. P. Sen, D. C. Smith, L. Schwinghamer, and J. J. Marleau, *J. Assoc. Offic. Anal. Chemists* **52**, 47 (1969).
445. J. Sander, *Arch. Hyg. Bakteriol.* **151**, 22 (1967).
446. J. Sander, F. Schweinsberg, and H. P. Menz, *Z. Physiol. Chem.* **349**, 1691 (1968).
447. L. S. DuPlessis, J. R. Nunn, and W. A. Roach, *Nature* **222**, 1198 (1969).
448. L. Pasternak, *Arzneimittel-Forsch.* **14**, 802 (1964).
449. L. Pasternak, *Naturwissenschaften* **49**, 381 (1962).
450. L. Pasternak, *Acta Biol. Med. Ger.* **10**, 436 (1963).
451. O. G. Fahmy and M. J. Fahmy, *Mutation Res.* **6**, 139 (1968).
452. O. G. Fahmy, M. J. Fahmy, J. Massasso, and M. Ondrej, *Mutation Res.* **3**, 201 (1966).
453. J. Veleminsky and T. Gichner, *Mutation Res.* **5**, 429 (1968).
454. E. Geisler, *Naturwissenschaften* **49**, 380 (1962).
455. O. N. Pogodina, *Cytologia* **8**, 503 (1966).
456. H. Marquardt, F. K. Zimmermann, and R. Schwaier, *Z. Vererbungslehre* **95**, 82 (1964).
457. H. Marquardt, F. K. Zimmermann, and R. Schwaier, *Naturwissenschaften* **50**, 625 (1963).
458. H. Marquardt, R. Schwaier, and F. Zimmermann, *Naturwissenschaften* **50**, 135 (1963).
459. H. V. Malling, *Mutation Res.* **3**, 537 (1966).
460 S. Udenfriend, C. T. Clark, J. Axelrod, and B. B. Brodie, *Cancer Res.* **24**, 1712 (1964).
461. D. F. Heath and J. A. E. Jarvis, *Analyst* **80**, 613 (1955).
462. D. Daiber and R. Preussmann, *Z. Anal. Chem.* **206**, 344 (1964).
463. P. Griess, *Chem. Ber.* **12**, 427 (1879).
464. G. Neurath, B. Pirmann, and M. Dunger, *Chem. Ber.* **97**, 1631 (1964).
465. O. L. Mayrhofer and K. Möhler, *Z. Lebensm.-Untersuch. Forsch.* **134**, 246 (1967).
466. H. J. Petrowitz, *Arzneimittel-Forsch.* **18**, 1486 (1968).
467. C. E. Looney, W. D. Philips, and E. L. Reilly, *J. Am. Chem. Soc.* **79**, 6136 (1957).
468. R. K. Harris and R. A. Spragg, *Chem. Commun.* No. 7, p. 362 (1967); *Chem. Abstr.* **67**, 38122n (1967).
469. E. A. Werner, *J. Chem. Soc.* **115**, 1093 (1919).
470. D. R. McCalla, A. Reuvers, and R. Kitai, *Can. J. Biochem.* **46**, 806 (1968).
471. E. R. Garrett, S. Goto, and J. F. Stubbins, *J. Pharm. Sci.* **54**, 119 (1965).
472. P. F. Swann, *Biochem. J.* **110**, 49 (1968).
473. H. Druckrey, R. Preussmann, D. Schmähl, and M. Müller, *Naturwissenschaften* **48**, 165 (1961).
474. M. G. Kelly, R. W. O'Gara, S. T. Yancey, and C. Botkin, *J. Natl. Cancer Inst.* **41**, 619 (1968).
475. E. M. Johnson and C. Lambert, *Teratology* **1**, 179 (1968).
476. T. von Kreybig, *Z. Krebsforsch.* **67**, 46 (1965).
477. F. K. Sanders and B. O. Burford, *Nature* **213**, 1171 (1967).
478. F. K. Sanders and B. O. Burford, *Nature* **220**, 448 (1968).
479. J. Rapoport, *Dokl. Akad. Nauk SSSR* **146**, 1418 (1962).
480. J. Rapoport, *Dokl. Akad. Nauk SSSR* **148**, 696 (1962).
481. T. Alderson, *Nature* **207**, 164 (1965).
482. H. O. Corwin, *Mutation Res.* **5**, 259 (1968).
483. F. K. Zimmermann, R. Schwaier, and U. von Laer, *Z. Vererbungslehre* **97**, 68 (1965).
484. R. Schwaier, *Z. Vererbungslehre* **97**, 55 (1965).
485. R. Schwaier, F. K. Zimmermann, and U. von Laer, *Z. Vererbungslehre* **97**, 72 (1965).
486. A. J. Müller, *Zuechter* **34**, 102 (1964).
487. A. J. Müller, *Naturwissenschaften* **52**, 213 (1965).
488. T. Gichner and J. Veleminsky, *Mutation Res.* **4**, 207 (1967).

489. A. Loveless and C. C. Hampton, *Mutation Res.* **7**, 1 (1967).
490. N. N. Zoz and S. I. Makarova, *Tsitologiya* **7**, 405 (1965).
491. A. Michaelis, J. Schöneich, and R. Rieger, *Chromosoma* **16**, 101 (1965).
492. E. Gläss and H. Marquardt, *Mol. Gen. Genet.* **101**, 307 (1968).
493. F. Jancik, B. Kacak, V. Vanilek, and M. Urublovska, *Chem. Listy* **52**, 909 (1958).
494. H. S. Rosenkranz, M. Bitoon, and R. M. Schmidt, *J. Natl. Cancer Inst.* **41**, 1099 (1968).
495. A. H. Blatt, *Org. Syn.* Coll. Vol. 3, 464 (1943).
496. R. Schoental, *Nature* **199**, 190 (1963).
497. R. Schoental, *Nature* **188**, 420 (1960).
498. R. Schoental, *Brit. J. Cancer* **16**, 92 (1962).
499. R. Schoental and D. J. Rive, *Biochem. J.* **97**, 466 (1965).
500. R. Schoental, *Nature* **215**, 535 (1967).
501. P. D. Lawley, *Nature* **218**, 580 (1968).
502. C. D. Gutsche and H. E. Johnson, *J. Am. Chem. Soc.* **77**, 109 (1955).
503. I. A. Rapoport, *Dokl. Akad. Nauk SSSR* **59**, 1183 (1948).
504. H. Henke, G. Höhne, and H. A. Künkel, *Biophysik* **1**, 418 (1964).
505. E. A. Adelberg, M. Mandel, and G. C. C. Chen, *Biochem. Biophys. Res. Commun.* **18**, 788 (1965).
506. A. Trams and H. A. Künkel, *Biophysik* **1**, 422 (1964).
507. F. K. Zimmermann and R. Schwaier, *Mol. Gen. Genet.* **100**, 63 (1967).
508. G. Zetterberg, *Exptl. Cell Res.* **20**, 659 (1960).
509. G. Zetterberg, *Hereditas* **47**, 295 (1961).
510. G. Zetterberg, *Hereditas* **48**, 371 (1962).
511. A. Abbondandolo and N. Loprieno, *Mutation Res.* **4**, 31 (1967).
512. R. Guglielminetti, S. Bonatti, and N. Loprieno, *Mutation Res.* **3**, 152 (1966).
513. N. Loprieno and C. H. Clarke, *Mutation Res.* **2**, 312 (1965).
514. N. Loprieno, *Mutation Res.* **1**, 469 (1964).
515. N. Loprieno, G. Zetterberg, R. Auglielminetti, and E. Michel, *Mutation Res.* **1**, 37 (1964).
516. H. Heslot, R. Ferrary, and J. Tempé, *Mutation Res.* **3**, 355 (1966).
517. M. J. Ord, *Exptl. Cell Res.* **53**, 73 (1968).
518. M. J. Ord, *Nature* **206**, 413 (1965).
519. B. A. Kihlman, *Radiation Botany* **1**, 35 (1961).
520. B. A. Kihlman, *Exptl. Cell Res.* **20**, 657 (1960).
521. C. J. Grant and H. Heslot, *Ann. Genet.* **8**, 98 (1965).
522. E. Gläss and H. Marquardt, *Z. Vererbungslehre* **98**, 361 (1966).
523. A. F. McKay and G. F. Wright, *J. Am. Chem. Soc.* **69**, 3028 (1946).
524. A. F. McKay, *J. Am. Chem. Soc.* **70**, 1974 (1948).
525. R. A. Henry, *J. Am. Chem. Soc.* **72**, 3287 (1950).
526. D. R. McCalla and A. Reuvers, *Can. J. Biochem.* **46**, 1411 (1960).
527. V. M. Craddock, *Biochem. J.* **106**, 921 (1968).
528. D. R. McCalla, *Biochim. Biophys. Acta* **155**, 114 (1968).
529. T. Sugimura, S. Fujimura, M. Nagao, T. Yokoshima, and M. Hasegawa, *Biochim. Biophys. Acta* **170**, 427 (1968).
530. G. P. Wheeler, *Cancer Res.* **22**, 651 (1962).
531. W. A. Skinner, H. F. Gram, M. O. Greene, J. Greenberg, and B. R. Baker, *J. Med. Pharm. Chem.* **2**, 299 (1960).
532. M. O. Greene and J. Greenberg, *Cancer Res.* **20**, 1166 (1966).
533. J. Leiter and M. A. Schneiderman, *Cancer Res.* **19**, 31 (1959).
534. A. Goldin, J. M. Venditti, and I. Kline, *Cancer Res.* **19**, Suppl., 429 (1959).

535. H. E. Skipper, F. M. Schabel, Jr., M. W. Trader, and J. R. Thompson, *Cancer Res.* **21**, 1154 (1963).
536. D. E. Hunt and R. F. Pittillo, *Appl. Microbiol.* **16**, 1879 (1968).
537. P. M. L. Siu, *Proc. Soc. Exptl. Biol. Med.* **129**, 753 (1968).
538. T. Sugimura and S. Fujimura, *Nature* **216**, 943 (1967).
539. R. Schoental, *Nature* **209**, 726 (1966).
540. H. Druckrey, R. Preussmann, S. Ivankovic, B. T. So, C. H. Schmidt, and J. Bucheler, *Z. Krebsforsch.* **68**, 87 (1966).
541. T. Sugimura, M. Nagao, and Y. Okada, *Nature* **210**, 962 (1966).
542. A. Terawaki and J. Greenberg, *Biochim. Biophys. Acta* **95**, 170 (1965).
543. J. D. Mandell and J. Greenberg, *Biochem. Biophys. Res. Commun.* **3**, 575 (1960).
544. J. D. Mandell, *Bacteriol. Proc.* p. 124 (1965).
545. S. Zamenhof, L. H. Heldenmuth, and P. J. Zamenhof, *Proc. Natl. Acad. Sci. U.S.* **55**, 50 (1960).
546. L. Silengo, D. Schlessinger, G. Mangiarotti, and D. Apirion, *Mutation Res.* **4**, 701 (1967).
547. D. Botstein and E. W. Jones, *J. Bacteriol.* **98**, 847 (1969).
548. E. Cerda-Olmedo, P. C. Hanawalt, and N. Guerola, *J. Mol. Biol.* **33**, 705 (1968).
549. I. Takahashi and R. A. Bernard, *Mutation Res.* **4**, 111 (1967).
550. A. Eisenstark, R. Eisenstark, and R. Van Sickle, *Mutation Res.* **2**, 1 (1965).
551. F. Lingens and O. Oltmans, *Z. Naturforsch.* **21b**, 660 (1966).
552. K. Nordström, *J. Gen. Microbiol.* **48**, 277 (1967).
553. N. Loprieno, R. Auglielminetta, A. Abbondandolo, and S. Bonatti, *Microbiol. Genet. Bull.* **23**, 21 (1965).
554. R. Megnet, *Mutation Res.* **2**, 328 (1965).
555. A. J. Müller and T. Gichner, *Nature* **201**, 1149 (1964).
556. M. V. R. Prasad, R. Krishnaswami, and M. S. Swaminathan, *Current Sci.* (*India*) **36**, 438 (1967).
557. L. S. Browning, *Genetics* **60**, 165 (1968).
558. L. S. Browning, *Mutation Res.* **8**, 157 (1969).
559. N. W. Gillham, *Genetics* **52**, 529 (1965).
560. D. R. McCalla, *Science* **148**, 497 (1965).
561. T. Gichner, A. Michaelis, and R. Rieger, *Biochem. Biophys. Res. Commun.* **11**, 120 (1963).
562. T. J. deBoer and H. J. Backer, *Org. Syn.* **34**, 96 (1954).
563. T. J. deBoer and H. J. Backer, *Rec. Trav. Chim.* **73**, 232 (1954).
564. A. O. Olson, *J. Chromatog.* **35**, 292 (1968).
565. D. R. McCalla, *J. Protozool.* **13**, 472 (1966).
566. H. Nashed and G. Garbur, *Z. Vererbungslehre* **98**, 106 (1966).
567. R. Schwaier, N. Nashed, and F. K. Zimmermann, *Mol. Gen. Genet.* **102**, 290 (1968).
568. L. I. Smith, *Chem. Rev.* **23**, 193 (1938).
569. A. Eistert, *Z. Angew. Chem.* **54**, 99 and 124 (1941).
570. F. G. Arndt, "Organic Analysis," Vol. 1, p. 197. Wiley (Interscience), New York, 1953.
571. R. Gomper, *Chem. Ber.* **93**, 187 and 198 (1960).
572. E. B. Lewinn, *Am. J. Med. Sci.* **218**, 556 (1949).
573. R. M. Watrous, *Brit. J. Ind. Med.* **4**, 111 (1947).
574. F. L. Rose, *in* "The Evaluation of Drug Toxicity" (A. L. Walpole and A. Spinks, eds.), p. 116. Little, Brown, Boston, Massachusetts, 1958.
575. I. J. Mizrahi and P. Emmelot, *Cancer Res.* **22**, 339 (1962).
576. H. Druckrey, *Acta, Unio Intern. Contra Cancrum* **19**, 510 (1963).
577. O. M. Friedman, *Biochim. Biophys. Acta* **23**, 215 (1957).

578. O. M. Friedman, G. N. Mahapatra, B. Dash, and R. Stevenson, *Biochim. Biophys. Acta* **103**, 286 (1965).
579. E. Kriek and P. Emmelot, *Biochim. Biophys. Acta* **91**, 59 (1964).
580. A. Holy and K. H. Scheit, *Biochim. Biophys. Acta* **123**, 430 (1966).
581. H. Meerwein, H. Rathjen, and H. Werner, *Chem. Ber.* **75**, 1610 (1942).
582. F. C. Palazzo, *Gazz. Chim. Ital.* **79**, 13 (1949).
583. W. A. Waters, "The Chemistry of Free Radicals." Oxford Univ. Press, London and New York, 1946.
584. K. A. Jensen, G. Kølmark, and M. Westergaard, *Hereditas* **35**, 521 (1949).
585. E. Cerda-Olmedo and P. C. Hanawalt, *Mol. Gen. Genet.* **101**, 191 (1968).
586. M. G. Whiting, *Econ. Botany* **17**, 270 (1963).
587. M. G. Whiting, *Federation Proc.* **23**, 1343 (1964).
588. H. Matsumoto, T. Nagahama, and H. O. Larson, *Biochem. J.* **95**, 13c (1965).
589. G. L. Laqueur, E. G. McDaniel, and H. Matsumoto, *J. Natl. Cancer Inst.* **39**, 355 (1967).
590. G. L. Laqueur, O. Mickelsen, M. G. Whiting, and L. T. Kurland, *J. Natl. Cancer Inst.* **31**, 919 (1963).
591. M. Spatz, E. G. McDaniel, and G. L. Laqueur, *Proc. Soc. Exptl. Biol. Med.* **121**, 417 (1966).
592. M. Spatz, D. W. E. Smith, E. G. McDaniel, and G. L. Laqueur, *Proc. Soc. Exptl. Biol. Med.* **124**, 691 (1967).
593. G. L. Laqueur and H. Matsumoto, *J. Natl. Cancer Inst.* **37**, 217 (1966).
594. M. Spatz and G. L. Laqueur, *Proc. Soc. Exptl. Biol. Med.* **127**, 281 (1968).
595. M. Spatz, W. J. Dougherty, and D. W. E. Smith, *Proc. Soc. Exptl. Biol. Med.* **124**, 476 (1967).
596. H. Matsumoto and H. H. Higa, *Biochem. J.* **98**, 20 (1966).
597. R. C. Shank and P. N. Magee, *Biochem. J.* **105**, 521 (1967).
598. J. A. Miller, *Federation Proc.* **23**, 1361 (1964).
599. P. N. Magee, V. M. Craddock, and P. F. Swann, *in* "Carcinogenesis: A Broad Critique," Symposium on Fundamental Cancer Research, p. 89. Wilkins & Wilkins, Baltimore, Maryland.
600. H. J. Teas and J. G. Dyson, *Proc. Soc. Exptl. Biol. Med.* **125**, 988 (1967).
601. D. W. E. Smith, *Science* **152**, 1273 (1966).
602. M. G. Gabridge, A. Denunzio, and M. S. Legator, *Science* **163**, 689 (1969).
603. H. J. Sax and K. Sax, *Science* **149**, 541 (1965).
604. K. Nishida, A. Kobayashi, and T. Nagahara, *Kagoshima Daigaku Nogakubu Gakujutsu Hokoku* **15**, 118 (1956).
605. H. Matsumoto and F. M. Strong, *Arch. Biochem. Biophys.* **101**, 298 (1963).
606. B. H. Korsch and N. V. Riggs, *Tetrahedron Letters* **10**, 523 (1964).
607. N. V. Riggs, *Chem. & Ind.* (*London*) p. 926 (1956).
608. D. K. Dastur and R. S. Palekar, *Nature* **210**, 841 (1966).
609. A. Kobayashi and H. Matsumoto, *Arch. Biochem. Biophys.* **110**, 373 (1965).
610. U. Weiss, *Federation Proc.* **23**, 1357 (1964).
611. W. W. Wells, M. G. Yang, W. Bolzer, and O. Mickelsen, *Anal. Biochem.* **25**, 325 (1968).
612. M. E. Campbell, O. Mickelsen, M. G. Yang, G. L. Laqueur, and J. C. Keresztesy, *J. Nutr.* **88**, 115 (1966).
613. L. C. J. Culvenor and L. W. Smith, *Australian J. Chem.* **16**, 1955 (1966).
614. M. E. Fowler, *Am. J. Vet. Med. Assoc.* **152**, 1131 (1968).
615. R. Schoental, *Bull. World Health Organ.* **29**, 823 (1963).
616. J. M. Watt and M. S. Breyer-Brandwijk, "The Medicinal and Poisonous Plants of Southern Africa," 2nd ed. Livingstone, Edinburgh and London, 1962.

617. R. Shoental, *Voeding* **16**, 268 (1955).
618. L. B. Bull and A. T. Dick, *J. Pathol. Bacteriol.* **78**, 483 (1959).
619. C. C. J. Culvenor, A. T. Dann, and A. T. Dick, *Nature* **195**, 570 (1962).
620. G. S. Christie, *Nature* **189**, 593 (1962).
621. C. C. J. Culvenor and D. T. Downing, *Conf. Biol. Effects Alkylating Agents, New York, 1969, Ann. N. Y. Acad. Sci.*, **163**, 837 (1970).
622. A. R. Mattocks, *Nature* **217**, 723 (1968).
623. R. Schoental, *Cancer Res.* **28**, 2237 (1968).
624. J. M. Barnes, *New Scientist*, June 20, p. 619 (1968).
625. F. L. Warren, *Prog. Chem. Org. Nat. Prod.* **12**, 198 (1955).
626. J. S. McKenzie, *J. Exptl. Biol. Med. Sci.* **36**, 11 (1958).
627. R. Schoental, *J. Pathol. Bacteriol.* **77**, 485 (1959).
628. L. B. Bull and A. T. Dick, *Australian J. Biol. Med. Sci.* **38**, 515 (1960).
629. C. R. Green and G. S. Christie, *Brit. J. Exptl. Pathol.* **42**, 369 (1961).
630. S. M. Kupchan, R. W. Doskoch, and P. W. Venevenhoren, *J. Pharm. Sci.* **53**, 343 (1964).
631. A. M. Clark, *Nature* **183**, 731 (1959).
632. A. M. Clark, *Z. Verebungslehre* **91**, 74 (1960).
633. L. M. Cook and A. C. E. Holt, *J. Genet.* **59**, 273 (1966).
634. A. M. Clark, *Z. Vererbungslehre* **94**, 115 (1963).
635. N. G. Brink, *Mutation Res.* **3**, 66 (1966).
636. N. G. Brink, *Mutation Res.* **8**, 139 (1969).
637. N. G. Brink, *Z. Vererbungslehre* **94**, 331 (1963).
638. T. Alderson and A. M. Clark, *Nature* **210**, 593 (1966).
639. S. Avanzi, *Caryologia* **14**, 251 (1961).
640. A. R. Mattocks, *Anal. Chem.* **39**, 443 (1967).
641. A. R. Mattocks, *Anal. Chem.* **40**, 1749 (1968).
642. A. R. Mattocks, *J. Chem. Soc.* p. 1975 (1964).
643. A. R. Mattocks, *J. Chromatog.* **27**, 505 (1967).
644. R. K. Sharma, G. S. Khajuria, and C. K. Atal, *J. Chromatog.* **19**, 434 (1965).
645. A. Klásek, V. Svárovsky, S. S. Ahmed, and F. Šantávy, *Collection Czech. Chem. Commun.* **33**, 1738 (1968).
646. A. Klásek, T. Reichstein, and F. Šantávy, *Helv. Chim. Acta* 1088 (1968).
647. A. H. Chalmers, C. C. J. Culvenor, and L. W. Smith, *J. Chromatog.* **20**, 270 (1965).
648. A. M. Pamukeu, *Ann. N. Y. Acad. Sci.* **108**, 938 (1963).
649. W. C. Evans, E. T. R. Evans, and L. E. Hughes, *Brit. Vet. J.* **110**, 295 (1954).
650. W. C. Evans, *Vet. Record* **76**, 365 (1964).
651. A. M. Pamukeu, *Acta, Unio Intern. Contra Cancrum* **18**, 625 (1962).
652. I. A. Evans and J. Mason, *Nature* **208**, 913 (1965).
653. I. A. Evans, *Cancer Res.* **28**, 2252 (1968).
654. Anonymous, *Sci. News Letter* **88**, 402 (1965).
655. T. Nakabayashi, *Bull. Agr. Chem. Soc. Japan* **19**, 104 (1955).
656. J. Kwasniewski, *Arch. Pharm.* **288**, 307 (1955).
657. B. L. Van Duuren, personal communication (1969).
658. J. N. Kaplanis, M. J. Thompson, W. E. Robbins, and B. M. Bryce, *Science* **157**, 1436 (1967).
659. N. R. Jones and R. A. Evans, *Biochem. J.* **46**, 38 (1950).
660. P. H. Weswig, A. M. Freed, and J. R. Haag, *J. Biol. Chem.* **165**, 737 (1946).
661. W. C. Evans, N. R. Jones, and R. A. Evans, *Biochem. J.* **46**, 38 (1950).
662. F. E. Moon and J. M. McKean, *Brit. Vet. J.* **109**, 321 (1953).
663. G. B. S. Heath and B. Wood, *J. Comp. Pathol.* **68**, 201 (1958).

664. I. A. Evans, A. J. Thomas, W. C. Evans, and C. M. Edwards, *Brit. Vet. J.* **114**, 253 (1958).
665. W. C. Evans, I. A. Evans, A. J. Thomas, J. E. Watkins, and A. G. Chamberlain, *Brit. Vet. J.* **114**, 180 (1958).
666. A. M. Pamukcu, S. K. Goksoy, and J. M. Price, *Cancer Res.* **27**, 917 (1967).
667. J. M. Price and A. M. Pamukcu, *Cancer Res.* **28**, 2247 (1968).

CHAPTER 8

Alkylating Agents II (Epoxides, Aldehydes, Lactones, Alkyl Sulfates, Alkane Sulfonic Esters, and Related Derivatives)

1. *Ethylene Oxide*

The general considerations of ethylene oxide that are germane to this review include toxicity (1–3), pharmacological and physiological properties (4, 5), and mode of action (6–8). The mutagenicity of ethylene oxide in *Drosophila* (9, 10), *Neurospora* (11–14), barley (15–18) [nonmutagenicity in barley (19)], and chromosome aberrations in maize (20), barley (21), and *Vicia* (22) have been reported.

Analyses of ethylene oxide have been achieved via titrimetric (23, 24) colorimetric (indirect analysis based ultimately on the determination of formaldehyde after hydrolysis and oxidation of the glycol) with sodium chromotropate (25, 26), acetyl acetone and ammonia (27) and phenylhydrazine (28), spectrophotofluorimetric (29), infrared (30), and gas chromatographic techniques (31–34).

Ethylene oxide is made on an industrial scale (approximately 2.5 billion pounds per year) by the action of alkali on ethylene chlorohydrin or by catalytic oxidation in air of ethylene, viz.:

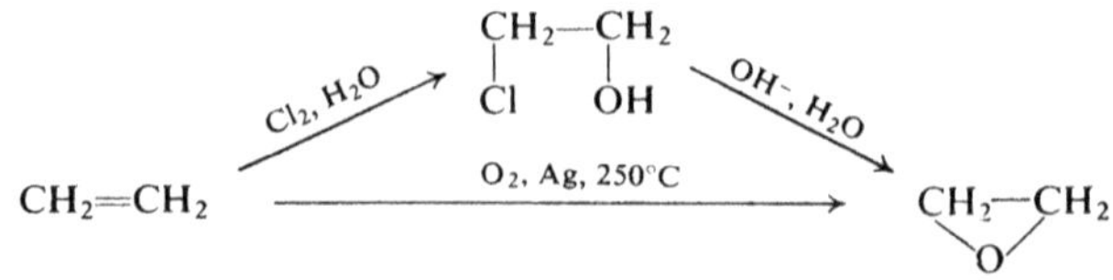

Epoxides such as ethylene oxide and propylene oxide owe their industrial importance to their high reactivity which is due to the ease of opening of the highly strained three-membered ring. Figure 8.1 illustrates the preparation of a number of commercially important compounds prepared from ethylene oxide. Other industrially important chemicals synthesized from ethylene oxide include monoethanolamine, acrylonitrile, and surface-active agents. In the latter instance, the most important class of nonionic surfactants con-

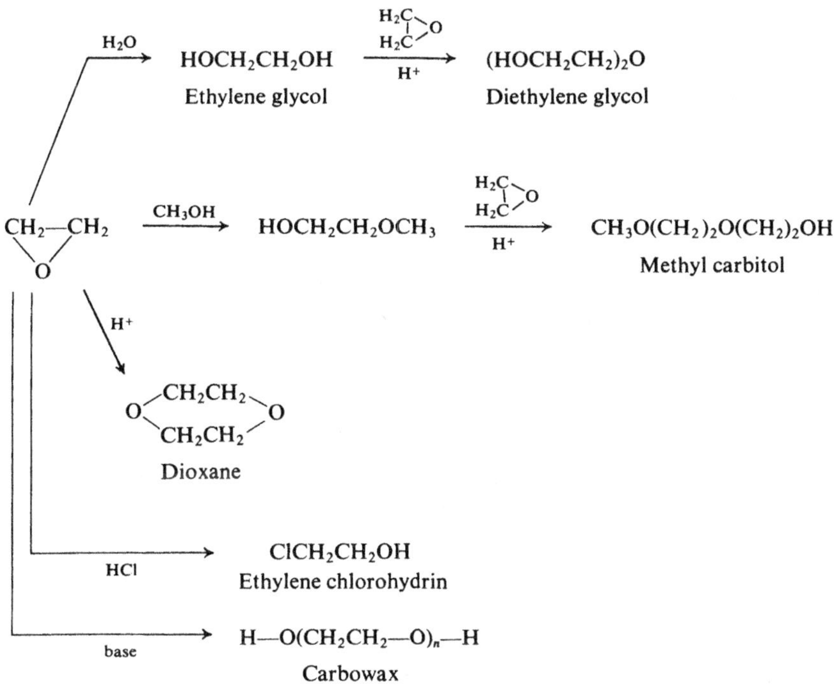

FIG. 8.1. Some commercially important derivatives of ethylene oxide.

sists of the reaction products of ethylene oxide or propylene oxide with compounds such as *p*-alkyl phenols, glycols, or fatty alcohols which contain a reactive hydrogen atom. The commercial adducts which are frequently termed oxyalkylates are mixtures with rather broad molecular weight. Ethylene oxide also has utility as a solvent and plasticizer in combination with other chemicals, and in the production of high-energy fuels, diverse plastics and textile auxiliaries, and hydroxyethylated cellulosic fibers and starch.

The area of utility which is most germane to our consideration (in terms of residues of toxic and mutagenic agents) is that of gas sterilization (35, 36) and fumigation. All of the compounds most actively employed in gaseous sterilization, e.g., ethylene oxide, propylene oxide, β-propiolactone, methyl bromide, and formaldehyde are alkylating agents. They are widely used in the sterilization of items which are labile to heat (e.g., plastic devices), radiation, or liquid chemical sterilization. The mode of action upon microbes appears to be a nonspecific alkylation of such chemical groups as —OH, —NH_2—, and —SH with the loss of a hydrogen atom and the production of an alkyl hydroxyethyl group.

The use of ethylene oxide for sterilization has thus raised a number of significant questions regarding (a) the possible entrapment and/or interaction of ethylene oxide in plastic, pharmaceuticals, or food which may then exert a toxic effect when placed in contact with living tissue, and (b) the effect of sorbed ethylene oxide on the possible changes in the physical and chemical properties of the medical plastics, per se. It had been earlier shown that plastic tubings which have been ethylene oxide sterilized can cause significant hemolysis when placed in contact with human blood (37, 38). For example, Bain and Lowenstein (38) reported that when mixed leukocyte cultures were incubated in disposable plastic tubes sterilized with ethylene oxide, the cell survival was severely affected by a toxic residue left on the plastics. The residue was dissipated only after 4 or 5 months storage at room temperature, and the cell survival returned to values obtained with ultraviolet sterilized control tubes.

O'Leary and Guess (39) in their study of the toxicogenic potential of medical plastics sterilized with ethylene oxide vapors presented data demonstrating the ability of ethylene oxide to remain entrapped in an open system, such as surgical tubing, gas washing bottles, plastic syringe, or plastic bottles, at various temperatures above its boiling point. In this same study the homolyzing ability of known amounts of ethylene oxide to that of freshly gas-sterilized plastic pharmaceutical products as well as the effects of ester type plasticizers upon the sorption of ethylene oxide into polyvinyl chloride products was shown.

Residual ethylene oxide could present a hazard in the sterilization of such devices as syringes. The residual gas, for example, might produce a toxicity from the following sequences of oxidation

Epoxide → glycol → glyoxal → glyoxalic acid → glycolic acid → oxalic acid

In addition, the effect of residual ethylene oxide (or its oxidation products) upon the syringe contents adds still another dimension to the potential hazard and novel toxicities that could result. This is especially true in the case of the rubber and polyvinyl components of many devices.

For example, the ethylene oxide reaction product [2-(2-hydroxyethylmercapto)benzothiazole] of the vulcanization accelerator 2-mercaptobenzothiazole was found to be more toxic than the precursor using cells in culture, mice, and rabbits (40).

Kulkarni (41) described the retention of ethylene oxide for a period of 1 week following the sterilization of plastic and rubber catheters. It was pointed out that ethylene oxide, if not removed, may be released later under use and may cause blood hemolysis, erythema, and edema of the tissues. Cunliffe and Wesley (42) have shown that ethylene chlorohydrin emanated from polyvinyl chloride tubing 6 days after ethylene oxide sterilization. It has also been demonstrated by Gunther (24) that high concentrations of ethylene oxide can be taken up by polyethylene, gum rubber, and plasticized polyvinyl chloride. This reemphasizes the general problem of gas entrapment occurring in plastics.

In the pharmaceutical field, ethylene oxide has been used as a sterilizing agent for antibiotics, e.g., penicillin in bulk and streptomycin (43). In the former case sterilization was performed without decomposition, while streptomycin showed a loss of 35% activity. The influence of ethylene oxide on various vitamins has been studied and a deleterious effect on some of the B vitamins has been reported (44, 45). The mechanism of the reaction of ethylene oxide with nicotinamide, nicotinic acid, histidine, methionine, and cysteine has been studied (46, 47). All of the reaction products elaborated involved hydroxyethylation of an atom with one or more lone pair of electrons, either nitrogen or sulfur. For example, imidazole (I) yielded 1,3-bis(2-hydroxyethyl)imidazolinium ion (II), and *N*-acetylmethionine was converted to *S*-(2-hydroxyethyl)-*N*-acetylmethionine. The double alkylation of the mercapto group of cysteine was found to result in a sulfonium derivative.

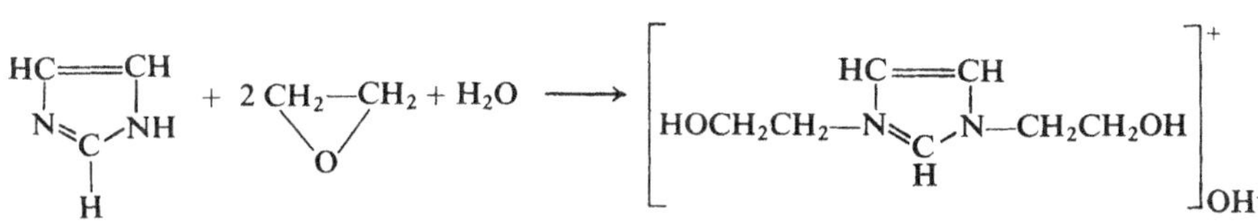

(I) (II)

Windmueller *et al.* (47) suggested that since water facilitates proton reaction (the reaction of ethylene oxide with tertiary nitrogen requires the presence of an available proton), this may explain the enhancing effect of small amounts of moisture on the effectiveness of ethylene oxide fumigation procedures. [This in turn may partially determine the rate of destruction of nutrients during the fumigation of foods (47).]

Adler (48) reported residual ethylene glycol residues (in some cases approximately 1%) in ethylene oxide-sterilized pharmaceuticals (e.g., cortisone, hydrocortisone, and prednisolone acetates).

Under existing United States food and drug regulations, seven different foodstuffs (cocoa, glazed fruits, gums, processed nut meats, dried prunes, processed spices, and starches) may be commercially treated with propylene oxide and three products (whole spices, black walnut meats, and cocoa) with ethylene oxide.

These epoxides are capable of eliminating or greatly diminishing the numbers of viable organisms present. Also, when properly used, they do not impart any after-odor or flavor-tint to foods by direct effect or as a consequence of the residues which result.

Residues are produced mainly by two reactions of epoxides, e.g., (a) their slow chemical combination with water to form the corresponding glycols and (b) their combination with the elements of hydrochloric acid to form the corresponding chlorohydrins. Since nominally dry materials contain moisture, it is apparent that glycol formation can occur unavoidably. Moreover, without the presence of some moisture, sterilization cannot be effected in this way.

Wesley, Rourke, and Darbishire (49) described the formation of persistent toxic chlorohydrins in foodstuffs by fumigation with ethylene oxide and with propylene oxide. Under conditions for effective fumigation (e.g., 750 ml/m^3 for 5 hours) with either oxide, these reagents were found to combine not only with moisture but also with chlorine from the natural inorganic chloride content of foodstuffs, forming the corresponding chlorohydrin. Concentrations of ethylene chlorohydrin up to about 1000 ppm were found in whole spices after commercial fumigation with ethylene oxide. These chlorohydrins are sufficiently involatile to be persistent under food processing conditions, and most importantly they are toxic substances by all accounts.

The isolation and determination of chlorohydrins in foods fumigated with ethylene oxide or with propylene oxide has been investigated by Ragelis *et al.* (50) using gas-liquid chromatography and infrared spectroscopy to determine and identify the chlorohydrin present. The levels of 2-chloroethanol found in five foods commercially fumigated with ethylene oxide ranged from 45 ppm for paprika to 110 ppm for pepper. Levels of 1-chloro-2-propanol found in six food products commercially treated with propylene oxide ranged from 4 ppm for cocoa to 47 ppm for glazed citron.

Heuser and Scudamore (33, 34) described the analysis of ethylene oxide as well as fumigant residues (ethylene chlorohydrin produced *in situ* by naturally occurring inorganic chloride with ethylene oxide) in wheat and flour.

Initial exposure of flour to ethylene oxide for periods of 1 to 6 hours as well as with periods of preliminary aerations from 45 minutes to 2 hours and main aeration from 3 to 4 hours resulted in recoveries of free ethylene oxide of 95.0% to 95.8% with a lower detection limit of about 0.3 ppm, hence, suggesting *in situ* residues of ethylene oxide reaction products of approximately 4% to 5% resulting from the partial aeration of flour with ethylene oxide.

Ethylene oxide has been used in the tobacco industry both to shorten the aging process (51) and to reduce the nicotine content in tobacco leaves (52). The reaction products of nicotine with gaseous ethylene oxide and the aspects of their pyrolysis has been reported by Obi *et al.* (53). It was found that *N*-hydroxyethyl nicotine was the main reaction product when tobacco leaves were treated with ethylene oxide. In the reaction with nicotine monoacetate two products were formed, e.g., *N'*-hydroxyethyl nicotine and *N*-hydroxyethyl nicotine in the weight ratio of 1 to 3.5.

Muramatsu (54) reported ethylene oxide residues in cigarette smoke from tobacco treated with oxyethylene docosanol as well as nontreated tobacco. Ethylene oxide was also found in the smoke of cigarettes that possess charcoal filters. Additional hydroxyethylation reactions have been reported for the reaction of ethylene oxide with cellulose, sugars, proteins, or free amino acid components in prunes (55). The hydroxyethyl radical is chemisorbed to form addition products with carboxyl, amino, sulfhydryl, hydroxyl, or phenolic groups present in the cells of insects, molds, bacteria, and yeasts with resultant lethal interference to their metabolism according to Sair (56).

2. *Propylene Oxide*

The general considerations of propylene oxide that are relevant include mode of action (57–59), toxicity (3, 60–62), carcinogenicity (63), and mutagenicity (for *Drosophila*) (9, 64). Analytic techniques for propylene oxide include titrimetry (23, 24), infrared spectroscopy (65), and gas chromatography (66).

Propylene oxide is synthesized commercially in a manner analogous to ethylene oxide, i.e., from propylene through the intermediate propylene chlorohydrin or by the direct oxidation of propylene with either air or oxygen. The yearly consumption of propylene oxide is approximately 700 million pounds and is largely used in the production of propylene glycols, mixed polyglycols, and various propylene glycol ethers and esters.

Large amounts are also used in the synthesis of hydroxypropyl celluloses and sugars, surfactants, urethan elastomers, isopropanol amine, and a variety of other derivatives that are useful in many applications, including textiles, cosmetics, pharmaceuticals, agricultural chemicals, petroleum, plastics, rubber, and paints. Propylene oxide, which is highly reactive chemically, being intermediate between ethylene oxide and butylene oxide, is also used as a fumigant herbicide, preservative, and in some cases as a solvent. (It is a powerful low-boiling solvent for hydrocarbons, cellulose nitrate, and acetate and vinyl chloride, and acetate.)

Propylene oxide is commonly employed as a gaseous decontaminant (35, 67, 68) to destroy specific groups of organisms such as fungi, yeasts, coliform, and salmonellae. Although its penetrating ability and microbiocidal properties are less than that of ethylene oxide, it is frequently used as a substitute for ethylene oxide and has been suggested for use as a soil sterilant.

Propylene oxide has also been used on a variety of foods such as dried fruits (68, 69), powdered, or flaked foods [e.g., cocoa, yeast powder, and cereal flakes (70)], to control spoilage. It has been shown (49, 50, 71) above that both ethylene and propylene oxide may react with inorganic chloride in foodstuffs to form chlorohydrins which in themselves are toxic.

Persistent residues of these were found because of their low volatility and relatively unreactive chemical nature. Recent studies (72, 73) involving the use of ethylene oxides and propylene oxides for the fumigation of flour have also revealed appreciable amounts of residual epoxide.

3. *Ethylene Chlorohydrin*

Ethylene chlorohydrin (2-chloroethanol) is prepared commercially by the reaction of hypochlorous acid and ethylene. It has been proposed as an effective agent in hastening the early sprouting of dormant potatoes (74, 75) and has also been investigated as a means of treating seeds for the inhibition of biological activity (76). It is also used for the separation of butadiene from hydrocarbon mixtures, in dewaxing and removing of naphthenes from mineral oil, in the refining of rosin, in the extraction of pine lignin, and as a solvent for cellulose acetate, ethers, and various resins. Ethylene chlorohydrin is a potential reaction product during sterilization and fumigation procedures (34, 49), as discussed earlier.

The metabolism of ethylene chlorohydrin in the rat has revealed the conversion of liver glutathione to *S*-carboxymethyl glutathione. It was suggested that the toxicity of ethylene chlorohydrin is due to its conversion to chloroacetaldehyde *in vivo* (77).

The chemical properties (78), toxicity (79–81), tissue reactions (82), and mutagenicity in *Klebsiella pneumoniae* (83) of ethylene chlorohydrin have been described.

Analysis of ethylene chlorohydrin has been achieved by titrimetry (84), infrared spectroscopy (85), and gas chromatography (31, 34, 86).

4. *Glycidol (2,3-Epoxy-1-propanol)*

Glycidol (2,3-epoxy-1-propanol) is made by the dehydrochlorination of glycerol and monochlorohydrin with caustic. It is used for the preparation of glycerol and glycidyl esters, ethers, and amines, which have utility as pharmaceutical intermediates, and in textile finishings, e.g., glycidol esters as water-repellant finishes (87, 88). Glycidol has been used as an antibacterial and antimycotic agent for food products (89). Its toxicity (90, 91), mutagenicity in *Drosophila* (9), *Neurospora* (92), and barley (16), as well as its chromosome breakage properties in tissue cultures of mouse embryonic skin and Crocker mouse Sarcoma 180, have been reported (93).

Glycidol has been analyzed by paper chromatographic techniques (94, 95).

5. *Epichlorohydrin*

Epichlorohydrin (1-chloro-2,3-epoxypropane) is made commercially from propylene or the reaction of alkalies upon dichlorohydrins and is employed extensively as a solvent for natural and synthetic resins, gums, cellulose esters and ethers, paints, varnishes, nail enamels and lacquers, and as a raw material for the manufacture of a number of glycerol and glycidol derivatives, in the manufacture of epoxy resins, as a stabilizer in chlorine-containing materials, and as an intermediate in the preparation of condensates with polyfunctional substances.

The patented uses of epichlorohydrin include its utility as a cross-linking agent in the crease-proofing of textiles (96), paper processing (97), water-proofing of materials (98), fire-resistant epoxy resins (99), and as a curing agent for aminoplast resins (100).

The toxicology (62, 101, 102), carcinogenic potential (103), and mutagenicity in *Drosophila* (9), *Neurospora* (92), *E. coli* (104, 105), and barley (106) have been reported.

Analytic procedures for epichlorohydrin include colorimetry (107–109), infrared (110) and nmr spectroscopy (111, 112), and paper chromatography (95).

6. *Di(2,3-epoxypropyl) Ether*

Di(2,3-epoxypropyl) ether (diglycidyl ether) is prepared through epoxidation of diallyl ether, then chlorohydrination followed by dehydrochlorination with caustic (113):

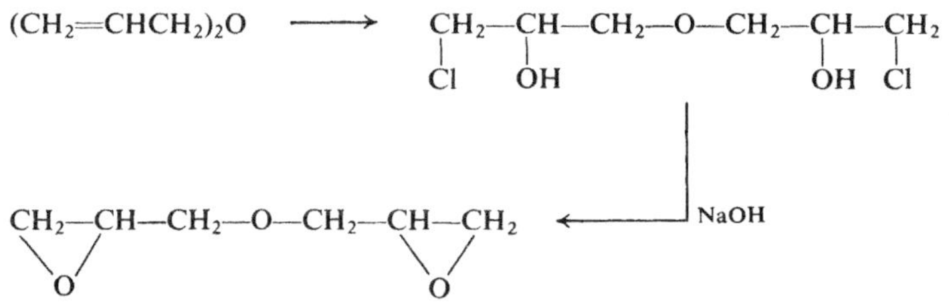

Its broad area of utility includes the preparation of trioxane copolymers (114), the curing of polysulfide polymers (115), preparation of thermoset resins (116), vulcanizable polyethers and acetals (117), anion-exchangers (118), polymers as flocculating agents (119), diluent in aromatic amine-cured epoxy adhesives (120), as a hardener in photographic emulsions (121), and for the removal of remnants of Ziegler-Natta catalysts in polymerization (122).

Its acute (90, 123) and chronic toxicity (90, 124) and mutagenicity in bacteriophage T2 (125) and *Neurospora* (14, 92), and its role as an inducer of chromosome aberrations in *Vicia faba* (126–132), in root tips of broad bean (133), and in root cells of *Tradescantia* (134) have all been described.

Di(2,3-epoxypropyl) ether has been determined by titrimetric (135) and paper chromatographic procedures (95).

7. *1,2:3,4-Diepoxybutane*

dl-Diepoxybutane (butadiene diepoxide; DEB) is prepared via the bromination of *cis*-2-butene-1,4-diol followed by the conversion to the epoxide with potassium hydroxide in ether (136); the *meso* form from 1,4-dihydroxy-2-butene or from 3,4-epoxy-1-butene (137).

DEB is used in the prevention of microbial spoilage (138), in curing of polymers, as a cross-linking agent for textile fibers, and as an intermediate in the preparation of erythritol and pharmaceuticals. DEB has been reported to inhibit Walker rat Carcinoma 256 (139), and has been evaluated for the treatment of Hodgkin's disease and lymphoreticulosarcoma (140). Experimentally it is an active radiomimetic substance producing skin cancers and sarcomas and depression of the hemopoietic system (141–144).

Its acute (145–146) and chronic toxicity (91, 146), effect on DNA (147), and interaction with amino acids (6, 148) have been reported. The mutagenicity of DEB in *Drosophila* (10, 149–153), *Neurospora* (11, 14, 154–159), *E. coli* (160), *Salmonella typhimurium* (161), *Saccharomyces cerevisiae* (162), barley (15, 163, 164), maize (165–167), tomato (168), *Vicia faba* (169), *Penicillium* (169), *Arabidopsis thaliana* (170), houseflies, *Musca domestica* L. (171), and mammals (172), and as an inducer of chromosome aberrations in *Vicia faba* (126, 129, 173–175), *Drosophila* (150), mammals (172), *Allium cepa* (176), and mouse lymphocytes *in vitro* (177) have been reported.

Analysis of diepoxybutane has been achieved by titrimetry (178, 179), infrared spectroscopy (136), and gas chromatography (136).

8. *Formaldehyde*

Formaldehyde is manufactured chiefly via passage of a mixture of methanol vapor and air over a silver, copper, or iron-molybdenum oxide catalyst. This process yields essentially pure formaldehyde containing some methanol and traces of formic acid as a primary product. About 14% of United States production of formaldehyde is obtained by the partial oxidation of the lower petroleum hydrocarbons which yields a mixture of lower aliphatic aldehydes, alcohols, and acids.

Since pure monomeric formaldehyde is a gas at ordinary temperatures and

TABLE 8.1

Some Uses of Formaldehyde and Its Derivatives

Use	Form
Agricultural	
Disinfection of seeds	Formaldehyde, paraformaldehyde
Prevention of scab (potato, wheat, barley, oats)	Formaldehyde, urea formaldehyde
Fertilizer	Urea formaldehyde
Fungicide	Formaldehyde, paraformaldehyde
Preservative	Formaldehyde
Textile	
Creaseproof, crushproof, flame-resistant, shrinkproof fabrics	*N*-Methylol derivatives
Paper	
Wet strength, shrink resistance, grease resistance, pigment binder	Formaldehyde and derivatives
Photography	
Developers	Formaldehyde-bisulfite compositions
Hardening and insolubilizing agents	Formaldehyde
Sterilizing agent	Formaldehyde, paraformaldehyde
Drugs	
Germicides	Mannich reaction products
Bactericide	Formaldehyde
Specialty chemicals	
Pentaerythritol, ethylene glycol, glycolic acid, methylol, hexamethylene tetramine	Formaldehyde

hence cannot be readily handled, it is marketed chiefly in the form of aqueous solutions containing 37% to 50% formaldehyde by weight. (The United States production capacity in terms of the 39% aqueous solution was estimated at 3.4 billion pounds for 1965.) Solutions of formaldehyde are also commercially available in the lower aliphatic alcohols, e.g., methanol, propanol, and butanol. It is also commercially utilized in the form of its solid, hydrated linear polymer, paraformaldehyde $[HO(CH_2O)_nH]$ and the cyclic trimer, *s*-trioxane (α-trioxymethylene; $OCH_2OCH_2OCH_2$).

The major uses of formaldehyde and its polymers are in the synthetic resin industry (e.g., in the production of thermosetting resins, oil-soluble resins, and adhesives, which accounts for 50% of the total production). The remainder is used in the manufacture of a broad spectrum of textiles, paper, fertilizer, miscellaneous products, and specialty chemicals.

TABLE 8.2

Some Uses of Formaldehyde[a]

Use	U.S. Production (1964) (37% solution) (millions lb.)
Resins	
Phenolic	530
Urea	550
Melamine	160
Acetal	105
	1345
Urea formaldehyde concentrates	
Industrial	200
Agricultural	75
	275
Special chemicals	
Hexamethylene tetramine	150
Pentaerythritol	220
Ethylene glycol	380
Sequestering agents	35
	785
Other uses	240
Total	2645

[a] *Chemical Week* **95**, 113 (1964).

Formaldehyde is extremely reactive and will react with practically every type of organic chemical (180).

In general, the major chemical reactions of formaldehyde with other compounds involve the formation of methylol ($—CH_2OH$) or methylene derivatives. Other typical reactions include alkoxy-, amido-, amino-, cyano-, halo-, sulfo-, and thiocyanomethylations. For example, reactions with amides and carbamates yield methylol derivatives, e.g., methylolureas and methylol carbamates which are used in the treatment of textiles (for crease-resistance).

The major commercial uses are listed in Tables 8.1 and 8.2. Industrial urea–formaldehyde concentrates are used in adhesives, coating compositions, and in the modification of textiles and paper products. The agricultural uses of these condensates include the control of potato scab and other plant diseases as well as in the production of ureaform fertilizers which gradually release nitrogen in the soil. In the textile industry formaldehyde is extensively employed alone and in the form of its *N*-methylol derivatives for the production of creaseproof, crushproof, flame resistant, and shrinkproof fabrics.

In addition to its use in the manufacture of urea and phenolic and melamine resins, formaldehyde is utilized for the preparation of resins by reaction of urethans, aniline, aromatic hydrocarbons, ketones, and other chemicals as well as in the production of resinous products from proteins such as casein, glue and soybean protein, and lignin.

The utility of formaldehyde in medicinal products is due to its capacity to modify and reduce the toxicity of viruses, venoms, and irritating pollens. The formaldehyde deactivation of viruses in vaccine preparation is ascribed to the chemical reaction with the nucleic acid moieties of these compounds.

Formaldehyde has been found widely spread in man's environment, e.g., in tobacco leaf (181), tobacco smoke (182), incinerator effluents (183, 184), and automobile (185) and diesel exhaust (186). Measurements of total aldehydes of automobile exhaust (187) and of atmospheric aldehydes (188) indicate that from 40% to 50% of the total aldehyde is formaldehyde. Another source of aldehydes such as formaldehyde and acetaldehyde is the thermal degradation of epoxy thermoplastic materials (189).

Neimann *et al.* (190) proposed the degradation sequence shown in Eqs. (1) to (8):

(1)

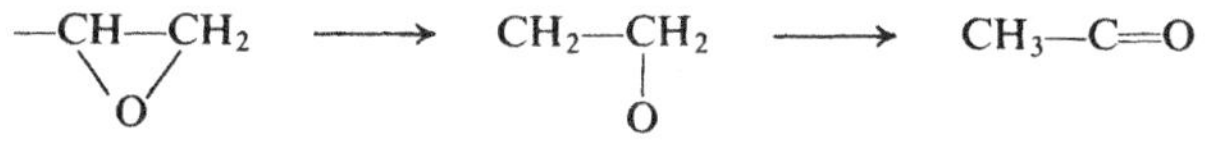
(2)

$$CH_3—C{=}O \longrightarrow CH_3 + CO \tag{3}$$

$$RO—CH_2 \longrightarrow R + HCHO \tag{4}$$

$$CH_3—C{=}O + RH \longrightarrow R + CH_3CHO \tag{5}$$

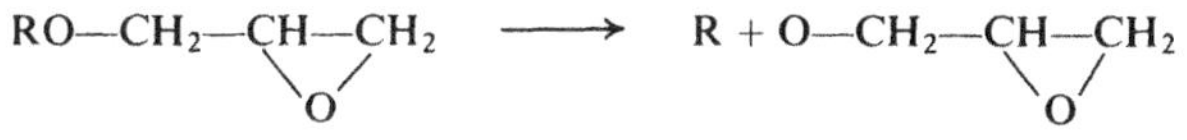
(6)

(7)

(8)

According to this view, acetaldehyde, formaldehyde, and acrolein will be the principal degradation products.

Lee (191, 192) has cited the primary formation of formaldehyde, acetaldehyde, acrolein, acetone, and propylene with methane, ethylene, and hydrogen as secondary products of the thermal degradation of amine-cured epoxide resins.

The problem of free formaldehyde resulting from textiles treated with formaldehyde resins (e.g., urea, melamine, and phenol) to improve crease- and shrinkage-resistance has been recently cited (193, 194). The amount of free formaldehyde has been shown to be between 0.027% and 0.075%; it persists beyond 48 hours and is responsible for increasing cases of contact dermatitis. Besides formaldehyde, the formation of other products obtained by the hydrolysis of the resins such as organic acids of the type $HOCH_2NHCOOH$ and a glycine–formaldehyde complex were postulated.

The chemistry (195) and toxicity (61, 101, 196, 197) as well as the reported carcinogenicity of formaldehyde have been described (198). Because of the wide use and distribution of formaldehyde as cited above, information on its reaction with dietary and body constituents and on its metabolic fate is of considerable importance. It is established that formaldehyde can react readily with proteins through various side-chain groups, as well as with certain peptide bonds (199), with the formation of hydroxymethylamide ($HOCH_2CONH_2$) and hydroxymethylamine (NH_2CH_2OH) by hydroxymethylation of amido (—$CONH_2$) and amino groups, respectively.

When administered orally or parenterally to animals, formaldehyde is rapidly converted to formic acid by formaldehyde dehydrogenase of the liver, erythrocytes, and other tissues; formic acid is then largely oxidized to CO_2 by catalase. A minor metabolic pathway is via the folic acid cycle leading to the urinary excretion of serine and methionine. Other metabolic investigations of formaldehyde in dog, rabbit, guinea pig, and cat (200, 201) have been described.

The mutagenicity of formaldehyde has been described most extensively for *Drosophila* (202–215) [with hydrogen peroxide (216)], and established for *Neurospora cassida* [also with hydrogen peroxide (19, 217)] and *E. coli* (218). Various theories of formaldehyde mutagenesis have been postulated, e.g., by Rapoport (219), Alderson (220), Auerbach (207), and Jensen and co-workers (19).

It would be of importance to determine *unequivocally* what special metabolic conditions transform foods treated with formaldehyde into an effective mutagen for *Drosophila*, as well as to elicit whether addition of formaldehyde to food produces mutations in mammalian germ cells. Casein treated with formaldehyde and subsequently washed has been shown by Auerbach (208) to be a potent mutagen in *Drosophila*. In many big breeding stations, the skimmed milk fed to animals is stabilized with formaldehyde (221), and this has caused concern about possible genetic effects on the breeding stock.

Formaldehyde has been analyzed by a variety of procedures including titrimetry (222), iodometry (223), colorimetry [with chromotropic acid (188, 224) and Schiff reagent (223)], spectrophotofluorimetry [with acetylacetone and ammonia (28, 225) and with 6-amino-1-naphthol-3-sulfonic acid (29)], and gas chromatography (226–228).

9. *Acetaldehyde*

The toxicity (197, 229–231), pharmacology (232), metabolism (233, 234), and mutagenicity (in *Drosophila*) (235) of acetaldehyde have been described. The analysis of acetaldehyde has been accomplished using polarographic (236), argentometric titration (237), mercurometric oxidation (238), infrared (239) and ultraviolet (240) spectrophotometric, and paper (241, 242) and gas-liquid (243–245) chromatographic techniques.

Acetaldehyde is produced commercially by (a) the vapor phase dehydrogenation or partial oxidation of ethanol, (b) high-temperature oxidation of saturated hydrocarbons, (c) the liquid phase hydration of acetylene, or (d) the liquid phase oxidation of ethylene. The amount of acetaldehyde produced from ethanol alone was 900 million pounds in the United States in 1960. It is an intermediate in the manufacture of a host of important products including acetic acid, acetic anhydride, butyl alcohol, butyraldehyde, chloral, pentaerythritol, peroxyacetic acid, acrylonitrile, cellulose acetate, vinyl acetate resins, and pyridine derivatives. It is the product of most hydrocarbon oxidations; it is a normal intermediate product in the respiration of higher plants; it occurs in traces in all ripe fruits and may form in wine and other alcoholic beverages after exposure to air. Acetaldehyde is an intermediate product in the metabolism of sugars in the body and hence occurs in traces in blood. It has been reported in fresh leaf tobacco (181, 246) as well as in tobacco smoke (239, 244, 247, 248).

Acetaldehyde, as well as propionaldehyde, butyraldehyde, isobutyraldehyde, isovaleraldehyde, and crotonaldehyde, has been identified in automobile and diesel exhaust by paper (249) as well as gas chromatography (250–254). Acetaldehyde is a highly reactive compound exhibiting the general reactions of aldehydes (e.g., under suitable conditions the oxygen or any hydrogen may be replaced) and hence undergoes a number of condensation, addition, and polymerization reactions.

Acetaldehyde has a number of industrial uses: the condensation products with phenol or urea are thermosetting resins; the reaction with aliphatic and aromatic amines yield Schiff bases which are used as accelerators and antioxidants in the rubber industry and for the production of butadiene and polyvinylacetal resins. Acetaldehyde has been used as a preservative for fruit and fish, as a denaturant for ethanol, in fuel compositions, for hardening gelatin, glue, and casein products, for the prevention of mold growth on leather, and as a solvent in the rubber, tanning, and paper industries.

10. *Acrolein*

Acrolein (acrylic aldehyde) is manufactured on a commercial scale by (a) the direct oxidation of propylene utilizing catalysts such as mixed oxides of bismuth and molybdenum, molybdenum and cobalt, and molybdenum cuprous oxide, and (b) cross-condensation of acetaldehyde with formaldehyde using lithium phosphate on activated alumina or sodium silicate on silica gel as catalyst. The extreme reactivity of acrolein is attributed to the conjugation of a carbonylic group with the vinyl group within its structure.

Relatively large quantities of acrolein are consumed in the manufacture of derivatives such as 1,2,6-hexanetriol, hydroxyadipaldehyde, and glutaldehyde via the intermediate acrolein dimer (3,4-dihydro-2-formyl-2*H*-pyran), viz.:

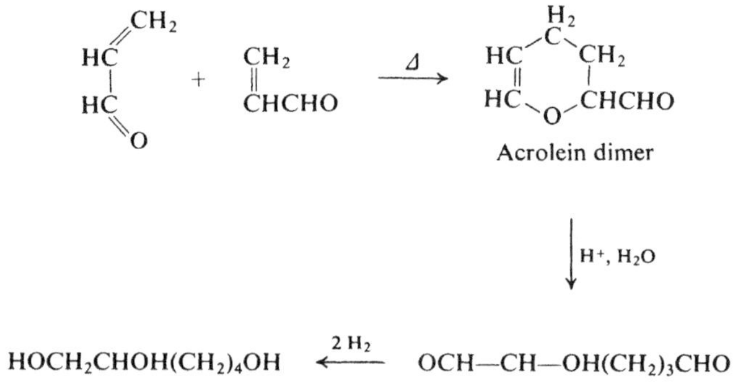

Acrolein dimer is valuable as a starting point for the synthesis of a variety of chemicals useful in textile finishing, paper treating, and the manufacture of rubber chemicals, pharmaceuticals, plasticizers, and synthetic resins.

Other important reactions of acrolein involve its ability to undergo a variety of polymerization reactions (homo-, co-, and graft polymerization) as well as to undergo reactions with ammonia and formaldehyde, respectively, to yield the industrially important derivatives acrylonitrile and pentaerythritol. One of the largest uses is in the production of methionine which is used in supplementing fowl, swine, and ruminant feeds. Epoxidation of acrolein with hydrogen peroxide yields glycidaldehyde which is extensively used as a cross-linking agent for textile treatment and leather tanning.

Polarographic (255), gas (250–253) and paper chromatographic (249), as well as colorimetric techniques (256, 257) have been used to identify and determine acrolein as an air pollutant in automobile exhaust and in the atmospheres of paint and varnish plants.

Acrolein has been identified in both tobacco leaf (247) and tobacco smoke (182, 258–261). It has also been found that the use of humectants such as glycerol in tobacco can serve as precursors of volatile aldehydes upon combustion (262). For example, pyrolysis of humectants at 600°C yielded 10 to 15 g of volatile carbonyls (acrolein, acetaldehyde, and acetone, calculated as acetaldehyde) per 100 g of tested polyol.

Acrolein (as well as formaldehyde, hydrocarbons, organic peroxides, formic acid, sulfur dioxide, ammonia, and nitrogen oxide) has been identified as a volatile contaminant in smog (263). The toxicity (197, 264, 265) of acrolein as well as its mutagenicity for *Drosophila* (235) have been described. Other analytic techniques that have been found useful for the detection and estimation of acrolein include thin-layer chromatography (266) and titrimetry (223).

11. *Chloral Hydrate*

Chloral hydrate (hydrated trichloroacetaldehyde) is prepared by (a) chlorination of a mixture of ethanol and acetaldehyde followed by hydration of the intermediate trichloroacetaldehyde or (b) the action of hypochlorous acid on trichloroethylene. Chloral hydrate is used primarily as a hypnotic and sedative. It also has utility in the cross-linking of nylon fibers (267) and in herbicidal formulations (268). The hypnotic action of chloral hydrate depends on its reduction to trichloroethanol in the liver (catalyzed by alcohol dehydrogenase) and other tissues including whole blood (269, 270). A variable fraction of chloral hydrate is oxidized in the liver and kidney by a DPN-dependent enzyme system to trichloroacetic acid (271). Chloral hydrate is also a metabolite of trichloroethylene (probably the first step leading to the formation of trichloroethanol, then conjugation and excretion of the latter as a glucuronide) (272). The toxicity (273), pharmacology (273, 274), and mutagenicity (275, 276) of chloral hydrate have been described. Analytic procedures for chloral hydrate include colorimetry (270, 277, 278), titrimetry (279), polarography (280), and ion-exchange column chromatography (281).

12. *Citronellal*

Citronellal (3,7-dimethyl-6-octenal) is the chief constituent of citronella oil (15%) and is also found in many other volatile oils, such as lemon, lemon grass, and melissa.

The structure of citronellal was elaborated by Naves (282), and it was synthesized from β-pinene by Webb (283). Citronellal possesses bactericidal (284, 285) and fungicidal action (286) and has been used as an ingredient in soaps, as a perfume in insectifuges, as a silkworm attractant (287), and for the potentiation of dihydrostreptomycin in the treatment of tuberculosis (288).

Citronellal is metabolized via cyclization and conversion to methane-3,8-diol, which is excreted as the glucuronide.

Rapoport (235) found that unsaturated aldehydes, including citronellal and acrolein, were more active mutagens toward *Drosophila* than acetaldehyde.

The analysis of citronellal has been achieved via UV spectroscopy (289) and paper (290, 291) and gas-liquid chromatography (292–295).

13. *Ketene*

Ketene is formed by pyrolysis of virtually any compound containing an acetyl group, e.g., acetone:

$$CH_3COCH_3 \xrightarrow{700^\circ C} CH_2{=}C{=}O + CH_4$$

or by the reaction of ozone on olefins such as propylene and 2-pentene as well as by the photooxidation of hydrocarbons (olefins) (296).

The major uses of ketene are in the manufacture of acetic anhydride and the dimerization to diketene, viz.:

Diketene is an important intermediate for the preparation of dihydroacetic acid, acetoacetic esters, acetoacetanilide, *N,N*-dialkylacetoacetamides, and cellulose esters which are used in the manufacture of fine chemicals, drugs, dyes, and insecticides.

Ketene has utility as a rodenticide (297), in textile finishing (298), in the acetylation of viscose rayon fiber (299), and as an additive for noncorrosive hydrocarbon fuels (300).

Ketene is a very useful acetylating agent for ROH and RNH_2 compounds.

It reacts rapidly, and since the reactions involve additions, there are no by-products to be separated:

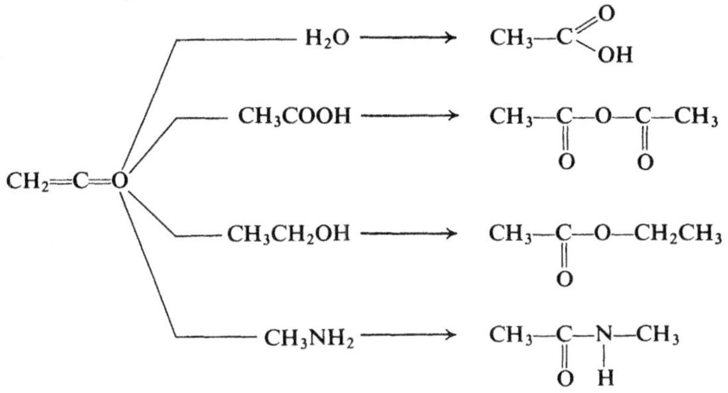

The general utility of ketene in organic synthesis has been reviewed by Lacey (301), and its thermal (302) and photo decomposition (303, 304), its toxicity (305–307), and colorimetric (308) and UV spectroscopic analysis (309) described.

Ketene has been found to be mutagenic in *Drosophila* (310) but nonmutagenic in *Neurospora* (19).

14. *β-Propiolactone*

β-Propiolactone (β-hydroxypropionic acid lactone: BPL) is produced commercially from formaldehyde and ketene in the presence of a catalyst such as zinc chloride, hydrated aluminum silicate, or zinc trifluoroacetate, viz.:

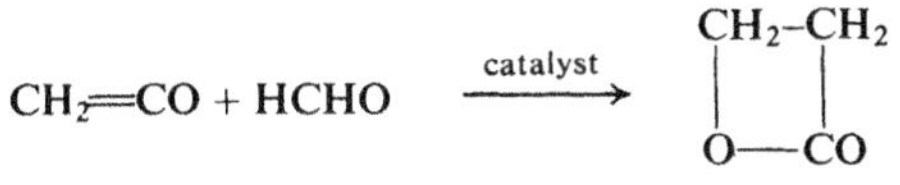

Commercial grade BPL (97%) (which can contain trace quantities of the reactants) is stable for several years when stored at 4°C; the half-life of aqueous solutions is approximately $3\frac{1}{2}$ hours at 25°C. BPL is slowly hydrolyzed by water and rapidly by aqueous alkalies to give hydracrylic acid (β-hydroxypropionic acid) and its salts, respectively.

β-Propiolactone possesses a broad spectrum of current and suggested industrial uses, e.g., in wood processing, protective coatings and impregnation of textiles, modification of flax cellulose, urethan foam manufacture, intermediate in the preparation of insecticides, plasticizers, medicinals, additive for leaded gasolines, and in the modifications of tobacco flavors.

BPL has been widely used as a sterilizing agent, and also as an aerosol for sterilizing rooms, plasma, arterial or homografts, or as a "toxoiding" agent in place of formalin. It has been shown to be an effective vapor phase decontaminant for the treatment of enclosed areas (it is effective against bacterial spores and vegetative cells as well as viruses). The virucidal, toxicological, and biological effects of BPL as well as its degradation products, hydracrylic acid and β-chloropropionic acid, have been described by Kelly and Hartman (311, 312). Because liquid BPL is a good organic solvent, problems are often encountered when it is used as a decontaminant, e.g., the dissolution of paints, lacquers, and waxes on surfaces in the tested area. The problem of BPL vapor decontamination has been reviewed by Hoffman *et al.* (313). It has been shown that BPL vapor does adsorb on surfaces and may require a relatively long time to desorb.

Chemically, BPL is highly reactive, due to the strained four-membered ring. It is an alkylating carcinogen (314), and undergoes rapid electrophilic reactions, e.g., with thiols forming thioethers. In the case of cysteine, the compound formed is *S*-2-carboxyethylcysteine (315).

The chemical properties and reactions (316), mechanisms of action (317–322), toxicity (323–325), carcinogenicity (314, 326–329), mutagenicity (330, 331) in *Vicia faba* (332, 333), *Neurospora* (332), *E. coli*, and *Serratia marcescens* (331, 334), nonmutagenicity in barley (163), and as causal agent in chromosome aberrations in *Vicia faba* (332–335), *Allium* (335), and *Neurospora* (332) of β-propiolactone have all been described. Procedures for the analysis of β-propiolactone include titrimetry (336, 337), infrared (338), and mass spectroscopy (339, 340).

15. *Ethylene Sulfide*

Ethylene sulfide is prepared via the reaction of ethylene oxide or ethylene chlorohydrin with potassium thiocyanate. Its areas of utility include gaseous sterilization (43), the preparation of vicinal dithiols (341), the grafting of thiol groups into cellulose chains (342), polymerization and use of polyethylene sulfide in rubber (343, 344), elastomer-resin compositions (345), urethan polymers (346), and the preparation of dextran mercaptoethyl ethers (347). Its toxicity (348) and mutagenicity in *Drosophila* (349) have been described. Ethylene sulfide has been analyzed by infrared, Raman (350), nmr (351, 352), and UV spectroscopy (353), and a review of the chemistry of ethylene sulfide and its derivatives was prepared by Reynolds and Fields (354).

16. *Dimethyl Sulfate*

Dimethyl sulfate (DMS) is obtained by distillation of methyl hydrogen sulfate which pyrolyzes as follows:

$$2\,CH_3HSO_4 \rightarrow H_2SO_4 + (CH_3)_2SO_4$$

DMS is hydrolyzed by water alone about five times as fast as diethyl sulfate (DES). In the hydrolysis of DMS, the first methyl group is removed much more rapidly than the second. If an appreciable concentration of methanol accumulates in the solution, a secondary reaction occurs as indicated:

$$(CH_3)_2SO_4 + H_2O \rightarrow CH_3OH + CH_3HSO_4$$
$$(CH_3)_2SO_4 + CH_3OH \rightarrow (CH_3)_2O + CH_3HSO_4$$

DMS has been extensively employed as an alkylating agent both in the laboratory and in industry. Its utility includes the methylation of cellulose (355), preparation of alkyl ethers of starch (356), preparation of alkyl lead compounds (357), solvent for the extraction of aromatic hydrocarbons (358), curing agent for furyl alcohol resins (359), and the polymerization of olefins (360). The chemical properties (361), toxicity (362, 363), mutagenicity in *Drosophila* (364, 365), *E. coli* (364), and *Neurospora* (366, 367), and induction of chromosome breakage in plant material (368) of dimethyl sulfate has been described. Dimethyl sulfate has been analyzed by titrimetric (369), colorimetric (370, 371), and by proton magnetic resonance spectroscopy (372).

17. *Diethyl Sulfate*

Diethyl sulfate (DES) is prepared commercially by the addition of ethylene to sulfuric acid, and is extensively used in a variety of ethylation processes in organic synthesis and in a number of commercial areas: stabilization of organophosphorus insecticides (373, 374), finishing of cellulosic yarns (375), etherification of starch (356), as catalyst in olefin polymerization (376), and acrolein-pentaerythritol resin formation (377).

The effect of temperature (0° to 40°C) on *in vitro* and *in vivo* reactions of diethyl sulfate (378) indicate that hydrolysis occurs in accordance with the equation:

$$\text{Step 1:} \quad C_2H_5OSO_2OC_2H_5 + H_2O \rightarrow C_2H_5OH + C_2H_5OSO_2O^- + H^+$$
$$\text{Step 2:} \quad C_2H_5OSO_2O^- + H_2O \rightarrow C_2H_5OH + SO_4^{=} + H^+$$

The *in vitro* hydrolysis was found to be dependent on temperature. It has been shown that the first ethyl group was removed much faster than the second and that the rate of hydrolysis of ethyl sulfuric acid (step 2) in the temperature range 0° to 40°C was found to be insignificant.

The chemical properties (361), toxicity (196, 379), mutagenicity in *Drosophila* (364, 365, 380–382), *E. coli* (383–385), bacteriophage T2 (386), *Aspergillus nidulans* (387), peas (388), wheat (389), barley (390, 391), *Neurospora* (366), and *S. pombe* (392) of diethyl sulfate have been described.

The analysis of diethyl sulfate has been achieved via titrimetric (393), colorimetric (370, 371), infrared (394) and Raman spectroscopy (395), and gas chromatographic techniques (396).

18. *Methyl Methanesulfonate*

Methyl methanesulfonate (MMS) is prepared industrially from sulfur trioxide and methane (397) and is used as a catalyst in polymerization, alkylation, and esterification reactions, and as a solvent.

The oncogenic (398) and leukemogenic (399, 400) effects of MMS (resembling x-rays and nitrogen mustard), its usefulness in cancer chemotherpay (401) and in the sterilization of houseflies (*Musca domestica*) (402), and its antifertility effects in mice (403, 404) and male rats (403, 404) have been reported.

The hydrolysis of MMS (405) and its metabolism in the rat (369, 406–409) have been described. The methylation proficiency of MMS with DNA (408, 409) and its mutagenicity in bacteriophage (386, 410), *Schizosaccharomyces pombe* (411), barley (412, 413), mice (414–418), postmeiotic cells of the rat (414), as well as its induction of chromosome breakage in barley (412) and *Bellevalia* (419) have been reported.

The chemistry of MMS (361) and its analysis by colorimetry (370, 371) and infrared (420, 421) and Raman spectroscopy (421) have been described.

19. *Ethyl Methanesulfonate*

Ethyl methanesulfonate (EMS) has been prepared by the reaction of methane sulfonic acid and ethylene in the presence of boron trifluoride (422). It has been used for the production of mutants with high griseofulvin production capacity in the griseofulvin fermentation process from *Penicillium urticae* (423) as well as in the production of several mutant lines of cotton seeds that proved economically valuable (424).

EMS has been shown to be a highly effective mutagen in a wide variety of organisms, e.g., in *Drosophila* (210, 364, 425–431), *Hadrobracon* (432), *E. coli* (104, 433–435), *B. subtilis* (436), *B. cereus* (437), *Corynebacterium* (437), bacteriophage (386, 438–440), *Neurospora* (367, 441–443), yeast (*Schizosaccharomyces pombe*) (397, 401, 444, 445), (*Saccharomyces cerevisiae*) (446, 447), barley (448–455), maize (456–457), wheat (389, 458, 459), *Vicia* (460), sweet clover (461), *Arabidopsis* (462–464), *Aspergillus nidulans* (387), *Nicotiana tabacum* (465), *Hordeum vulgare* (466), *Nigella damascena* seeds (467), Chinese hamster cells *in vitro* (468); it is also effective in inducing dominant lethal mutations in mice (481) and rats (414).

EMS has induced chromosome aberrations in barley (413, 469), wheat (469), *Vicia faba* (470, 471), *Crepis capillaris* seeds (472), cultured human cells

(473), and in mice (474). The carcinogenicity of EMS in neonatal mice (475) as well as its sterilizing capability in the housefly (*Musca domestica*) (476) have also been described.

Analysis of EMS and MMS have been achieved by infrared and Raman spectroscopy (421).

20. *Myleran*

Myleran [1,4-di(methanesulfonoxy)butane; busulfan] is prepared by the reaction of 1,4-butanediol and methanesulfonyl chloride in the presence of a

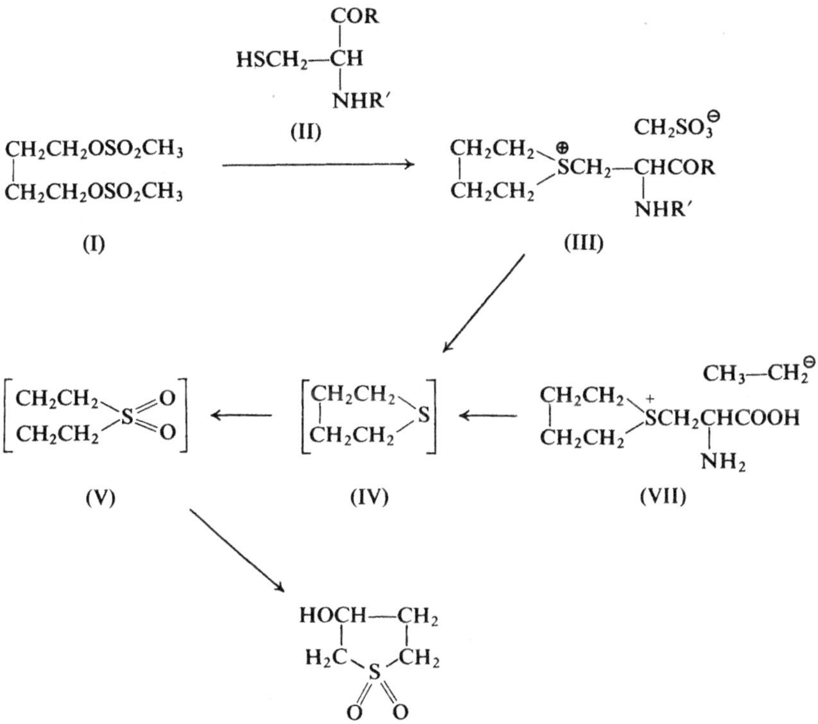

FIG. 8.2. Metabolism of Myleran. R = OH, amino acid, peptide or protein residue; R′ = H, amino acid, peptide or protein residue.

base such as pyridine or dimethylaniline. Myleran is a clinically useful chemotherapeutic agent active against chronic myeloid leukemia.

The utility of Myleran (477–479), its acute and chronic toxicity (480–481), the synthesis of tritiated (482, 483), ^{35}S- and ^{14}C-Myleran (484), and the metabolism of labeled and unlabeled Myleran in the rat (484–489), mouse (485), rabbit (485), and human (490, 491) have been described.

Roberts and Warwick (321, 492) have identified the major metabolites of Myleran and have proposed the route of metabolism as shown in Fig. 8.2. Myleran [Fig. 8.2(I)] reacts with cysteine or a cysteinyl moiety (II) to form a cyclic sulfonium ion (III) which undergoes decomposition to tetrahydrothiophene (IV), which is converted to tetrahydrothiophene-1,1-dioxide (V) and then to 3-hydroxytetrahydrothiophene-1,1-dioxide (VI) (which was found in the urine of rats, mice, and rabbits). While (III), (IV), and (V) were not isolated, it was shown that *S*-β-L-alanyltetrahydrothiophenium mesylate (VII) and (IV) and (V) were metabolized almost entirely to (VI).

The comparative effects of Myleran on spermatogenesis (493), its implication in human teratogenesis (494), its mutagenicity in mice (172, 414), *Drosophila* (151, 431, 495–597), barley (16, 17, 450, 498), and the monocellular alga *Chlamydomonas eugametos* (499–500), and its induction of chromosomal aberrations in human lymphocytes (501, 502), barley (503), *Vicia faba* (504, 505), and *Hordeum sativum* (506, 507), and mouse spermatogenesis (172) have been reported.

Myleran has been analyzed by colorimetric (508), gravimetric (509), and paper chromatographic techniques (321, 485, 489, 490, 510).

21. L-*Threitol-1,4-bismethanesulfonate*

DL-, L-, or D-Threitol-1,4-bismethanesulfonate (TBMS) was prepared by Feit (511) from (a) treatment of the corresponding DL-, L-, D, or *meso*-1,4-dibromo-2,3-butanediol with silver methanesulfonate or (b) from DL-, L-, D-, or *meso*-1,2:3,4-diepoxybutane reacted with methanesulfonic acid.

The antitumor activity and pharmacology of this difunctional alkylating agent has been reported (511–513). TBMS is stable at low pH (5.5 to 6) and, under some experimental conditions, can be transformed into diepoxybutane among other products (the rate of this transformation increases with the alkalinity of the solution) (514).

Davis and Ross (515) have shown that β-hydroxymethanesulfonates (I) can react in the following sequence:

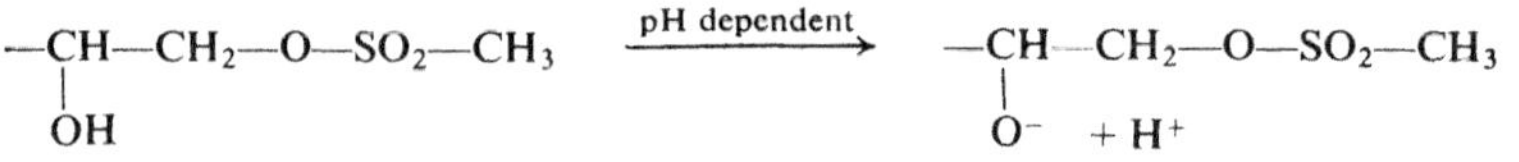

(I) (II) (a)

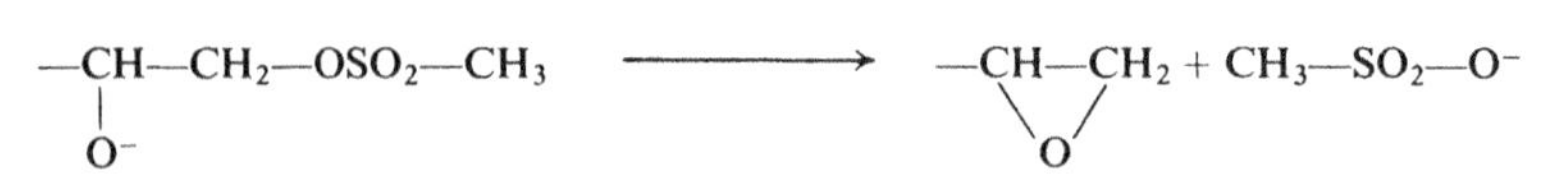

(III) (b)

(IV) (c)

An epoxide (III) is formed which subsequently hydrolyzes into a glycol (IV). The first reaction is accelerated with increasing pH owing to the increasing proportion of the ionized form II of the β-hydroxyl which displaces the methanesulfonate group. It was demonstrated that conversions of TBMS to the epoxide at 37°C (pH 7.5) could occur before effective hydrolysis to the glycol took place (515).

Matagne (514) recently studied the activity of L-TBMS on barley chromosomes in relation to chemical transformation during treatment. The frequency of induced chromosomal aberrations was dependent on pH and consequently on the transformation rate of TBMS into epoxides during treatment. The hydrolysis of TBMS was postulated to occur as follows:

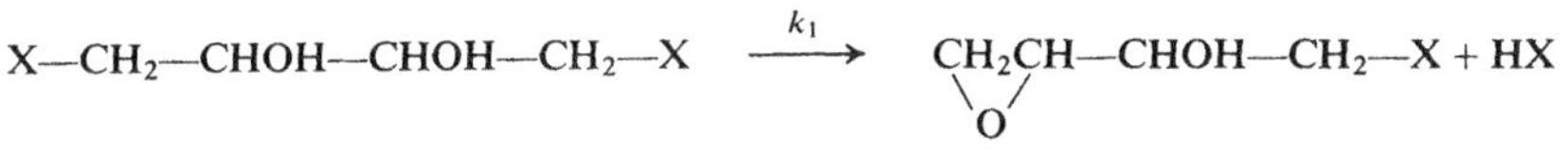

(V) (VI)

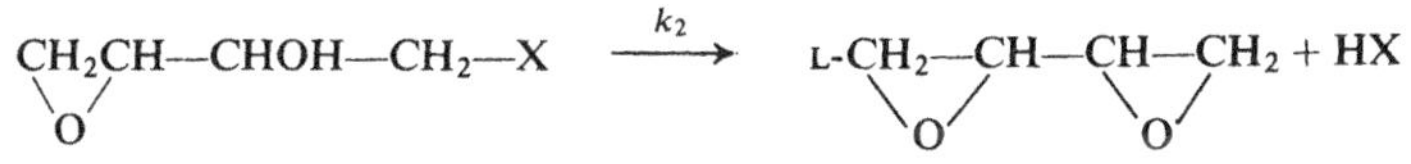

(VII)

($X = CH_2—SO_2—O—$)

Hence, two new alkylating agents, e.g., a monoepoxide (VI) and diepoxybutane (DEB) (VII), should be successively formed, but since k_2 is probably greater than k_1 the monoepoxide should never accumulate.

L-TBMS has been shown to be more active than the D- or *meso*-form (L > D > *meso*) in inducing chromosomal aberrations in plants, e.g., *Arabidopsis thaliana* (516), *Vicia faba* (517, 518), and *Nigella damascena* (517, 518). Other chromosomal aberration studies in *Allium cepa* (176) and *Vicia faba* (519, 520) with L-TBMS have been described.

REFERENCES

1. A. M. Thiess, *Arch. Toxikol.* **20**, 127 (1963).
2. R. E. Joyner, *Arch. Environ. Health* **8**, 700 (1964).
3. K. H. Jacobson, E. B. Hackley, and L. Feinsilver, *A.M.A. Arch. Ind. Health* **13**, 237 (1956).

4. R. L. Stehle, W. Bourne, and E. Lozinsky, *Arch. Exptl. Pathol. Pharmakol.* **104**, 82 (1924).
5. A. Haddow, *in* "The Physiopathology of Cancer" (F. Hamburger and H. Fishman, eds.), Vol. II, p. 411, Harper (Hoeber), New York, 1953.
6. H. L. Fraenkel-Conrat, *J. Biol. Chem.* **154**, 337 (1944).
7. J. N. Brönsted and M. Kilpatrick, *J. Am. Chem. Soc.* **51**, 428 (1929).
8. P. Alexander and K. A. Stacey, *Ann. N. Y. Acad. Sci.* **68**, 1225 (1958).
9. I. A. Rapoport, *Dokl. Akad. Nauk SSSR* **60**, 469 (1948).
10. M. J. Bird, *J. Genet.* **50**, 480 (1952).
11. H. G. Kølmark and B. J. Kilbey, *Mol. Gen. Genet.* **101**, 89 (1960).
12. H. G. Kølmark, B. J. Kilbey, and S. Kondo, *Proc. 11th Intern. Consr. Genet., The Hague, 1963* Vol. I, p. 61. Pergamon Press, Oxford, 1965.
13. H. G. Kølmark and B. J. Kilbey, *Mol. Gen. Genet.* **101**, 185 (1968).
14. G. Kølmark and M. Westergaard, *Hereditas* **39**, 209 (1953).
15. L. Ehrenberg and A. Gustafsson, *Hereditas* **43**, 595 (1957).
16. L. Ehrenberg, *Abhandl. Deut. Akad. Wiss. Berlin, Kl. Med.* **1**, 124 (1960).
17. D. Wettstein, A. Gustafsson, and L. Ehrenberg, *Arbeitsgemeinschaft Forsch. Landes Nordrhein-Westfalen* **73**, 73 (1959).
18. K. Šulovská, D. Lindgren, G. Eriksson, and L. Ehrenberg, *Hereditas* **62**, 264 (1969).
19. K. A. Jensen, I. Kirk, G. Kølmark, and M. Westergaard, *Cold Spring Harbor Symp. Quant. Biol.* **16**, 245 (1951).
20. A. C. Faberge, *Genetics* **40**, 571 (1955).
21. J. Moutschen-Dahmen, M. Moutschen-Dahmen, and L. Ehrenberg, *Hereditas* **60**, 267 (1968).
22. A. Loveless, *Heredity* **6**, Suppl., 293 (1953).
23. O. F. Lubatti, *J. Soc. Chem. Ind. (London)* **63**, 133 (1944).
24. D. Gunther, *Anal. Chem.* **37**, 1172 (1965).
25. F. E. Critchfield and J. B. Johnson, *Anal. Chem.* **29**, 797 (1957).
26. N. E. Bolton and N. H. Ketcham, *Arch. Environ. Health* **8**, 711 (1964).
27. J. C. Gage, *Analyst* **82**, 587 (1957).
28. S. Belman, *Anal. Chim. Acta* **29**, 129 (1963).
29. E. Sawicki, T. W. Stanley, and J. Pfaff, *Anal. Chim. Acta* **28**, 156 (1963).
30. E. Kröller, *Deut. Lebensm.-Rundschau* **62**, 227 (1966).
31. S. Ben-Yehoshua and P. Krinsky, *J. Gas Chromatog.* **6**, 350 (1958).
32. B. Berck, *J. Agr. Food Chem.* **13**, 373 (1965).
33. S. G. Heuser and K. A. Scudamore, *Analyst* **93**, 252–258 (1968).
34. S. G. Heuser and K. A. Scudamore, *Chem. & Ind. (London)* p. 1557 (1967).
35. C. W. Bruch, *Ann. Rev. Microbiol.* **15**, 245 (1961).
36. C. R. Phillips and S. Kaye, *Am. J. Hyg.* **50**, 270 (1949).
37. C. Clarke, W. Davidson, and J. B. Johnson, *Australia (South) New Zealand J. Surg.* **36**, 53 (1966).
38. B. Bain and L. Lowenstein, *Med. Res. Council Can. Rept.* **MBT 1664** (1967).
39. R. K. O'Leary and W. L. Guess, *J. Pharm. Sci.* **57**, 12 (1968).
40. W. L. Guess and R. K. O'Leary, *Toxicol. Appl. Pharmacol.* **14**, 221 (1969).
41. R. K. Kulkarni, D. Bartak, D. K. Ousterhout, and F. Leonard, *J. Biomed. Mater. Res.* **2**, 165 (1968).
42. A. C. Cunliffe and F. Wesley, *Brit. Med. J.* **2**, 575 (1967).
43. S. Kaye, *Am. J. Hyg.* **50**, 289 (1949).
44. E. A. Hawk and O. Mickelsen, *Science* **121**, 442 (1955).
45. N. Diding, *Pharm. Acta Helv.* **35**, 582 (1960).

46. H. G. Windmueller, C. J. Ackerman, H. Blackerman, and O. Mickelsen, *J. Biol. Chem.* **234,** 889 (1959).
47. H. G. Windmueller, C. J. Ackerman, and R. W. Engel, *J. Biol. Chem.* **234,** 895 (1959).
48. N. Adler, *J. Pharm. Sci.* **54,** 735 (1965).
49. F. Wesley, B. Rourke, and O. Darbishire, *J. Food Sci.* **30,** 1037 (1965).
50. E. P. Ragelis, B. S. Fisher, and B. A. Klimeck, *J. Assoc. Offic. Anal. Chemists* **51,** 709 (1968).
51. P. M. Gross and L. F. Dixon, U.S. Patent 1,962,145 (1934); *Chem. Abstr.* **28,** 4824. (1934).
52. S. M. Samfield, E. E. Locklan, and B. A. Brock, U.S. Patent 2,760,495 (1956); *Chem. Abstr.* **51,** 677a (1957).
53. Y. Obi, Y. Shimada, K. Takahashi, K. Nishida, and T. Kisaki, *Tobacco* **166,** 26 (1968).
54. M. Muramatsu, *Japan. Monopolycent. Res. Inst. Sci. Papers* **110,** 217 (1968).
55. H. T. Gordon, W. W. Thronburg, and L. M. Werum, *J. Agr. Food Chem.* **7,** 196 (1959).
56. L. Sair, personal communication in B. Berck, *Occupational Health Rev.* **18,** 16 (1966).
57. H. Frankel-Conrat, *J. Biol. Chem.* **154,** 227 (1944).
58. W. C. J. Ross, *Advan. Cancer Res.* **1,** 429 (1953).
59. P. Alexander, *Advan. Cancer Res.* **2,** 4 (1954).
60. V. K. Rowe, R. L. Hollingsworth, F. Oyen, D. D. McCollister, and H. C. Spenser, *A.M.A. Arch. Ind. Health* **13,** 228 (1956).
61. H. F. Smyth, Jr., J. Seaton, and L. Fischer, *J. Ind. Hyg. Toxicol.* **23,** 259 (1941).
62. H. F. Smyth, Jr. and C. P. Carpenter, *J. Ind. Hyg. Toxicol.* **30,** 63 (1948).
63. A. L. Walpole, *Ann. N. Y. Acad. Sci.* **68,** 750 (1958).
64. A. Schalet, *Drosophila Inform. Serv.* **28,** 155 (1954).
65. O. D. Shreve, M. R. Heether, H. B. Knight, and D. Swern, *Anal. Chem.* **23,** 277 (1951).
66. R. Mestres and C. Barrois, *Soc. Pharm. Montpelier* **24,** 47 (1964).
67. C. R. Phillips, *Am. J. Hyg.* **50,** 288 (1949).
68. R. Whelton, H. J. Phaff, E. M. Mrak, and C. D. Fisher, *Food Ind.* **18,** 23 (1946).
69. F. S. Mury, M. W. Miller, and J. E. Brekke, *Food Technol.* **14,** 113 (1960).
70. C. W. Bruch and M. G. Koesterer, *J. Food Sci.* **26,** 428 (1961).
71. E. P. Ragelis, B. S. Fisher, and B. A. Klimeck, *J. Assoc. Offic. Anal. Chemists* **49,** 963 (1966).
72. C. Vojnovich and V. F. Pfeifer, *Cereal Sci. Today* **12,** 54–60 (1967).
73. L. Wiseblatt, *Cereal Chem.* **44,** 269 (1967).
74. F. E. Denny, *J. Soc. Chem. Ind.* (*London*) **47,** 239 (1928).
75. F. E. Denny, *Contrib. Boyce Thompson Inst.* **14,** 1 (1945).
76. M. Z. Condon, F. R. Andrews, M. G. Lambou, and A. M. Altschul, *Science* **105,** 525 (1947).
77. M. K. Johnson, *Biochem. Pharmacol.* **16,** 185 (1967).
78. E. T. Huntress, "The Preparation, Properties, Chemical Behavior and Identification of Organic Chlorine Compounds," p. 705. Wiley, New York, 1948.
79. A. M. Ambrose, *A.M.A. Arch. Ind. Health* **2,** 591 (1950).
80. M. W. Goldblatt and W. E. Chiesman, *Brit. J. Ind. Med.* **1,** 207 (1944).
81. S. Carson and B. L. Oser, *Tyxicol. Appl. Pharmacol.* **14,** 633 (1969).
82. W. L. Guess, *Toxicol. Appl. Pharmacol.* **14,** 659 (1969).
83. C. E. Voogd and P. V. D. Vet, *Experientia* **25,** 85 (1969).
84. K. Uhrig, *Ind. Eng. Chem., Anal. Ed.* **18,** 469 (1946).
85. J. Y. T. Chen and J. H. Gould, *J. Assoc. Offic. Anal. Chemists* **51,** 878 (1968).
86. J. S. Pagington, *J. Chromatog.* **36,** 528 (1968).

87. Deutsche Hydrierwerke G.m.b.H., German Patent 874,289 (1953); *Chem. Abstr.* **51**, 10003 (1957).
88. E. I. duPont deNemours & Co., Netherlands Appl. 6,608,665 (1966); *Chem. Abstr.* **66**, 29979z (1967).
89. M. W. Minkler, T. W. Findley, and R. S. Geister, U.S. Patent 2,890,119 (1959); *Chem. Abstr.* **53**, 16413a (1959).
90. C. H. Hine, J. K. Kodama, J. S. Wellington, M. K. Dunlap, and H. H. Anderson, *A.M.A. Arch. Ind. Health* **14**, 250 (1956).
91. J. L. Kodama, R. J. Guzman, M. K. Dunlap, G. S. Loquuam, R. Lima, and C. H. Hine, *Arch. Environ. Health* **2**, 50 (1961).
92. G. Kølmark and N. H. Giles, *Genetics* **40**, 890 (1955).
93. J. J. Biesele, F. S. Philips, J. B. Thiersch, J. H. Burchenal, S. M. Buckley, and C. C. Stock, *Nature* **166**, 1112 (1950).
94. J. Dyr and J. Mostek, *Kvasny Prumysl* **4**, 121 (1958); *Chem. Abstr.* **52**, 19011 (1958).
95. W. Schafer, W. Nuck, and H. Jahn, *J. Prakt. Chem.* [4] **11**, 1–10 and 11–19 (1960); *Chem. Abstr.* **55**, 3304 (1961).
96. Deering Milliken Res. Corp., British Patent 855,547 (1960); *Chem. Abstr.* **55**, P17034h (1961).
97. Hercules Powder Co., British Patent 871,205 (1961); *Chem. Abstr.* **55**, P27884d (1961).
98. P. Michel, Belgium Patent 555,772 (1957); *Chem. Abstr.* **53**, P19348h (1959).
99. G. Faerber, German Patent 1,002,945 (1957); *Chem. Abstr.* **53**, P23088a (1959).
100. L. F. Burnstein, U.S. Patent 2,810,700 (1957); *Chem. Abstr.* **52**, P2458f (1958).
101. C. P. Carpenter, H. F. Smyth, Jr., and V. C. Pozzani, *J. Ind. Hyg. Toxicol.* **31**, 343 (1949).
102. C. Schultz, *Deut. Med. Wochschr.* **89**, 1342 (1964).
103. *Anonymous*, *Food Cosmet. Toxicol.* **2**, 240 (1964).
104. A. Loveless and S. Howarth, *Nature* **184**, 1780 (1959).
105. B. S. Strauss and S. Okubo, *J. Bacteriol.* **79**, 464 (1960).
106. L. Ehrenberg, unpublished, cited in A. Loveless, "Genetic and Allied Effects of Alkylating Agents," p. 78. Pennsylvania State Univ. Press, University Park, Pennsylvania, 1966.
107. B. S. Boikina and E. A. Peregud, *Gigiena Truda i Prof. Zabolevaniya* **4**, 40 (1960); *Chem. Abstr.* **55**, 8886c (1961).
108. J. W. Daniel and J. C. Gage, *Analyst* **81**, 594 (1956).
109. E. A. Peregud and B. S. Boikina, *Gigiena i Sanit.* **25**, 71 (1960); *Chem. Abstr.* **54**, 25425h (1960).
110. R. F. Goodu and D. A. Delker, *Anal. Chem.* **30**, 2013 (1958).
111. H. A. Christ, P. Diehl, H. R. Schneider, and H. A. Dahn, *Helv. Chim. Acta* **44**, 865 (1961).
112. G. Aruldhas and V. U. Nayar, *Indian J. Pure Appl. Phys.* **4**, 361 (1960).
113. H. G. Eggert, W. Dietrich, and H. Rath, German Patent 956,678 (1957); *Chem. Abstr.* **53**, 22016a (1959).
114. E. Fischer and C. Schott, German Patent 1,233,140 (1967); *Chem. Abstr.* **66**, 86120m (1967).
115. E. M. Fettes and J. A. Gannon, U.S. Patent 2,789,958 (1957); *Chem. Abstr.* **51**, 17239 (1957).
116. G. W. Strother, Jr., U.S. Patent 3,303,144 (1967); *Chem. Abstr.* **66**, 66173c (1967).
117. H. F. Mark and S. M. Atlas, French Patent 1,438,201 (1966); *Chem. Abstr.* **66**, 29874m (1967).

118. V. V. Zykova, M. V. Vittikh, and L. O. Mareeva, *Izv. Akad. Nauk Kaz. SSR, Ser. Khim.* **16**, 54 (1966); *Chem. Abstr.* **66**, 46858b (1967).
119. T. L. Ashby, U.S. Patent 3,349,053 (1967); *Chem. Abstr.* **67**, 117700c (1967).
120. C. A. May and A. C. Nixon, *J. Chem. Eng. Data* **6**, 290 (1961); *Chem. Abstr.* **55**, 20502 (1961).
121. W. Himmelmann and O. Wahl, German Patent 1,095,113 (1960); *Chem. Abstr.* **55**, 20742c (1961).
122. J. Kresta, L. Ambroz, and K. Obrucova, *Chem. Prumysl* **17**, 25 (1967); *Chem. Abstr.* **66**, 116219v (1967).
123. C. H. Hine, J. K. Kodama, R. J. Guzman, M. K. Dunlap, R. Lima, and G. S. Loquuam, *Arch. Environ. Health* **2**, 31 (1961).
124. C. J. McCammon, P. Kotin, and H. L. Falk, *Proc. Am. Assoc. Cancer Res.* **2**, 229 (1957).
125. A. Loveless and C. C. Stock, *Proc. Roy. Soc.* **B150**, 497 (1959).
126. S. H. Revell, *Heredity* **6**, Suppl., 107 (1953).
127. S. H. Revell, *Brit. Empire Cancer Campaign Ann. Rept.* **30**, 42 (1953).
128. S. H. Revell, *Heredity*, Suppl., p. 107 (1953).
129. C. D. Darlington and J. McLeish, *Nature* **167**, 407 (1951).
130. J. McLeish, *Heredity* **6**, Suppl., 125 (1953).
131. J. McLeish, *Heredity* **8**, 385 (1954).
132. B. A. Kihlman, *J. Biophys. Biochem. Cytol.* **2**, 543 (1956).
133. S. H. Revell, *Ann. N. Y. Acad. Sci.* **68**, 902 (1958).
134. G. R. Lane, *in* "Radiobiology Symposium" (Z. M. Bacq and P. Alexander, eds.), p. 265. Academic Press, New York, 1954.
135. J. L. Jungnickel, E. D. Peters, A. Polgar, and F. T. Weiss, *in* "Organic Analysis" (J. M. Mitchell *et al.*, eds.), Vol. I, p. 127. Wiley (Interscience), New York, 1953.
136. B. L. Van Duuren, N. Nelson, L. Orris, E. D. Palmes, and F. L. Schmitt, *J. Natl. Cancer Inst.* **31**, 41 (1963).
137. W. T. Beech, *J. Chem. Soc.* p. 2483 (1951).
138. H. D. Michener and T. C. Lewis, U.S. Patent 2,934,439 (1960); *Chem. Abstr.* **54**, 14498i (1960).
139. F. L. Rose, J. A. Hendry, and A. L. Walpole, *Nature* **165**, 993 (1950).
140. J. Bichel, A. Stenderup, and K. A. Jensen, *Acta, Unio. Intern. Contra Cancrum* **16**, 503 (1960).
141. J. A. Hendry, R. F. Homer, F. L. Rose, and A. L. Walpole, *Brit. J. Pharmacol.* **6**, 235 (1951).
142. B. L. Van Duuren, L. Orris, and N. Nelson, *J. Natl. Cancer Inst.* **35**, 707 (1965).
143. P. Kotin and H. L. Falk, *Radiation Res.* Suppl. 3, 193 (1963).
144. A. Haddow, *Brit. Med. Bull.* **14**, 79 (1958).
145. C. S. Weil, N. Condra, C. Haun, and J. A. Striegel, *Am. Ind. Hyg. Assoc. J.* **24**, 305 (1963).
146. C. H. Hine and V. K. Rowe, "Industrial Hygiene and Toxicology," Vol. II, p. 1601. Wiley (Interscience), New York, 1967.
147. M. S. Melzer, *Biochim. Biophys. Acta* **138**, 613 (1967).
148. P. Alexander and S. F. Cousons, *Biochem. Pharmacol.* **1**, 25 (1958).
149. O. G. Fahmy and M. J. Fahmy, *Proc. 10th Intern. Congr. Genet., Montreal, 1958* Vol. 2. p. 78. Univ. of Toronto Press, Toronto, 1959.
150. M. J. Bird and O. G. Fahmy, *Proc. Roy. Soc.* **B140**, 556 (1953).
151. O. G. Fahmy and M. J. Fahmy, *J. Genet.* **54**, 146 (1956).
152. Y. Nakao and C. Auerbach, *Z. Vererbungslehre* **92**, 457 (1961).

153. O. G. Fahmy and M. J. Bird, *Heredity* **6**, Suppl., 149 (1953).
154. G. Kølmark and M. Westergaard, *J. Biophys. Biochem. Cytol.* **2**, 543 (1956).
155. H. G. Kølmark and B. J. Kilbey, *Z. Vererbungslehre* **93**, 356 (1962).
156. C. Auerbach and D. Ramsey, *Japan. J. Genet.* **43**, 1 (1968).
157. G. Kølmark and N. H. Giles, *Genetics* **38**, 674 (1953).
158. G. Kølmark and N. H. Giles, *Hereditas* **39**, 209 (1953).
159. M. Westergaard, *Abhandl. Deut. Akad. Wiss. Berlin Kl. Med.* **1**, 30 (1960).
160. C. H. Clarke, *Mutation Res.* **8**, 35 (1969).
161. Z. Hartman, *Carnegie Inst. Wash. Publ.* **612**, 107 (1956).
162. J. Kilbey, *Mutation Res.* **8**, 73 (1969).
163. H. Heslot and R. Ferrary, *Ann. Inst. Natl. Rech. Agron.* **44**, 1 (1958).
164. L. Ehrenberg, A. Gustafsson, and U. Lundquist, *Acta Chem. Scand.* **10**, 492 (1956).
165. J. D. Kreizinger, *Genetics* **45**, 143 (1960).
166. A. Bianchi and M. Contin, *J. Heredity* **53**, 277 (1962).
167. A. Bianchi, *Proc. 11th Intern. Congr. Genet., The Hague, 1963* Vol. 1, p. 95. Pergamon Press, Oxford, 1965.
168. G. Emery, *Science* **131**, 1732 (1960).
169. J. Moutschen-Dahmen, M. Moutschen-Dahmen, and R. Loppes, *Nature* **199**, 406 (1963).
170. R. Matagne, *Bull. Soc. Roy. Botan. Belg.* **102**, 239 (1969).
171. L. E. LaChance, M. Degrugillier, and A. P. Leverich, *Mutation Res.* **7**, 63 (1969).
172. J. Moutschen, *Genetics* **46**, 291 (1961).
173. N. S. Cohn, *Nature* **192**, 1093 (1961).
174. N. S. Cohn, *Exptl. Cell Res.* **24**, 569 (1961).
175. A. C. Faberge, *Genetics* **40**, 171 (1955).
176. R. Matagne, *Radiation Botany* **8**, 489 (1968).
177. J. Nemenzo and C. H. Hine, *Abstr. 8th Ann. Meeting Soc. Toxicol., Williamsburg, 1969, Toxicol. Appl. Pharmacol.* **14**, 653 (1969).
178. J. Mitchell, I. M. Kolthoff, E. S. Proskaver, and A. Weissberger, eds., "Organic Analysis," Vol. I, p. 127. Wiley (Interscience), New York, 1953.
179. D. J. Swann, *Anal. Chem.* **26**, 878 (1954).
180. Kirk-Othmer, "Encyclopedia of Chemical Technology," 2nd ed., Vol. 10, p. 77. Wiley (Interscience), New York, 1966.
181. W. G. J. Shaw, R. L. Stephens, and J. A. Weybrew, *Tobacco Sci.* **4**, 179 (1960).
182. J. R. Newsome, V. Norman, and C. H. Keith, *Tobacco Sci.* **9**, 102 (1965).
183. R. L. Stenburg, R. P. Hangebrauck, D. J. von Lehmden, and A. H. Rose, Jr., *J. Air Pollution Control Assoc.* **11**, 376 (1961).
184. R. L. Stenburg, R. P. Hangebrauck, D. J. von Lehmden, and A. H. Rose, Jr., *J. Air. Pollution Control Assoc.* **12**, 83 (1962).
185. A. P. Altshuller, I. R. Cohen, M. E. Meyer, and A. F. Wartburg, Jr., *Anal. Chim. Acta* **25**, 101 (1961).
186. K. W. Wilson, *Anal. Chem.* **30**, 1127 (1960).
187. C. F. Ellis, *U.S., Bur. Mines, Rept. Invest.* **5822** (1961).
188. A. P. Altshuller and S. P. McPherson, *J. Air Pollution Control Assoc.* **13**, 109 (1963).
189. J. M. Stuart and D. A. Smith, *J. Appl. Polymer Sci.* **9**, 3195 (1965).
190. M. B. Neimann, B. M. Kovarskay, C. I. Golubenkova, A. S. Strizhkova, I. I. Levantovskaya, and M. S. Akutin, *J. Polymer Sci.* **56**, 383 (1962).
191. L. H. Lee, *Proc. Battelle Symp. Thermal Stability Polymers, Columbus, Ohio*, 1963 PFL.
192. L. H. Lee, *Preprints Am. Chem. Soc. Meeting, New York* **5**, No. 2, 453 (1963).
193. L. Berrens, E. Young, and L. H. Janson, *Brit. J. Dermatol.* **76**, 110 (1964).

194. A. A. Fisher, N. B. Kanof, and E. M. Biondi, *Arch. Dermatol.* **86**, 753 (1962).
195. J. F. Walker, "Formaldehyde," 3rd ed. Reinhold, New York, 1964.
196. C. P. Carpenter and H. F. Smyth, Jr., *Am. J. Ophthalmol.* **29**, 1363 (1946).
197. E. Skog, *Acta Pharmacol. Toxicol.* **6**, 299 (1950).
198. T. Matsunaga, T. Soejima, Y. Iwata, and F. Watanabe, *Gann* **45**, 451 (1954).
199. L. J. Saidel, J. S. Saltzman, and W. H. Elfring, *Nature* **207**, 169 (1965).
200. G. Malorney, N. Rietbrock, and M. Schneider, *Arch. Exptl. Pathol. Pharmakol.* **250**, 419 (1965).
201. N. Rietbrock, *Arch. Exptl. Pathol. Pharmakol.* **251**, 189 (1965).
202. I. A. Rapoport, *Dokl. Akad. Nauk SSSR* **54**, 65 (1946).
203. I. A. Rapoport, *Zh. Obshch. Biol.* **8**, 359 (1947).
204. I. A. Rapoport, *Dokl. Akad. Nauk SSSR* **56**, 537 (1947).
205. I. A. Rapoport, *Dokl. Akad. Nauk SSSR* **51**, 713 (1948).
206. W. D. Kaplan, *Science* **108**, 43 (1948).
207. C. Auerbach, *Science* **110**, 419 (1949).
208. C. Auerbach, *Hereditas* **37**, 1 (1951).
209. T. Alderson, *Proc. 11th Intern. Congr. Genet., The Hague, 1963* Vol. 1, p. 65. Pergamon Press, Oxford, 1965.
210. T. Alderson, *Nature* **207**, 164 (1965).
211. C. Auerbach, *Am. Naturalist* **86**, 330 (1952).
212. A. F. E. Khishin, *Mutation Res.* **1**, 202 (1964).
213. W. J. Burdette, *Cancer Res.* **11**, 241 (1951).
214. C. Auerbach, *Z. Vererbungslehre* **81**, 621 (1956).
215. C. Auerbach, *Nature* **210**, 104 (1966).
216. F. H. Sobels, *Am. Naturalist* **88**, 109 (1954).
217. F. H. Dickey, G. H. Cleland, and C. Lote, *Proc. Natl. Acad. Sci. U.S.* **35**, 581 (1949).
218. M. Demerec, G. Bertani, and J. Flint, *Am. Naturalist* **85**, 119 (1951).
219. I. A. Rapoport, *Dokl. Akad. Nauk SSSR* **59**, 1183 (1948).
220. T. Alderson, *Nature* **187**, 485 (1960).
221. Anonymous, *Food Cosmet. Toxicol.* **4**, 99 (1966).
222. S. Ebel, M. Brueggemann, and E. Rosswog, *Deut. Apotheker-Ztg.* **107**, 1718 (1967).
223. M. B. Jacobs, "The Analytical Chemistry of Industrial Poisons, Hazards and Solvents," 2nd ed., Vol. I, p. 524. Wiley (Interscience), New York, 1949.
224. E. Sawicki, T. R. Hauser, and S. McPherson, *Anal. Chem.* **34**, 1460 (1962).
225. T. Nash, *Biochem. J.* **55**, 416 (1953).
226. F. Onuška, J. Janák, S. Duras, and M. Kremarova, *J. Chromatog.* **40**, 209 (1969).
227. R. S. Mann and K. W. Hahn, *Anal. Chem.* **39**, 1314 (1967).
228. R. E. Leonard, *J. Gas Chromatog.* **4**, 142 (1966).
229. H. F. Smyth, Jr., *Am. Ind. Hyg. Assoc. Quart.* **17**, 311 (1956).
230. E. Asmussen, *Acta Pharmacol.* **4**, 311 (1948).
231. H. Handovsky, *Compt. Rend.* **123**, 1242 (1936).
232. E. Asmussen, J. Hald, and V. Larsen, *Acta Pharmacol.* **4**, 311 (1948).
233. E. Jacobsen, *Pharmacol. Rev.* **4**, 107 (1952).
234. R. T. Williams, "Detoxification Mechanisms," 2nd ed. Wiley, New York, 1959.
235. I. A. Rapoport, *Dokl. Akad. Nauk SSSR* **61**, 713 (1948).
236. S. Sandler and Y. H. Chung, *Anal. Chem.* **30**, 1252 (1958).
237. H. Siegel and F. T. Weiss, *Anal. Chem.* **26**, 917 (1954).
238. J. E. Ruch and J. B. Johnson, *Anal. Chem.* **28**, 69 (1956).
239. J. D. Mold and M. T. McRae, *Tobacco Sci.* **1**, 173 (1957).
240. J. H. Ross, *Anal. Chem.* **25**, 1288 (1953).

241. R. Ellis, A. M. Gaddis, and G. T. Currie, *Anal. Chem.* **30**, 475 (1958).
242. D. A. Buyske, L. H. Owen, P. Wilder, and M. E. Hobbs, *Anal. Chem.* **28**, 910 (1956).
243. R. Stevens, *Anal. Chem.* **33**, 1126 (1961).
244. R. M. Irby and E. S. Harlow, *Tobacco* **148**, 21 (1959).
245. R. N. Baker, A. L. Alenty, and J. F. Zack, Jr., *J. Chromatog. Sci.* **7**, 312 (1969).
246. C. Neuberg and M. Kobel, *Biochem. Z.* **179**, 459 (1926).
247. R. A. W. Johnstone and J. R. Plimmer, *Chem. Rev.* **59**, 885 (1959).
248. R. M. Irby and E. S. Harlow, *133rd Am. Chem. Soc. Meeting, San Francisco, 1958* p. D–54.
249. E. D. Barber and J. P. Lodge, *Anal. Chem.* **35**, 348 (1963).
250. E. J. Hughes and R. W. Hurn, *J. Air Pollution Control Assoc.* **10**, 367 (1960).
251. C. F. Ellis, R. F. Kendall, and B. H. Eccleston, *Anal. Chem.* **37**, 511 (1965).
252. P. K. Mueller, M. F. Fraccia, and F. J. Schuette, *152nd Natl. Meeting Am. Chem. Soc., New York, 1966* p. Y–63.
253. C. J. Hughes, R. W. Hurn, and F. G. Edwards, *Gas Chromatog., 2nd Intern. Symp., East Lansing, Mich., 1959* p. 171. Academic Press, New York, 1961.
254. R. H. Linnel and W. E. Scott, *Arch. Environ. Health* **5**, 616 (1962).
255. D. M. Coulson, *Anal. Chim. Acta* **19**, 284 (1958).
256. I. R. Cohen and A. P. Altshuller, *Anal. Chem.* **33**, 726 (1961).
257. A. P. Altshuller, *Intern. J. Air Water Pollution* **6**, 169 (1962).
258. K. Grob, *Beitr. Tabakforsch.* **3**, 243 (1965).
259. K. Grob. *J. Gas Chromatog.* **3**, 52 (1965).
260. A. H. Lawrence, L. A. Lyerly, and G. W. Young, *Tobacco Sci.* **8**, 150 (1964).
261. J. T. Williamson and D. R. Allman, *Beitr. Tabakforsch.* **3**, 590 (1966).
262. T. Doihara, U. Kobashi, S. Sugawara, and Y. Kaburaki, *Sci. Papers, Central Res. Inst., Japan, Monopoly Corp.* No. 106, pp. 129 and 141 (1964).
263. R. D. Cadle and H. S. Johnston, *Proc. 2nd Natl. Air Pollution Symp., Pasadena, Calif., 1952* p. 28.
264. H. F. Smyth, Jr., C. P. Carpenter, and C. S. Weil, *Arch. Ind. Hyg. Occupational Med.* **4**, 119 (1951).
265. N. Iwanoff, *Arch. Hyg.* **73**, 307 (1910).
266. K. Onoe, *J. Chem. Soc. Japan, Pure Chem. Sect.* **73**, 337 (1952).
267. Badische Anilin and Soda Fabrik, A-G, Netherlands Appl. 6,607,766 (1966); *Chem. Abstr.* **67**, 22848t (1967).
268. H. Krueger, East German Patent 54,836 (1967); *Chem. Abstr.* **67**, 107587b (1967).
269. T. C. Butler, *J. Pharmacol. Exptl. Therap.* **95**, 360 (1949).
270. P. J. Friedman and J. R. Cooper, *Anal. Chem.* **30**, 1674 (1958).
271. J. R. Cooper and P. J. Friedman, *Biochem. Pharmacol.* **1**, 76 (1958).
272. G. Scansetti, A. F. Rubino, and G. Trompeo, *Med. Lavoro* **50**, 743 (1959); *Chem. Abstr.* **54**, 15666a (1960).
273. L. S. Goodman and A. Gilman, "The Pharmacological Basis of Therapeutics," 3rd ed., p. 132. Macmillan, New York, 1967.
274. F. J. Mackay and J. R. Cooper, *J. Pharmacol. Exptl. Therap.* **135**, 271 (1962).
275. A. Barthelmess, *Arzneimittel-Forsch.* **6**, 157 (1956).
276. A. Goldstein, *in* "Mutations" (W. J. Schull, ed.), p. 172. Univ. of Michigan Press, Ann Arbor, Michigan, 1960.
277. B. E. Cabana and P. K. Gessner, *Anal. Chem.* **39**, 1449 (1967).
278. A. E. Mayer, *Arzneimittel-Forsch.* **7**, 194 (1957).
279. N. Stanciv and U. Stoicescu, *Farmacia* (*Bucharest*) **4**, 313 (1956); *Chem. Abstr.* **52**, 8459f (1958).
280. J. Barlot and C. Albisson, *Chim. Anal.* (*Paris*) **38**, 313 (1956).

281. I. K. Tsitovich and E. A. Kuzmenko, *Zh. Anal. Khim.* **22**, 603 (1967); *Chem. Abstr.* **67**, 501565 (1967).
282. Y. R. Naves, *Bull. Soc. Chim. France* p. 505 (1951).
283. R. L. Webb, U.S. Patent 2,902,495 (1959); *Chem. Abstr.* **54**, P2404g (1960).
284. W. Kellner and W. Kober, *Arzneimittel-Forsch.* **6**, 768 (1956).
285. J. C. Maruzzella and E. Bramnick, *Soap, Perfumery Cosmetics* **34**, 743 (1961).
286. J. C. Maruzzella, J. S. Chiaramonte, and M. M. Carofalo, *J. Pharm. Sci.* **50**, 665 (1961).
287. T. Hirao and S. Ishikawa, *Nippon Sanshigaku Zasshi* **33**, 277 (1964); *Chem. Abstr.* **67**, 9109h (1967).
288. L. Kátó and B. Gozsy, *Arch. Intern. Pharmacodyn.* **117**, 52 (1958).
289. B. Hampel, *Z. Anal. Chem.* **170**, 56 (1959).
290. K. Hayashi and Y. Hashimoto, *Pharm. Bull. (Tokyo)* **5**, 74 (1957).
291. R. R. Paris and M. Godon, *Ann. Pharm. Franc.* **19**, 86 (1961).
292. A. H. M. Varelzakker and H. J. Van Zutphen, *Z. Lebensm.-Untersuch. -Forsch.* **115**, 222 (1961).
293. B. Starcher and G. P. Marletta, *Rass. Chim.* **19**, 99 (1967); *Chem. Abstr.* **67**, 93895s (1967).
294. E. von Rudloff, *Can. J. Chem.* **38**, 931 (1960).
295. E. Bayer, G. Kupfer, and K. H. Reuther, *Z. Anal. Chem.* **164**, 1 (1958).
296. E. A. Schuck and G. J. Doyle, *Air Pollution Found. (Los Angeles), Rept.* **29**, 110 (1959); *Chem. Abstr.* **54**, 4119g (1960).
297. Farbenfabriken Bayer Akt., British Patent 862,866 (1961); *Chem. Abstr.* **55**, 15823b (1961).
298. D. D. Gagliardi, *Am. Dyestuff Reptr.* **50**, 34 (1961).
299. M. Takigawa and I. Kanda, *Toho Reiyon Kenkyu Hokoku* **3**, 30 (1956); *Chem. Abstr.* **53**, 7602 (1959).
300. J. E. Fields and G. W. Zopf., Jr., U.S. Patent 2,291,843 (1960).
301. R. N. Lacey, *in* "Advances in Organic Chem. Methods Results" (R. A. Raphael, E. C. Taylor, and H. Wynberg, eds.), p. 213. Wiley (Interscience), New York, 1960.
302. J. R. Young, *J. Chem. Soc.* p. 2909 (1958).
303. G. B. Kistiakowsk and P. H. Kydd, *J. Am. Chem. Soc.* **79**, 4825 (1957).
304. W. G. Paterson and H. Gesser, *Can. J. Chem.* **35**, 1137 (1957).
305. H. A. Wooster, C. C. Lushbaugh, and C. E. Redemann, *J. Ind. Hyg. Toxicol.* **29**, 56 (1947).
306. J. F. Treon, H. E. Sigmon, K. V. Kitzmiller, F. F. Heyroth, W. J. Younker, and J. Cholak, *J. Ind. Hyg. Toxicol.* **31**, 209 (1949).
307. G. R. Cameron and A. Neuberger, *J. Pathol. Bacteriol.* **45**, 653 (1937).
308. R. M. Mendenhall, *Am. Ind. Hyg. Assoc. J.* **21**, 201 (1960).
309. A. J. Harrison and J. S. Lake, *J. Phys. Chem.* **63**, 1489 (1959).
310. I. A. Rapoport, *Dokl. Akad. Nauk SSSR* **58**, 119 (1947).
311. A. R. Kelly and F. W. Hartman, *Federation Proc.* **10**, 361 (1951).
312. A. R. Kelly and F. W. Hartman, *Federation Proc.* **11**, 419 (1952).
313. R. K. Hoffman, J. M. Buchanan, and D. R. Spiner, *Appl. Microbiol.* **14**, 989 (1966).
314. F. Dickens, *in* "Drugs with Lactone Groups as Potential Carcinogens" (R. Truhaut, ed.), UICC Monograph, Vol. 7, p. 144. Springer, Berlin, 1967.
315. F. Dickens and H. E. H. Jones, *Brit. J. Cancer* **15**, 85 (1961).
316. G. Mackell, *Ind. Chemist* **36**, 13 (1960).
317. R. K. Boutwell and N. H. Colburn, *Conf. Biol. Alkylating Agents, New York*, 1969, *N.Y. Acad. Sci. Ann.* **163**, 751 (1969).
318. N. H. Colburn and R. K. Boutwell, *Cancer Res.* **28**, 642 (1968).

319. W. C. J. Ross, *Ann. N. Y. Acad. Sci.* **68**, 669 (1958).
320. B. L. Van Duuren and B. M. Goldschmidt, *J. Med. Chem.* **9**, 77 (1966).
321. J. J. Roberts and G. P. Warwick, *Biochem. Pharmacol.* **6**, 217 (1961).
322. J. J. Roberts and G. P. Warwick, *Biochem. J.* **87**, 14P (1963).
323. R. K. Hoffman and B. Warshowsky, *Appl. Microbiol.* **6**, 358 (1958).
324. D. W. Fassett, *Ind. Hyg. Toxicol.* **4**, 1823 (1967).
325. B. L. Van Duuren, L. Langseth, L. Orris, M. Baden, and M. Kuschner, *J. Natl. Cancer Inst.* **39**, 1213 (1967).
326. E. D. Palmes, L. Orris, and N. Nelson, *J. Am. Ind. Hyg. Assoc.* **23**, 257 (1962).
327. F. Dickens, *Brit. Med. Bull.* **20**, 96 (1964).
328. F. Dickens, H. E. H. Jones, and H. B. Waynforth, *Brit. J. Cancer* **20**, 134 (1966).
329. F. Dickens and H. E. H. Jones, *Brit. J. Cancer* **19**, 392 (1965).
330. L. J. Lilly, *Nature* **207**, 433 (1965).
331. F. Mukai, S. Belman, W. Troll, and I. Hawryluk, *Proc. Am. Assoc. Cancer Res.* **8**, 49 (1967).
332. H. H. Smith and A. M. Srb, *Science* **114**, 490 (1951).
333. C. P. Swanson and T. Merz, *Science* **129**, 1364 (1959).
334. R. W. Kaplan, *Naturwissenschaften* **49**, 457 (1962).
335. H. H. Smith and T. A. Lofty, *Am. J. Botany* **42**, 750 (1955).
336. H. Roth, *Mikrochim. Acta* **6**, 766 (1958).
337. W. P. Tyler and D. W. Beesing, *Anal. Chem.* **24**, 1511 (1952).
338. H. K. Hall, Jr. and R. Zbinden, *J. Am. Chem. Soc.* **80**, 6428 (1958).
339. E. Honkanen, T. Moisio, and P. Karvonen, *Acta Chem. Scand.* **19**, 370 (1965).
340. W. H. McFadden, E. A. Day, and M. J. Diamond, *Anal. Chem.* **37**, 89 (1965).
341. E. M. Meade and F. N. Woodward, *J. Chem. Soc.* p. 1894 (1948).
342. G. Champetier and F. Hennequin-Lucas, *Compt. Rend.* **252**, 2785 (1961).
343. J. K. Stille and J. A. Empen, *J. Polymer Sci.* **15**, 273 (1967).
344. H. W. McKinney, U.S. Patent 2,962,457 (1960); *Chem. Abstr.* **55**, 6009 (1961).
345. H. Romeyn, Jr. and J. F. Petras, German Patent 925,314 (1955); *Chem. Abstr.* **52**, 2444a (1958).
346. U. S. Etlis, A. P. Sineokov, and G. A. Razundev, U.S.S.R. Patent 176,397 (1965); *Chem. Abstr.* **64**, 113976 (1966).
347. O. A. Zavada, A. D. Virnik, K. P. Khomyakov, and Z. A. Rogovin, *Khim. Prirodn. Soedin.* **2**, 437 (1966); *Chem. Abstr.* **67**, 12790d (1967).
348. J. R. Brown and E. Mastromatteo, *J. Am. Ind. Hyg. Assoc.* **25**, 560 (1964).
349. I. A. Rapoport, *Byul. Mosk. Obshchestva Ispytatelei Prirody, Otd. Biol.* **67**, 109 (1962).
350. H. W. Thompson and W. T. Cave, *Trans. Faraday Soc.* **47**, 951 (1951).
351. F. S. Mortimer, *J. Mol. Spectry.* **5**, 199 (1960).
352. A. E. Maciel and G. B. Savitsky, *J. Phys. Chem.* **69**, 3925 (1965).
353. L. B. Clark and W. T. Simpson, *J. Chem. Phys.* **43**, 3666 (1965).
354. D. D. Reynolds and D. L. Fields, *Chem. Heterocyclic Comp.* **19**, 576 (1964).
355. D. Kumar and R. Choudhury, *Makromol. Chem.* **25**, 217 (1958).
356. R. W. Kerr, U.S. Patent 2,858,305 (1958); *Chem. Abstr.* **53**, 4786i (1959).
357. I. T. Krohn, U.S. Patent 2,727,053 (1955); *Chem. Abstr.* **50**, 10761 (1956).
358. A. W. Francis, U.S. Patent 2,776,327 (1957); *Chem. Abstr.* **51**, 5403 (1957).
359. A. Schmidt, German Patent 911,659 (1954); *Chem. Abstr.* **52**, P10651g (1958).
360. W. E. Hanford and R. M. Joyce, Jr., U.S. Patent 2,478,390 (1949); *Chem. Abstr.* **44**, 1126 (1950).
361. C. M. Suter, "Organic Chemistry of Sulfur," p. 62. Wiley, New York, 1944.
362. C. Weber, *Arch. Exptl. Pathol. Pharmakol.* **17**, 113 (1902).

363. L. T. Fairhall, "Industrial Toxicology," 2nd ed. Williams & Williams, Baltimore, Maryland, 1957.
364. T. Alderson, *Nature* **203**, 1404 (1964).
365. M. Pelecanos and T. Alderson, *Mutation Res.* **1**, 173 (1964).
366. H. V. Malling, *Mutation Res.* **2**, 320 (1965).
367. G. Kølmark, *Compt. Rend. Trav. Lab. Carlsberg., Ser. Physiol.* **26**, 205 (1956); *Chem. Abstr.* **50**, 15715 (1956).
368. A. Loveless and W. C. J. Ross, *Nature* **166**, 1113 (1950).
369. P. F. Swann, *Biochem. J.* **110**, 49 (1968).
370. J. Epstein, R. W. Rosenthal, and P. J. Ess, *Anal. Chem.* **27**, 1435 (1955).
371. O. Klatt, A. C. Griffin, and J. S. Stehlin, Jr., *Proc. Soc. Exptl. Biol. Med.* **104**, 629 (1960).
372. L. H. Mayer, A. Saika, and H. S. Gutowsky, *J. Am. Chem. Soc.* **75**, 4567 (1953).
373. K. Wedemeyer and D. Delfs, U.S. Patent 2,770,567 (1956); *Chem. Abstr.* **51**, 9078d (1957).
374. Monsanto Chemical Co., British Patent 937,762 (1963); *Chem. Abstr.* **60**, 9844a (1964).
375. G. L. Deniston, German Patent 1,051,238 (1959); *Chem. Abstr.* **55**, 1020b (1961).
376. H. W. Coover, Jr., U.S. Patent 3,026,311 (1962); *Chem. Abstr.* **56**, 15678f (1962).
377. H. R. Guest, J. T. Adams, and B. W. Kiff, U.S. Patent 2,970,985 (1961); *Chem. Abstr.* **55**, 12940g (1961).
378. R. E. Heiner, C. F. Konzak, R. A. Nilan, and H. Bartels, *Nature* **194**, 788 (1962).
379. H. F. Smyth, Jr., C. P. Carpenter, and C. S. Weil, *J. Ind. Hyg. Toxicol.* **31**, 60 (1949).
380. I. A. Rapoport, *Dokl.—Biol. Sci. Sect. (English Transl.)* **141**, No. 6, 1476 (1961).
381. T. Alderson and M. Pelecanos, *Mutation Res.* **1**, 182 (1964).
382. M. Pelachos, *Nature* **210**, 1294 (1965).
383. T. Alderson, *Brit. Empire Cancer Campaign Ann. Rept.* p. 416 (1963).
384. B. Strauss and S. Okubo, *J. Bacteriol.* **79**, 464 (1960).
385. S. Zamenhof, G. Leidy, E. Hahn, and H. Alexander, *J. Bacteriol.* **72**, 1 (1956).
386. A. Loveless, *Proc. Roy. Soc.* **B150**, 497 (1959).
387. T. Alderson and A. M. Clark, *Nature* **210**, 593 (1966).
388. L. M. Monti, *Mutation Res.* **5**, 187 (1968).
389. F. D'Amoto, G. T. Scarascia, L. M. Monti, and A. Bozzini, *Radiation Botany* **2**, 217 (1962).
390. R. E. Heiner, C. F. Konzak, R. A. Nilan, and R. R. Legault, *Proc. Natl. Acad. Sci. U.S.* **46**, 1215 (1960).
391. H. Heslot, *Abhandl. Deut. Akad. Wiss. Berlin, Kl. Med.* **1**, 106 (1960); *Chem. Abstr.* **55**, 46541 (1961).
392. H. Heslot, *Abhandl. Deut. Akad. Wiss. Berlin, Kl. Med.* **1**, 98 (1960).
393. E. Schulek and J. Laszlovsky, *Acta Pharm. Hung.* **28**, 89 (1958); *Chem. Abstr.* **55**, 12269e (1961).
394. S. Detoni and D. Hadzi, *Spectrochim. Acta* Suppl., p. 601 (1957); *Chem. Abstr.* **54**, 4154 (1960).
395. N. Krishnamurthy and R. S. Krishnan, *Proc. Indian Acad. Sci.* **A57**, 352 (1963).
396. G. G. Gallo and L. Chiesa, *Farmaco (Pavia), Ed. Prat.* **18**, 206 (1963); *Chem. Abstr.* **59**, 8545 (1963).
397. J. C. Snyder and A. V. Grosse, U.S. Patent 2,493,038 (1950); *Chem. Abstr.* **44**, 4021h (1950).
398. N. K. Clapp, A. W. Craig, and R. E. Toya, Sr., *Science* **161**, 913 (1968).
399. J. W. Conklin, A. C. Upton, K. W. Christenberry, and T. P. McDonald, *Radiation Res.* **19**, 156 (1963).
400. A. C. Upton, *Natl. Cancer Inst. Monograph* **22**, 329 (1966).

401. A. J. Bateman, R. L. Peters, J. G. Hazen, and J. L. Steinfeld, *Cancer Chemotherapy Rept.* **50**, 675 (1966).
402. J. B. Kissam, J. A. Wilson, and J. B. Hays, *J. Econ. Entomol.* **60**, 1130 (1967).
403. H. Jackson, *Agents Affecting Fertility* p. 62 (1965).
404. H. Jackson, B. W. Fox, and A. W. Craig, *J. Reprod. Fertility* **2**, 447 (1961).
405. E. Hartman and R. E. Robertson, *Can. J. Chem.* **38**, 2033 (1960).
406. D. J. Pillinger, B.W. Fox, and A. W. Craig, *in* "Isotopes in Experimental Pharmacology" (L. J. Roth, ed.), p. 415. Univ. of Chicago Press, Chicago, Illinois, 1965.
407. E. A. Barnsley, *Biochem. J.* **106**, 18P (1968).
408. P. D. Lawley, *Nature* **218**, 580 (1968).
409. P. N. Magee, *in* "Molekulare Biologie des Malignen Wachstums" (H. Holzer and A. W. Holldorf, eds.), p. 79. Springer, Berlin, 1966.
410. E. Bautz and E. Freese, *Proc. Natl. Acad. Sci. U.S.* **46**, 1585 (1960).
411. N. Loprieno, *Mutation Res.* **3**, 486 (1966).
412. A. T. Natarjan and M. C. Ramanna, *Nature* **211**, 1099 (1966).
413. R. N. Rao and A. T. Natarajan, *Mutation Res.* **2**, 132 (1965).
414. M. Partington and H. Jackson, *Genet. Res.* **4**, 333 (1963).
415. M. Partington and A. J. Bateman, *Heredity* **19**, 191 (1964).
416. S. S. Epstein, *Toxicol. Appl. Pharmacol.* **14**, 652 (1969).
417. S. S. Epstein and H. Shafner, *Nature* **219**, 385 (1968).
418. U. H. Ehling, R. B. Cumming, and H. V. Malling, *Mutation Res.* **5**, 417 (1968).
419. E. Glaess and H. Marquardt, *Z. Vererbungslehre* **98**, 167 (1966); *Chem. Abstr.* **66**, 26660c (1967).
420. D. E. Freeman and A. N. Hambly, *Australian J. Chem.* **10**, 239 (1957).
421. A. Simon, H. Kiregsmann, and H. Dutz, *Chem. Ber.* **89**, 2378 (1956).
422. C. E. Johnson and A. P. Lieu, U.S. Patent 2,665,293 (1954); *Chem. Abstr.* **48**, 12790 (1954).
423. G. Nordmann, R. Thren, and G. Baerwald, East German patent 53,273 (1967); *Chem. Abstr.* **67**, 10385b (1967).
424. Sh. I. Ibragimov and R. I. Koval'Chuk, *Tr. Mosk. Obschestva Ispytatelei Prirody* **23**, 179 (1966); *Chem. Abstr.* **67**, 72670z (1967).
425. O. G. Fahmy and M. J. Fahmy, *Nature* **180**, 31 (1957).
426. J. L. Epler, *Genetics* **54**, 31 (1966).
427. J. K. Lim and L. A. Snyder, *Mutation Res.* **6**, 129 (1968).
428. J. B. Jenkins, *Mutation Res.* **4**, 90 (1967).
429. I. A. Rapoport, *Dokl. Vses. Akad. Sel'skoKhoz. Nauk* **12**, 12 (1947).
430. O. G. Fahmy and M. J. Fahmy, *Genetics* **46**, 447 (1961).
431. O. G. Fahmy and M. J. Fahmy, *Genetics* **46**, 1111 (1961).
432. E. A. Löbbecke and R. C. von Borstel, *Genetics* **47**, 853 (1962).
433. N. M. Schwartz, *Genetics* **48**, 1357 (1963).
434. B. S. Strauss, *Nature* **191**, 730 (1961).
435. W. G. Verly, H. Barbason, J. Dusart, and A. Petispas-Dewandre, *Biochim. Biophys. Acta* **145**, 752 (1967).
436. J. Corban, *Mol. Gen. Genet.* **103**, 42 (1968).
437. J. Nečašek, P. Pikalek, and J. Drobnik, *Mutation Res.* **4**, 409 (1967).
438. M. Osborn, S. Person, S. Philips, and F. Funk, *J. Mol. Biol.* **26**, 437 (1967).
439. D. R. Krieg, *Genetics* **48**, 561 (1963).
440. A. Loveless, *Nature* **181**, 1212 (1958).
441. H. V. Malling and F. J. DeSerres, *Mutation Res.* **6**, 181 (1968).
442. M. Westergaard, *Experientia* **13**, 224 (1957).

443. A. Nasim and C. Auerbach, *Mutation Res.* **4**, 1 (1967).
444. A. Nasim, *Mutation Res.* **4**, 753 (1967).
445. H. Heslot, *Abhandl. Deut. Akad. Wiss. Berlin. Kl. Med.* **2**, 193 (1961).
446. G. Lindegren, Y. L. Hwang, Y. Oshima, and C. C. Lindegren, *Can. J. Genet. Cytol.* **7**, 491 (1965).
447. F. Lingens and O. Oltmans, *Z. Naturforsch.* **19b**, 1058 (1964).
448. R. A. Nilan, "The Cytology and Genetics of Barley, 1951–1952." Washington State Univ. Press, Pullman, Washington, 1964.
449. E. A. Favret, *Hereditas* **46**, 622 (1960).
450. H. Heslot, R. Ferrary, R. Levy, and C. Monard, *Compt. Rend.* **248**, 729 (1959).
451. L. A. Ehrenberg, A. Austafsson, and U. Lundquist, *Hereditas* **47**, 243 (1961).
452. L. Ehrenberg, *Abhandl. Deut. Akad. Wiss. Berlin, Kl. Med.* **1**, 124 (1960).
453. H. Gaul, *Naturwissenschaften* **49**, 1 (1962).
454. C. F. Konzak, R. A. Nilan, J. R. Harle, and R. E. Heiner, *Brookhaven Symp. Biol.* **14**, 128 (1961).
455. L. Ehrenberg, U. Lundquist, S. Osterman, and B. Sparrman, *Hereditas* **56**, 277 (1966).
456. M. G. Neuffer and G. Ficsor, *Science* **139**, 1296 (1963).
457. E. Amano and H. H. Smith, *Mutation Res.* **2**, 344 (1965).
458. H. K. Shama Rao and E. R. Sears, *Mutation Res.* **1**, 387 (1964).
459. J. Mackey, *Wheat Inform. Serv.* **14**, 9 (1962).
460. M. R. Ghatnekar, *Caryologia* **17**, 219 (1964).
461. A. Kleinhofs, H. J. Gorz, and F. A. Hoskins, *Crop. Sci.* **8**, 631 (1968).
462. C. R. Bhatia, *Mutation Res.* **4**, 375 (1967).
463. G. Röbbelen, *Naturwissenschaften* **49**, 65 (1962).
464. A. D. McKelvie, *Radiation Botany* **3**, 105 (1963).
465. H. Dulieu, *Mutation Res.* **4**, 177 (1967).
466. T. J. Arnason, L. M. El-Sadek, and J. L. Mirocha, *Can. J. Genet. Cytol.* **8**, 746 (1966).
467. J. Gilot, M. Moutschen-Dahmen, and J. Moutschen-Dahmen, *Experientia* **23**, 673 (1967).
468. E. H. Y. Chu and H. V. Malling, *Proc. Natl. Acad. Sci. U.S.* **61**, 1306 (1968).
469. M. S. Swaminathan, V. L. Chopra, and S. Bhaskaran, *Indian J. Genet.* **22**, 192 (1962).
470. A. T. Natarajan and M. D. Upadhya, *Chromosoma* **15**, 156 (1964).
471. J. Moutschen-Dahmen, A. Moes, and J. Gilot, *Experientia* **20**, 494 (1964).
472. V. V. Shevchenko, *Genetika* **4**, 24 (1968).
473. T. H. Chang and F. T. Elequin, *Mutation Res.* **4**, 83 (1967).
474. B. M. Cattanach, C. E. Pollard, and J. H. Isaacson, *Mutation Res.* **6**, 297 (1968).
475. M. A. Walters, F. J. C. Roe, B. C. V. Mitchley, and A. Walsh, *Brit. J. Cancer* **21**, 367 (1967).
476. J. B. Kissam and S. B. Hays, *J. Econ. Entomol.* **59**, 748 (1966).
477. A. Haddow and G. M. Timmis, *Lancet* **1**, 207 (1953).
478. V. A. Chernov, *Med. Prom. SSSR* **12**, 60 (1958).
479. D. A. G. Galton, *Advan. Cancer Res.* **4**, 73 (1956).
480. S. Ishikawa, H. Yokotani, F. Watanabe, Y. Aramaki, and K. Kajiwara, *Takeda Kenkyusho Nempo* **16**, 64 (1957).
481. A. I. Tareeva and A. I. Yakovleva, *Khim. i. Med.* **13**, 34 (1960); *Chem. Abstr.* **55**, 10679a (1961).
482. G. Koch, *Bull. Soc. Chim. Belges* **68**, 59 (1959).
483. G. Koch, W. G. Verly, and Z. M. Bacq, *Arch. Intern. Physiol.* **67**, 117 (1959).
484. E. G. Trams, R. A. Salvador, G. Maengwyn-Davies, and V. DeQuattro, *Proc. Am. Assoc. Cancer Res.* **2**, 256 (1957).

485. B. W. Fox, A. W. Craig, and H. Jackson, *Biochem. Pharmacol.* **5**, 27 (1960).
486. J. J. Roberts and G. P. Warwick, *Nature* **183**, 1509 (1959).
487. J. J. Roberts and G. P. Warwick, *Biochem. Pharmacol.* **1**, 60 (1958).
488. L. A. Elson, *Biochem. Pharmacol.* **1**, 39 (1958).
489. E. G. Trams, M. V. Nadkarni, V. DeQuattro, G. D. Maengwyn-Davies, and P. K. Smith, *Biochem. Pharmacol.* **2**, 7 (1959).
490. M. V. Nadkarni, E. G. Trams, and P. K. Smith, *Cancer Res.* **19**, 713 (1959).
491. M. V. Nadkarni, E. G. Trams, and P. K. Smith, *Proc. Am. Assoc. Cancer Rev.* **2**, 235 (1957).
492. J. J. Roberts and G. P. Warwick, *Nature* **184**, 1288 (1959).
493. H. Jackson, A. W. Craig, and B. W. Fox, *Acta, Unio. Intern. Contra Cancrum* **16**, 611 (1964).
494. I. Diamond, M. A. Anderson, and S. R. McCreadie, *Pediatrics* **25**, 85 (1960).
495. G. Röhrborn, *Z. Vererbungslehre* **90**, 116 (1959).
496. G. Röhrborn, *Z. Vererbungslehre* **90**, 457 (1959).
497. O. G. Fahmy and M. J. Fahmy, *Nature* **177**, 996 (1956).
498. A. T. Natarajan and M. C. Ramanna, *Nature* **211**, 1102 (1966).
499. W. G. Verly, A. Dewandre, J. Moutschen-Dahmen, and M. Moutschen-Dahmen, *J. Mol. Biol.* **6**, 175 (1963).
500. W. G. Verly, *Rev. Franc. Etudes Clin. Biol.* **9**, 878 (1964).
501. E. Gebhart, *Humangenetik* **7** (1970) (in press).
502. E. Gebhart, *Mutation Res.* **7**, 254 (1969).
503. A. T. Natarajan and M. S. Ramanna, *Nature* **211**, 100 (1966).
504. A. Michaelis and R. Rieger, *Zuechter* **30**, 150 (1960).
505. R. Rieger and A. Michaelis, *Kulturpflanze* **8**, 230 (1960).
506. J. Moutschen and M. Moutschen-Dahmen, *Hereditas* **44**, 415 (1958).
507. J. Moutschen and M. Moutschen-Dahmen, *Experientia* **15**, 320 (1959).
508. R. Truhaut, E. Delacoux, G. Brule, and C. Bohuon, *Clin. Chim. Acta* **8**, 235 (1963).
509. A. K. Ruzhentseva and I. A. Starostina, *Khim. Farm. Inst. im Ordzhon. Kidzke* **3**, 31 (1960); *Chem. Abstr.* **55**, 10797 (1961).
510. C. T. Peng, *J. Pharmacol. Exptl. Therap.* **120**, 229 (1957).
511. P. W. Feit, *J. Med. Chem.* **7**, 14 (1964).
512. E. R. White, *Cancer Chemotherapy Rept.* **24**, 95 (1962).
513. R. Jones, W. B. Kesler, H. E. Lessner, and L. Rare, *Cancer Chemotherapy Rept.* **10**, 99 (1960).
514. R. Matagne, *Mutation Res.* **4**, 621 (1967).
515. W. Davis and W. C. J. Ross, *Biochem. Pharmacol.* **12**, 915 (1963).
516. R. Matagne, *Mutation Res.* **7**, 241 (1969).
517. J. Moutschen, R. Matagne, and J. Gilot, *Nature* **210**, 762 (1966).
518. J. Moutschen, R. Matagne, and J. Gilot, *Bull. Soc. Roy. Botan. Belg.* **100**, 11 (1967).
519. J. Moutschen and M. Reekmans, *Caryologia* **17**, 495 (1964).
520. J. Moutschen, *Cellule* **65**, 163 (1965).

CHAPTER 9

Drugs, Food Additives, Pesticides, and Miscellaneous Mutagens

1. *Acridines*

The preparation, physical, chemical (1), and biological properties (2–6), and chemotherapeutic utility (7) of the acridines have been reviewed. The acridine dyes generally available for medicinal use are proflavine (3,6-diaminoacridine) and acriflavine (a mixture of 3,6-diamino-10-methylacridinium chloride and 3,6-diaminoacridine). Acridine derivatives exert a bactericidal and bacteriostatic action (8, 9) upon a variety of organisms, with the greatest activity shown against gram-positive organisms in an alkaline medium. The acridine dyes have been used primarily as topical antiseptics in concentrations ranging from 1:10,000 to 1:1000. For veterinary use, acridine dyes are occasionally administered IV in acute mastitis and septicemic infections. Acridines are actively bacteriostatic only in their ionized form; the ionization must be of a cationic nature, and the degree of ionization must be greater than 50% at pH 7.3. Their mode of action is further described elsewhere.

a. Proflavine. The induction of mutations in phage T2 (10) and in bacteriophage T4 by the photodynamic action of proflavine (11–14), as well as related dyes (e.g., acridine orange, acriflavine, acridine yellow, and 9-aminoacridine) (13), has been described. Proflavine has also been reported to induce chromosome breakage in human cells (15).

b. Acriflavine. The mutagenicity of acriflavine in *Drosophila* (16), *E. coli* (16, 17), yeast (18, 19), phage CHI of *Serratia marscescens* (20), its enhancement of both the lethal and mutagenic effects of ultraviolet radiation, and prevention of host cell reactivation in bacteria (21) have been reported.

c. Acridine Orange. The influence of acridine orange and acridine yellow on the initiation of skin carcinogenesis (22), the binding of acridine orange to proteins (23–25), nucleic acid (6), DNA (13, 26), polysaccharides (27, 28), its mutagenicity in *E. coli* (13, 29), *Allium cepa* (30), as well as its antimutagenic effect in *E. coli* (29) have been described.

d. 5-Aminoacridine. 5-Aminoacridine is mutagenic in *Drosophila* (31), phage T4 (13), and *Saccharomyces cerevisiae* (32) [during meiosis (33), and antimutagenic during mitosis (33)].

e. Acridine Mustard. Acridine mustard {2-methoxy-6-chloro-9-[3-(ethyl-2-chloroethyl)aminopropylamino]acridine dihydrochloride} (ICR-170) was prepared by Peck *et al.* (34, 35), and has been shown to be a potent antineoplastic agent for ascites tumor in mice (36). By comparative studies it has been found that both the acridine ring and the nitrogen mustard side chain (the alkylating part of the ICR-170 molecule) are essential for its antineoplastic activity.

ICR-170 has been found to be mutagenic in *Drosophila* (37–41), *Salmonella* (42), and *Neurospora* (43–45), and was found to induce chromosome aberrations in *Vicia faba* (46).

The analysis of the acridines has been performed by spectroscopy (47), paper (47), column (48), partition, and ion-exchange chromatography (48).

2. *Antibiotics*

a. Mitomycin C. Mitomycin C is an antibiotic first isolated by Horta *et al.* (49) from the broth of *Streptomyces caespitosus*. It is distinguished from other mitomycin fractions by its thermal stability, high melting point, ultraviolet absorption peak, and solubility in organic solvents.

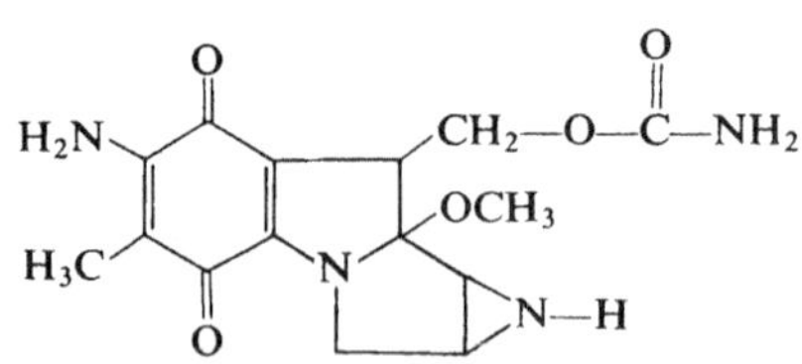

Mitomycin C

Although mitomycin C has strong activity against bacteria and viruses, its primary medical use is as an antineoplastic agent in the treatment of Hodgkin's disease. The drug has also produced responses in experimental animals in a wide range of solid tumors such as lung, breast, colon, stomach, pancreas, and osteogenic sarcoma (50). However, mitomycin C disappears rapidly from the plasma and there is no evidence of specific tissue localization (51). Mitomycin C also possesses a broad range of activity against experimental tumors, including Ehrlich ascites, tumor, Ridgway and Wagner osteogenic sarcomas, Friend virus leukemia, and Glioma 26.

Mitomycin C can be considered as a derivative of urethan and of ethylenimine. It is biologically inactive in its natural state, but it becomes a mono- and bifunctional alkylating agent upon chemical or enzymic reduction.

The isolation, chemistry, and elaboration of structure (52, 53) of mitomycin C, its pharmacology and toxicology (54), clinical trials (50, 55), mode of action (56–58), and cellular effects (59, 60) have been described.

Mitomycin C is mutagenic in *Drosophila* (61, 62) and Habrobracon sperm (63), and induces chromosome aberrations in cultured human leukocytes (64–66) and *Vicia faba* (67).

b. Streptonigrin. Streptonigrin is isolated from culture filtrates of *Streptomyces floccules* (68). Salient features of its isolation and characterization (68), structure (69), antitumor spectrum (70–72), chemotherapy in cancer (73–76), as well as its analysis by ultraviolet and infrared spectroscopy (68) and paper chromatography (68) have been described.

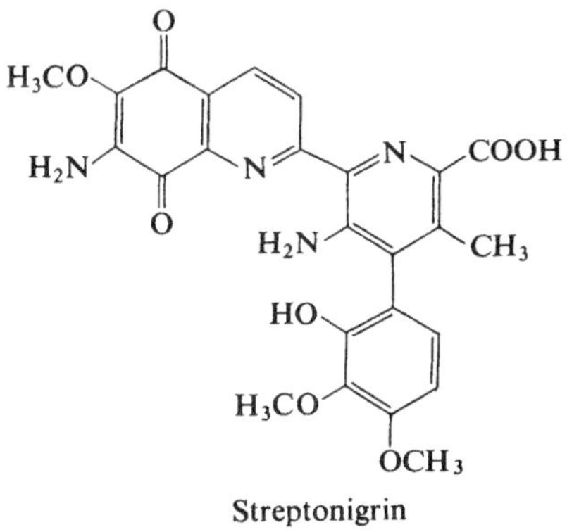

Streptonigrin

It is unstable if brought for 15 minutes into a pH range of 2.0 to 7.0; beyond pH 8.0 substantial decomposition occurs. Solutions of streptonigrin are photosensitive, especially at pH 7.0 and higher.

Streptonigrin has been shown to be an extremely potent and selective inhibitor of bacterial DNA synthesis (77); its reactions with DNA (78, 79) and comparison of cytological effects with mitomycin C have been described (80). It has been shown to be mutagenic in the ascomycete *Ophiostoma multiannulatum* (81) and to induce chromosome aberrations in cultured human leukocytes (82, 83) and in *Vicia faba* (80, 84).

c. Streptozotocin. Streptozotocin is a broad spectrum antibiotic isolated from *Streptomyces achromogenes.* It is active against a variety of gram-positive and gram-negative organisms both *in vitro* and *in vivo*, and is not cross-resistant with ten commercially available antibiotics.

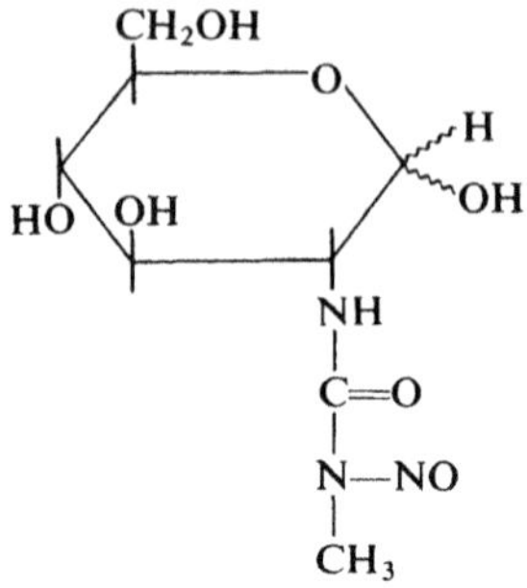

Streptozotocin

Its isolation, structure, and chemistry (85, 86), as well as its assay, stability, and bacterial studies have been described (87–89).

At room temperature, fully active samples are present after 30 days, while at 4°C streptozotocin is stable for periods of up to 6 months. Streptozotocin exhibits maximum stability at pH 4, with stability decreasing rapidly at either higher or lower pH.

The phage-inducing capacity (90, 91), its antitumor activity (against Ehrlich carcinoma, Sarcoma 51784, and Walker 256 in the mouse and rat) (92), and its diabetogenic action (92, 93) have all been described. Streptozotocin has been shown to be a mutagen for the histidine loci in several different classes of *Salmonella typhimurium* mutants (94).

The possibility that streptozotocin acts as an alkylating agent via the formation of diazomethane is strongly supported by the presence of the *N*-nitrosomethylamide portion of its structure (treatment of the antibiotic with alkali, 2 *N* aqueous NaOH, at 0°C yields diazomethane). Also, the action spectrum of streptozotocin in terms of mutant classes affected and the degree of this response are similar to that of nitrosoguanidine. The microbial mutagenicity of streptozotocin in animal-mediated assays has most recently been demonstrated by Legator *et al.* (95, 96). In the host-mediated system, however, streptozotocin behaves quite differently from nitrosoguanidine.

The increase in mutant frequency obtained by the host-mediated system is detected at lower concentrations, increases more rapidly with increasing concentration, and persists longer with streptozotocin than with nitrosoguanidine. The former is also more toxic for microorganisms in concentrations which elicit a mutagenic response. It has also been shown that with streptozotocin, microbial mutagenic activity is present in blood and urine and is detectable soon after injection. Nitrosoguanidine and dimethyl nitrosamine are negative in a blood-plate test.

The analysis of streptozotocin has been accomplished by polarography (97), spectrophotometry (98) (based on a reaction of the *N*-nitroso group), and paper chromatography (99).

d. Patulin. Patulin [4-hydroxy-4*H*-furo[3,2-*c*]pyran-2(6*H*)one] is an α,β-unsaturated lactone antibiotic derived from the metabolism of several species of *Aspergillus* and *Penicillium* (e.g., *A. clavatus, A. claviforme, P. patulum, P. expansum, P. meliniis, P. Leucopus*). Its isolation (100–103), synthesis (104), and biosynthesis (105) have been described.

Some of these fungal species are likely contaminants of foods. For example, *P. expansum*, the common storage rot of fruit, *A. clavatus, A. terreus, P. cyclopium*, and *P. urticae*, isolated from flour (106), and *Byssochlamys nivea*, and the heat-resistant fruit juice contaminant identified by Keuhn (107) as the Gymnoascus species of Karow and Foster (108) have all been shown to produce patulin (109, 110). Patulin is also produced by fungi in apples (111) and by field crops (103), and has been isolated by Ukai *et al.* (112) from a *Penicillium* which infected a malt feed responsible for the death of cows.

The potential health hazard of the presence of patulin in foods or animal feeds has been stressed by Kraybill and Shimkin (113), and Mayer and Legator (114). The stability of patulin in fruit juices and flour was investigated by Scott and Somers (115). Patulin has been found to be stable for several weeks over the pH range of 3.3 to 6.3 and was slowly inactivated at pH 6.8 (116), and stable in grape and apple juice but not in orange juice or the flours [partially due to reaction with thiols (117)]. The reaction of patulin with thiols at a pH as low as 4.5 has been reported previously (118). It is decomposed by glutathione at pH's as low as 2.3 and 3.0. It has been shown that if high concentrations of patulin are initially present in fruit juices of low thiol content, appreciable concentrations may remain in the processed juice.

The toxicity (109) and carcinogenicity (119) as well as the inhibition of cell division, nuclear division, or both in bacteria (120), plants (121), and in cell culture (122) of patulin have been described. It has been recently shown to induce petite mutations in *Saccharomyces cerevisiae* (114) and chromosome aberrations in avian eggs during mitosis (123) and in human leukocyte cell culture (124).

The isolation and detection of patulin has been accomplished by paper and thin-layer chromatographic techniques (103).

e. Phleomycin. Phleomycin, a water-soluble, copper-containing antibiotic complex, was isolated from the culture filtrate of *Streptomyces verticillus* (125). It is unstable in acidic solution and stable in neutral and weakly alkaline solutions. It possesses antitumor activity against mouse Sarcoma 180, Adenocarcinoma 755, and Ehrlich ascites carcinoma (126–128). Details of the isolation of phleomycin (125), its antitumor properties (126–129), toxicity (125, 127), pharmacology (130), fungicidal (131) and bactericidal properties (125, 127), as well as mechanism of action (132–135) have been described.

Phleomycin has been shown to selectively inhibit DNA synthesis in *E. coli* (136) and HeLa cells (136), as well as to induce chromosome aberrations in cultured human leukocytes (137), mouse ovum (138), HeLa cells (139, 140), and *Vicia faba* (141, 142).

f. Azaserine. Azaserine (diazoacetyl-L-serine) is an antibiotic isolated from the broth culture filtrate of a strain of *Streptomyces* by Bartz *et al.* (143). Its synthesis (144), pharmacology and toxicity (145), tumor inhibition (146), metabolism (147, 148), and teratogenesis in rat (149, 150) and chick embryo (151), and its amebicidal (152) and bacterial action (153) and analysis (154) have been described.

Azaserine is a glutamine antagonist which inhibits purine synthesis in some bacteria and causes accumulation of formyl glycinamide ribotide in *E. coli* (155, 156).

Azaserine has been shown to be mutagenic in *E. coli* (17, 57, 157–159) and T2 phage (160), and to induce chromosomal aberrations in root tips of *Allium cepa* (161), *Vicia faba* (80, 162), and *Tradescantia patudosa* (163), and to inhibit larval development in *Musca domestica* (164).

g. Daunomycin. Daunomycin is a new antibiotic with antitumor activity which has been isolated from *Streptomyces peucetius.* [An identical antibiotic, rubidomycin, has been isolated from cultures of *Streptomyces coeruleorubidus* (165).] It has structural similarities with a number of anthracycline antibiotics, e.g., rhodomycin, cinerubins, pyrromycins, and rhutilantins (166–168).

Daunomycin is a glycoside with an aglycon chromophore (daunomycinone) linked to an amino sugar (daunosamine) (169, 170).

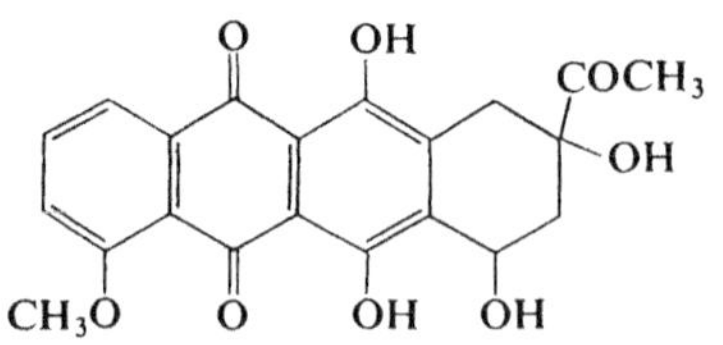

Daunomycinone

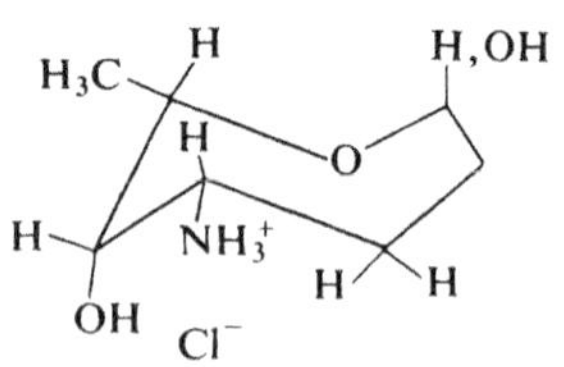

Daunosamine

The aglycon is a chromophore of a polyoxyanthraquinone type.

Daunomycin has recently been shown to be effective in acute leukemias and in certain solid tumors in children (171–173), and in addition has exhibited a marked inhibitory effect in a variety of animal transplantable tumors in solid or ascites forms (e.g., Ehrlich carcinoma, Sarcoma 180, Walker carcinoma, OGG myeloma, and MC sarcoma) and has increased the average survival time of the tumor-bearing animals (174).

The isolation (175), structure (169, 170), mode of action (56), mitotic activity (176–178), binding of DNA *in vivo* and *in vitro* (56), and inhibition of both DNA and RNA synthesis (179) have been described.

Daunomycin inhibits the cell growth, decreases mitotic index, and produces strong chromosomal aberrations on normal and tumor cells growing *in vitro* (resting cells show nuclear and nucleolar alterations) (176, 179–182).

3. *Ethidium Bromide*

Ethidium bromide (homidium bromide; 2,7-diamino-9-phenyl-10-ethylphenathridium bromide), synthesized by Watkins (183), is an intercalating dye with trypanocidal utility against *Trypanosoma congolese* and *T. vivax* infections in cattle.

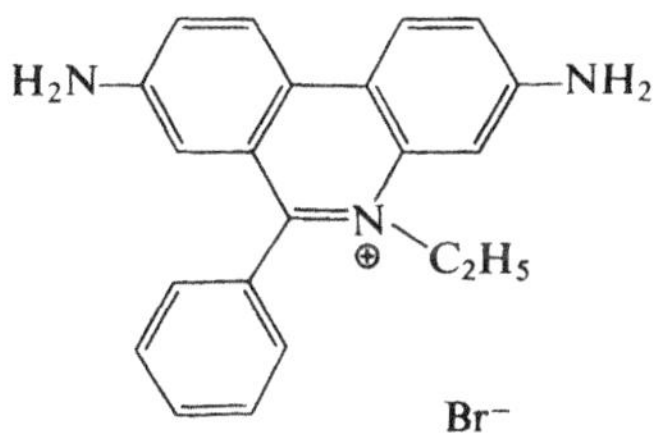

Ethidium bromide

Ethidium bromide inhibits the synthesis of nucleic acid by living organisms (184, 185) and also by cell-free enzyme systems (186, 187). It is probable that the effects of ethidium bromide on nucleic acid synthesis *in vivo* and *in vitro* are direct consequences of its physical binding to DNA and RNA (188, 189).

The induction of mutations of yeast mitochondria by ethidium bromide has been reported by Slonimski *et al.* (190).

4. *8-Hydroxyquinoline (8-Quinolinol)*

8-Hydroxyquinoline (oxine) is prepared by the Skraup reaction (heating *o*-aminophenol with *o*-nitrophenol, glycerol, and sulfuric acid). It has been used for over 60 years as an antibacterial agent (191) and fungistat (192–194). The bactericidal activity of oxine and its derivatives has been attributed to their ability to chelate trace metals which are essential to the cell growth of the microorganism. Oxine and its derivatives have additional utility as seed protectants against soil fungi (195), in insecticide emulsion concentrates (196), as preservatives for organic coatings and mildew inhibition (197, 198), and in the prevention of egg spoilage (199).

The properties of oxine and its derivatives have been reviewed by Hollingshead (200), and its carcinogenicity (201, 202) and mutagenicity in *Aspergillus oryzae* (203) and *Aspergillus niger* (204) and its chromosome aberrations in *Bromus inermis* (205) root tips have been described.

Gebhart studied the action of 8-hydroxyquinoline on human chromosomes *in vitro* and found that it causes gaps and fragmentations of chromatids and chromosomes. It has long been known that 8-hydroxyquinoline forms chelates with a variety of metal ions, and it was suggested by Gebhart (206) that this could thus indicate how the agent affects chromosomes.

The analysis of oxine has been achieved via potentiometric titration (207), polarography (208), infrared (209) and electron spin resonance spectroscopy (210), and paper chromatography (211–213).

5. *Dithranol*

Dithranol (1,8,9-trihydroxyanthracene) is prepared from chrysazin (1,8-dihydroxyanthraquinone) by reduction with hydrogen and a nickel catalyst (214). Dithranol is a parasiticide that is used in the treatment of psoriasis (215), ringworm infections, and other chronic dermatoses. The metabolic interrelations between dithranol and 1,8-dihydroxyanthraquinone have been described by Ippen and Montag (216). Following oral administration of 1,8-dihydroxyanthraquinone in humans, dithranol was identified in the feces. Conversely, after external application of dithranol, 1,8-dihydroxyanthraquinone was found in the urine, indicating the probable existence of an oxidation-reduction system of these two compounds in the human body.

The mechanism of action and cytostatic effect of dithranol in psoriasis has been proposed by Krebs and Schaltegger (217) and Swanbeck (218–220). In the latter *in vitro* study it was shown spectroscopically and by viscometry that dithranol interacts with DNA forming a complex with ~1 dithranol molecule per 8 base pairs of DNA, and that dithranol-induced inhibition of DNA synthesis may be due to the inability of the DNA–dithranol complex to function as a template for *de novo* synthesis of DNA.

Dithranol has been shown to have a tumor-promoting property in mice following pretreatment with 7,12-dimethylbenzanthracene (DMBA) (221). This is of interest because this agent has been used in dermatological practice often after earlier treatment with coal tar. Gillberg *et al.* (222) have recently shown that dithranol induces petite mutants in yeast *Saccharomyces cerevisiae*.

Dithranol has been analyzed by ultraviolet spectroscopy (223) and colorimetry (224).

6. *Miracil D*

Miracil D[1-(2-diethylaminoethylamino)-4-methylthiaxanthone; Lucanthone] is prepared by the reaction of 1-chloro-4-methylthiaxanthone with 2-diethylaminoethylamine (225, 226).

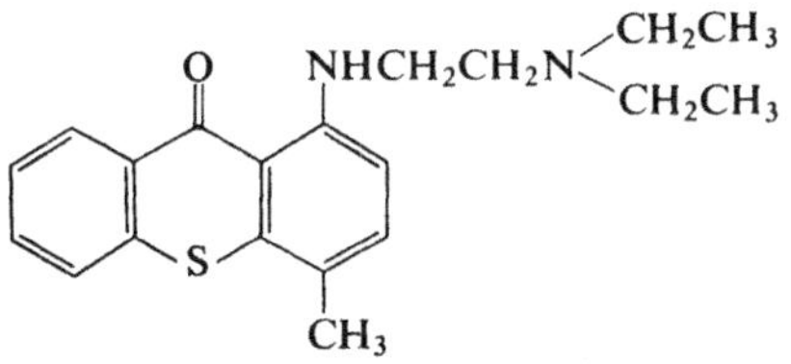

Miracil D

Miracil D (a 10-thiaxanthenone) shares certain structural features with actinomycin D and the acridines. It has carcinostatic potency against a variety of transplantable neoplasms in rodents (227, 228) and is an effective agent in the treatment of schistosomiasis in man (229). In *E. coli* (230) miracil D stops growth, causes complete inhibition of RNA synthesis and partial inhibition of DNA and protein synthesis, and blocks the induction of β-galactosidase. In a subcellular system miracil D inhibits DNA-directed RNA polymerase (230) and in mouse leukemia L1210 the inhibition of DNA synthesis is more pronounced than that of RNA or protein synthesis (231). It has been found to be mutagenic in *Drosophila* (232) and to induce achromatic lesions and chromatid breaks in human leukocyte chromosomes *in vitro* (233).

7. *Steroid Diamines*

The isolation and characterization (234) of the alkaloidal malouetine from *Malouetia bequaertiana* E. Woodson of the Apocynaceae family of plants (found primarily in tropic areas), the elaboration of its curarizing properties (235), as well as its partial synthesis from the alkaloid funtumafrine B (236) have been described.

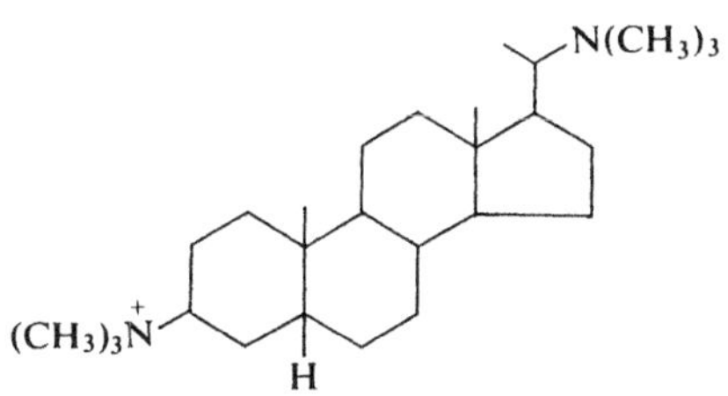

Malouetine

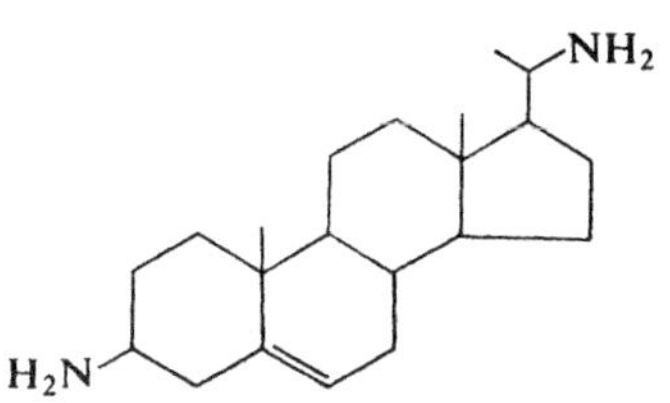

Irehdiamine

The steroid diamines, malouetine [5α-pregnan-3β,20α-ylenebis(trimethylammonium iodide); Mal] and irehdiamine (pregn-5-ene-3β,20α-diamine), have been demonstrated by Mahler and co-workers to exhibit strong chemical interactions with nucleic acids (237, 238). Both malouetine and irehdiamine have been shown by Mahler and Baylor (239) to be effective inhibitors of phage replication (T2H and T4D) and mutagens, and it was postulated that DNA was the primary target of diamine action *in vivo*.

8. *Nitrofurazone*

Nitrofurazone (5-nitro-2-furaldehyde semicarbazone) is prepared from 2-formyl-5-nitrofuran, and is used as a local antibacterial agent in veterinary medicine for the treatment of surface bacterial infections of the skin, mastitis, and swine enteritis. It is also used as a feed additive in the United States in doses of 0.0055% to prevent avian coccidiosis and as a food preservative in Japan.

Tests on nitrofuran in solution and during the storage and processing of foods showed that it is degraded principally by reduction with thiol-containing substances (240). Destruction by cysteine was enhanced by increases in pH and temperature and by the presence of ferrous ions.

The chemistry and mode of action of nitrofurazone (241, 242), effect on tumor growth (243, 244), carcinogenic activity (245), effect on DNA synthesis and phage induction (246), cross-resistance studies with *E. coli* (247, 248), and its mutagenicity (249) and nonmutagenicity in *E. coli* (157) have been described.

9. *Lysergic Acid Diethylamide*

Lysergic acid diethylamide (LSD) was first prepared by Hoffman (250) in 1938, who accidentally noticed its hallucinogenic effect in 1943. It is the most potent of the presently known hallucinogens (ingestion of 100 μg or even less is sufficient to cause symptoms of the same intensity as elicited by 500 mg of mescaline).

Three of the hallucinogens named in the Federal Register as "abuse drugs" are ergot alkaloids, namely, LSD, lysergic acid, and ergine (lysergic acid amide). Ergine is present in very small quantities in ergot, and is the hallucinogenic substance of morning-glory seeds. Lysergic acid, which is nonhallucinogenic, is obtained by the alkaline hydrolysis of crude ergot. Its abuse potential exists because it is the necessary starting material in the synthesis of LSD and other related derivatives.

Of the four possible isomers of lysergic acid diethylamide (i.e., *d*-, *d*-iso-, *l*-, and *l*-iso-) only *N*,*N*-diethyl-*d*-lysergamide is reported to have significant hallucinogenic effect (251).

The absorption, distribution, and elimination of LSD has been extensively studied (252–260).

The final distribution of LSD in animals is independent of the route of administration: 20% of the blood LSD is found in the red cells and 40% to 70% is bound to plasma proteins. Corresponding to this, only about one-fifth of the total blood concentration is found in the spinal fluid. The highest concentration is present in the liver and in the kidneys and only low concentrations are found in the central nervous system. The concentration of LSD in the organs, measured by radioactivity, decreases rapidly, the rate varying according to species, e.g., in plasma the half-time is 100 minutes for monkeys, 130 minutes for cats, and 7 minutes for mice.

LSD *in vitro* is transformed in the liver to 2-hydroxy LSD. *In vivo* most of the LSD is metabolized to four or more different compounds (and excreted primarily in the bile), none of which is identical to 2-hydroxy LSD, nor are any of the metabolites hallucinogenic.

The genetic effect of LSD in animals has been reported in the teratogenic studies of mice (261, 262), rats (263), hamsters (264), rabbits (265), and monkeys (266). Administration of the drug to pregnant females early in gestation has been accompanied by malformations of the central nervous system, liver, and other viscera, and has also been associated with a high incidence of fetal death.

An autoradiographic study on the placental transfer and tissue distribution of ^{14}C-LSD in mice (267) revealed that in the early stage of pregnancy 2.5% and in the later stage 0.5% of the radioactive dose passed the placental barrier into the fetus in 5 minutes (over 70% of this fetal radioactivity was *unchanged* ^{14}C-LSD). Information concerning possible teratogenic effects in humans is less available (268) and less convincing (269). However, the possible teratogenic effect on two human embryos has been suggested (269, 270).

There is a considerable body of conflicting literature in regard to the chromosome breaking potential of LSD. For example, chromosome breakage has been reported both *in vitro* and *in vivo* in human somatic cells (266, 271–274) and germinal cells in mice (275, 276).

In contrast to the above observation, other reports indicated that the number of chromosome abnormalities in leukocytes from persons exposed to LSD was not significantly higher (277–279) nor were there any significant chromosomal aberrations in root tips of *Vicia faba*, cell cultures of Chinese hamster, or *in vitro* cultures of human leukocytes treated with LSD (280).

The question of mutagenicity of LSD appears to be equally conflicting. Although mutagenic effects in *Drosophila* were noted by Browning (281) and Vann (282), Grace *et al.* (283) were unable to detect any mutagenic or chromosome effects of LSD in *Drosophila*, and Zetterberg (284) reported that LSD had no significance on the frequency of back mutations to *Ophiostoma multianulatum*.

The chemistry of LSD has been reviewed (251, 285, 286), as well as its pharmacology (287), clinical and therapeutical aspects (288–290), abuse (291, 292), and possible relation to the induction of leukemia described (293).

More recently, circular dichroism experiments were reported which indicate that LSD interacts directly with purified calf thymus DNA, probably by intercalation, causing conformational changes in the DNA. It was suggested that this observed interaction may serve as a model for the interaction of LSD on the chromosomal material of intact cells (294).

Analysis of LSD has been performed by fluorimetry (253, 295, 296), ultraviolet (297, 298), infrared (298), and mass spectroscopy (299), paper (300, 301), thin-layer (298, 302, 303), gas (304, 305), and column chromatography (298, 306), and bioassay (254, 307).

10. *Caffeine and 8-Ethoxycaffeine*

Caffeine, theophylline, and theobromine are three closely related alkaloids that occur in plants widely distributed throughout the world. Caffeine is found in the extensively used beverages such as coffee, tea, cocoa, mate, as well as in some "soft drinks" (particularly the cola-flavored drinks made from the nuts of the tree *Cola acuminata* which contain about 2% caffeine).

In addition to its consumption in beverages, caffeine has broad medical utility, e.g., with antihistamines to combat motion sickness, with analgesics or ergot alkaloids for relief of tension and migraine headaches, as a stimulant, and as a coronary artery dilator. The pharmacology (308) and metabolism of caffeine have been described (309–312).

Caffeine is rapidly and essentially completely absorbed from the gastrointestinal tract of man, then distributed in various tissues in approximate proportions to their water content; it passes rapidly into the central nervous system. It is almost entirely metabolized in man (only about 1% being excreted in the urine). The rate of biotransformation is fairly uniform, with the average half-life being 3.5 hours (15% metabolized per hour). Although a considerable amount of caffeine accumulates in the body of moderately heavy coffee drinkers during the day, no day-to-day accumulation of the drug was found in the study of Axelrod and Reichenthal (309). Cornish and Christman (311) found methylated uric acids as well as 1-methyl-, 7-methyl-, and 1,7-dimethylxanthine (*p*-xanthine) as urinary excretion products following caffeine ingestion. Schmidt and Schoyerer (312) in similar studies detected caffeine, theobromine, *p*-xanthine, and 1-methylxanthine in the urine of subjects given 300 mg caffeine daily.

In a more recent study of the metabolism of xanthine alkaloids in man by

Warren (310) it was found that habitual consumers of beverages containing caffeine require about 7 days of abstention before caffeine completely disappears from their blood. In such "decaffeinated" subjects, administration of 500 mg of caffeine orally leads to the appearance of caffeine and *p*-xanthine in both red cells and plasma within 3 hours and its disappearance from the blood within 24 hours. The first step in the metabolism of caffeine in man would appear to be the removal of the 3-methyl group and the formation of *p*-xanthine (1,7-dimethylxanthine).

Evidence exists that caffeine is mutagenic in *Drosophila* (313–317), bacteria (*E. coli*) (17, 319–324), a fungus (*Ophiostoma multiannulatum*) (325–328), human cells in tissue culture (315, 329, 330), plants (onion-tip root) (331), and bacteriophage T5 (332), and that it produces chromosome aberrations in *Drosophila* (317, 318) and human cells (315, 329–331), in tissue culture (315, 329–331, 333), and in mice (333). The lack of mutagenicity of caffeine in *Drosophila* (317) and inconclusive results in mice (334, 335) have also been reported.

The teratogenic effect in mice of caffeine in high doses (336) as well as the ability of caffeine to penetrate the preimplantation blastocyst in the rabbit have been reported (337). Although Goldstein and Warren (338) have shown that caffeine is transferred from the mother to the 7-week-old human embryo, there is no evidence that caffeine can penetrate the blastocyst before implantation.

Caffeine has been analyzed by nonaqueous titrimetry (339, 340), iodometry (340), argentimetry (311), colorimetry (341), spectrophotometry (309, 311, 342, 343), and paper (344, 345), thin-layer (303, 346, 347), and gas chromatographic techniques (348, 349).

8-Ethoxycaffeine. 8-Ethoxycaffeine (8-EOC) is prepared from the reaction of 8-chlorocaffeine, ethanol, and sodium (350). The mechanism of action of 8-EOC in regard to its inhibition of nucleic acid synthesis and induction of chromosome aberrations has been described by Ts'o *et al.* (351) and Kihlman (352–354).

The induction of chromosome aberrations in cultured human leukocytes (355), fungi (328), *Vicia faba* (84, 328, 356, 357), and *Allium* (358) has been described.

11. *Cyclamate and Cyclohexylamine*

Cyclamic acid (cyclohexanesulfamic acid) is prepared by the sulfuration of cyclohexylamine or via the treatment of cyclohexylamine and triammonium nitrilosulfate. The sodium and calcium salts are prepared from cyclohexyl ammonium *N*-cyclohexylsulfamate by treatment with the respective hydroxides.

Calcium and sodium cyclamates are extensively used alone or in combinations with saccharin as noncaloric sweeteners in a wide variety of food products such as beverages, cereals, and bakery products. Ten parts of cyclamates are usually mixed with one part of saccharin; the mixture is 60–100 times sweeter than sugar.

The available supply of cyclamate in the United States increased from about 5 million pounds in 1963 to 15 million pounds in 1967. Prior to the banning of cyclamates (due to its induction of bladder tumors in the rat in chronic feeding studies) in October, 1969 by the United States, and shortly thereafter by the United Kingdom, Sweden, Denmark, Germany, Finland, and Canada, the projected consumption of cyclamates in 1970 was 21 million pounds. About 70% of the usage of non-nutritive sweeteners is in soft-drink products. In 1963, 4% of the soft drinks were marketed with cyclamate compared to 15% in 1968 (359). It is estimated that some individuals may consume as much as 40 ounces of soft drinks, hence accounting for a maximum daily intake of 4 gm of cyclamate per day (80 mg/kg for a 50 kg human).

The metabolism of cyclamate has been intensively studied in rats (360, 361), dogs, rabbits (360), and man (362–366). Cyclohexylamine has been found in the urine of man following oral administration (362–367). Human excretion of cyclohexylamine after daily ingestion of soda containing 512 mg of cyclamate ranged from a trace to 28% of the dose on 3 to 5 days after ingestion. Other metabolites present in human urine included cyclohexanone, cyclohexanol, and *N*-hydroxycyclohexylamine (367). The conversion of cyclamate to cyclohexylamine in animals has also been reported following administration of cyclamate in saccharin mixtures (368).

Evidence of breakdown of cyclamate to cyclohexene in canned foods due to the presence of nitrite has been reported by Higuchi *et al.* (369). In view of the formation of monochlorocyclohexane, cyclohexanone, and cyclohexanol in addition to cyclohexene in the nitrous acid assay of cyclamate (370) and the formation of cyclohexylamine on hydrolysis of cyclamate by organic acid (371), there is a possibility that such breakdown products may also be formed during the processing of foods sweetened with cyclamate. The properties and food application (372–374) and acute and chronic toxicity (375–378) of cyclamates have been reported.

Bajusz (379) found that oral administration of calcium cyclamate to hamsters induces myocardial lesions accompanied by coronary sclerosis as well as soft tissue calcification in other organs.

Cyclamate induced chromosome breakage in onion-root tips (380), in human leukocytes *in vitro* (381), and in both leukocyte and monolayer cultures from human skin and carcinoma of the larynx (382). The synergistic radiomimetic effects of caffeine, alcohol, and Sucaryl (which contain 8% sodium cyclamate and 0.8% saccharin) in onion-root tips (380) have been described.

The analytic chemistry of cyclamate has been reviewed by Richardson (383), and colorimetric (384, 385) (following conversion to cyclohexylamine), titrimetric (370, 386), and paper (387), thin-layer (388–390), and gas chromatographic (360, 361, 391) procedures have been described.

Cyclohexylamine. It has been estimated that about 30% of all people who consume cyclamate convert it to cyclohexylamine. So far, there is no positive evidence where this conversion takes place (it may occur in the gut by intestinal flora or in the liver or both). Cyclohexylamine is considerably more toxic than the cyclamates (LD_{50} in rats is about 300 mg/kg, 20 to 40 times less than for cyclamates). It was found that when male rats are given an abnormally high dose of 250 mg of the amine per kilogram of body weight, about 17% of their germinal cells show chromosome breaks.

Cyclohexylamine also induces chromosome breaks *in vitro* in rodent cells in culture as well as *in vivo* in rat spermatogonia (392, 393). Cyclohexylamine has also been shown to be an indirect sympathomimetic agent with actions much like those of tyramine and/or amphetamine (394–397). The toxicity (376, 398–400) of cyclohexylamine, its role in the metabolism of cyclamate (362–367), and its analysis (383, 384) have been described.

12. *EDTA*

Ethylenediaminetetraacetic acid (EDTA; Versene) and its alkali metal salts are prepared via the reaction of ethylenediamine, hydrogen cyanide, and base. The free acid is less stable than its salts and tends to decarboxylate when heated to temperatures of 150°C. EDTA has been employed as a therapeutic agent in the treatment of lead (401), mercury (402), manganese (403), and radioactive metal poisoning (404), and is an anticoagulant in blood used for transfusion (405) and in other instances where control and investigation of metal ion concentration is desired (406).

EDTA and its alkali salts (e.g., disodium and calcium disodium salt) are widely used as sequestrants in food systems. Table 9.1 illustrates the foods in which EDTA can be employed. Sequestrants such as EDTA and its derivatives are not considered antioxidants in the classic sense in which hindered phenols such as BHT and BHA arrest oxidation by chain termination, or like ascorbates by scavenging oxygen. They are of value in antioxidant systems due to their property of forming poorly dissociable chelate complexes with trace quantities (0.1 to 5 ppm) of divalent and trivalent metals such as copper and iron in fats and oils.

By chelating these metals, pro-oxidant catalytic effects are eliminated or minimized and higher efficiency can be derived from the antioxidant. In most cases the combined effect of sequestrants and antioxidants is synergistic

TABLE 9.1

Foods in Which EDTA Can Be Employed[a,b]

Food	Use	Limitations (ppm) Calcium disodium EDTA	Disodium EDTA
Banana (dried component of ready to eat cereal)	Promote color retention	NA[c]	315
Beans (canned kidney)	Promote color retention	NA	165
Beans (processed dry pinto)	Promote color retention	800	NA
Carbonated beverages (canned)	Improve flavor retention, retard can corrosion, increase stability of some FD&C colors	33	NA
Chick peas (cooked, canned)	Promote color retention	NA	165
Clams (cooked, canned)	Promote color retention	340	NA
Crabmeat (cooked, canned)	Retard struvite formation, promote color retention	275	NA
Dressings (nonstandardized)	Promote spice flavor retention, protect against rancidity	75	75
Fermented malt beverages	Prevent gushing or wildness	75	NA
French dressings[d]	Promote flavor retention, protect against rancidity, prevent formation of gummy ring on glass containers	75	NA
Mayonnaise[d]	Promote spice flavor retention, protect against rancidity	75	NA
Oleomargarine[d]	Promote flavor retention	75	NA
Pecan pie filling	Promote color retention	100	NA
Potato salad	Improve keeping qualities	100	NA
Potatoes (frozen, white)	Promote color retention	NA	100
Salad dressing[d]	Promote spice flavor retention, protect against rancidity	75	NA
Sandwich spread	Protect spice flavor retention, protect against rancidity, keep vegetables crisp	100	100
Sauces	Promote spice flavor retention, protect against rancidity	75	75
Shrimp (cooked, canned)[e]	Retard struvite formation, promote color retention	250	NA
Spice extractives in soluble carriers	Promote color and flavor retention	60 60	NA NA
Vitamin solutions containing B_{12}	Stabilizer for B_{12}	NA	150

[a] T. E. Furia, *Food Technol.* **18**, 50 (1964).
[b] As of November 1964.
[c] NA = not allowed.
[d] FDA Standards of Identity apply to these products.
[e] Use of calcium disodium EDTA is allowed under temporary permits.

and their combined utilization in foods containing fats and oils is reflected by superior initial products with extended shelf-life.

Other areas of utility of EDTA and its derivatives include the stabilization of acrylonitrile for polymerization (407), the purification of antibiotics (dihydrostreptomycin) (408), pesticidal compositions (e.g., hydrazine salt of EDTA) (409), plant nutrients (410), plant growth regulation (411), water treatment (412), stabilization of hydrazine fuel compositions (413), and in bactericidal (414) and germicidal (415) compositions.

The chemistry (416, 417), pharmacology (418), toxicity (419, 420), metabolism [in humans (421), in rats (422)], effect on chromosome radiation damages and induction of chromosome aberrations (423), breakage of chromosomes of *Tradescantia paludosa* (424, 425), production of mitotic abnormalities in onion roots (426), chromosomal changes in *Drosophila* (427, 428), *Vicia* (429), and *Hordeium* (429), and the synergistic effect in the production of chromosome aberrations in *Vicia* (430) have been described for EDTA.

Analytic applications of EDTA (431) as well as its analysis by titrimetric (432), potentiometric (433), complexometric, and argentometric titrations (434), polarographic (435), ultraviolet (436) and nmr spectroscopic (437) and paper chromatographic (421, 438) techniques have been described.

13. *Allylisothiocyanate*

The natural source of allylisothiocyanate is sinigrin, one of the major glucosinolates in a variety of plants such as *Brassica oleraceae* (cabbages, kale, brussel sprouts, cauliflower, broccoli, and kohlrabi), *B. carinata* (Ethiopian rapeseed), *B. juncea* (Indian or brown mustard), *B. nigra* (black mustard), *Sinapsis alba* (white mustrad), *S. arvensis* (Charlock), and *Amoracia rustican* (horseradish).

Allylisothiocyanate is obtained from sinigrin via enzymic hydrolysis as shown:

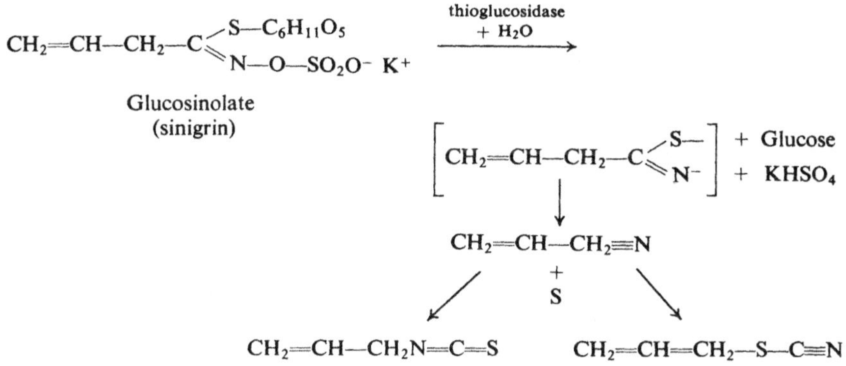

Allylisothiocyanate is used as a food additive (modifier, body) of hot sauces, relish flavors, salad dressings, and synthetic mustard. It is also used medically as a counterirritant, it possesses antihelminthic (439) and additional fungistatic properties (440, 441), and has utility as a meat preservative (442) and plant-growth regulator (443).

The toxicity (444), cytotoxicity (445, 446), mutagenicity in *Drosophila* (447–450) and *Ophiostoma* (451), as well as its induction of chromosome

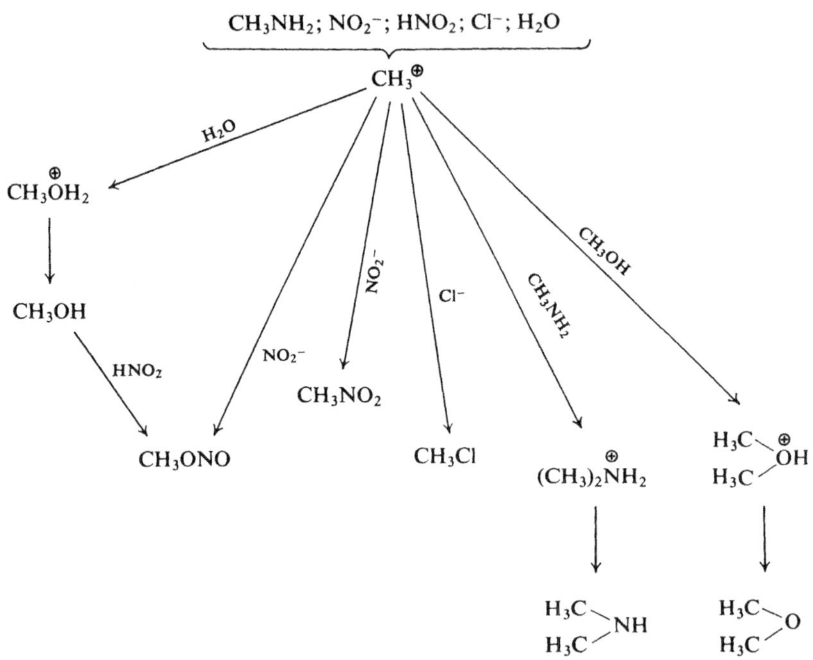

FIG. 9.1. Reaction of methylamine with nitrous acid and carbonium ion interactions.

aberrations in *Drosophila* (447) and *Allium* (452) for allylisothiocyanate have been described.

Its analysis has been accomplished by colorimetry (453–455), iodometry (456), argentometry (456), thin-layer (457) and gas–liquid chromatography (458), and nmr spectroscopy (459).

14. *Nitrites and Nitrous Acid*

Sodium nitrite is used as a preservative in meat, fish, and cheese (460). Nitrites are included in curing mixtures for meats to develop and fix the color (nitrites decompose to nitric oxide which then reacts with heme pigments to form nitrosomyoglobin). Nitrates are generally believed to be a reservoir of nitrite by reduction. Although nitrites and nitrates have been demonstrated to possess antimicrobial action *in vitro*, they seldom provide sterilization. As with other microbial inhibitors, the effectiveness of nitrite, a weak acid salt, is related to pH (461). The possible formation of nitrosamines from amines present in or derived from the diet occurs by reaction with nitrous acid at pH 4. In man, gastric juice attains a pH of 1.1. Such high concentrations of hydrogen ions give rise to the nitrosyl cation NO^+, a highly reactive nitrosating agent. The presence of meat, myoglobin, or hemoglobin serve to neutralize this cation by reaction with Fe^{++}. Fish and cheese do not provide this neutralization, hence permitting the nitrosation of the present amines to possible carcinogenic nitrosamines.

The toxic effects of nitrite in infants (462–465), ruminants (465), rats (466), mice (467), and chickens (468) have been reported. Nitrite has been shown to oxidize hemoglobin and reduce GSH in the red blood cell to methemoglobin and oxidized GSH, respectively (469, 470) [although the latter cellular GSH reaction with nitrite has been questioned (471)].

Solutions of nitrous acid are unstable and evolve oxides of nitrogen at concentrations greater than 0.05 *M*. The action of nitrous acid on aliphatic amines has been reviewed by Austin (472). Methylamine was found to yield methanol, methyl nitrite, methyl chloride, nitromethane, methyl nitrolic acid, and evolved gases (nitrogen and carbon dioxide). The multiplicity of the products suggested the initial formation of a carbonium ion. Figure 9.1 illustrates the sequence of the reaction of methylamine with nitrous acid as well as the interaction of the carbonium ion with a number of reactants.

The mutagenicity of nitrous acid toward transforming-DNA (473–476), *E. coli* (477, 478), *Salmonella typhimurium* (477, 479, 480), *Neurospora* (45, 481, 482), *Saccharomyces cerevisiae* (483–487), *Schizosaccharomyces pombe* (488, 489), phage [S13 and ϕX174 (490), T4 (491, 492), T2 (493, 494)], *Aspergillus nidulans* (495), *A. niger* and *A. amstelodami* (496), and tobacco mosaic virus have been reported (497–500).

The analysis of nitrous acid has been primarily performed by colorimetric techniques (501–504).

15. *Maleic Hydrazide*

Maleic hydrazide (1,2-dihydro-3,6-pyridazinedione; MH) is prepared commercially via the reaction of hydrazine with maleic anhydride, and has extensive application as a herbicide, fungicide, growth inhibitor, and growth regulator. Its utility includes the prevention of sucker production in tobacco plants, growth control of weeds, grass, and foliage, inhibition of sprouting of potatoes, onions, and stored root crops, and the protection of citrus seedlings against frost damage. The diethanolamine and sodium salts of maleic hydrazide are used in commercial formulations because of their enhanced solubility.

The growth-restricting effects of maleic hydrazide consist of destroying the apical dominance of shoots, stopping root growth, and destroying germ cells. The action of maleic hydrazide is analogous to that of x-rays: it is specifically directed to growing tissues and therefore mitotic cells. Studies on the effect of maleic hydrazide on the enzymes of intact cells suggest that maleic hydrazide affects the enzymes requiring sulfhydryl groups for activity (505, 506).

Among the possible routes of metabolism and degradation for maleic hydrazide are the following:

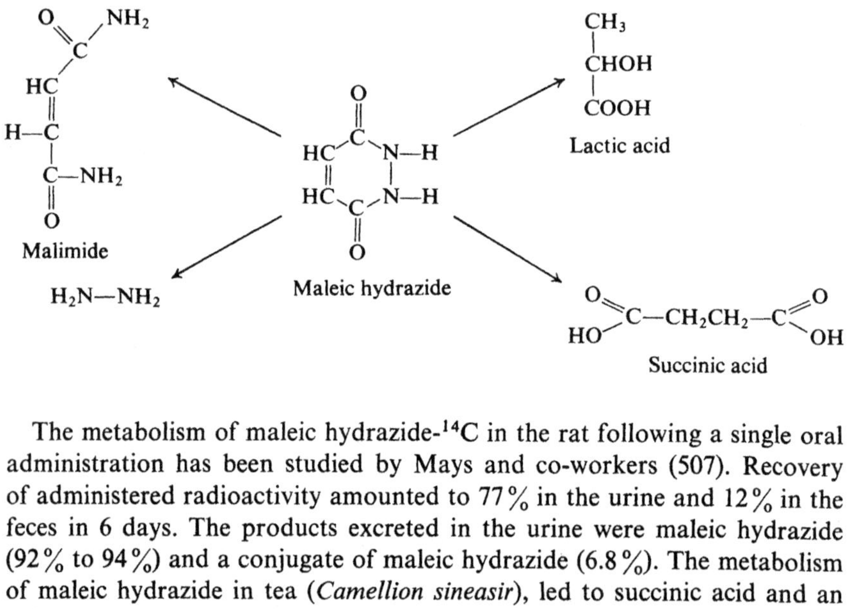

The metabolism of maleic hydrazide-^{14}C in the rat following a single oral administration has been studied by Mays and co-workers (507). Recovery of administered radioactivity amounted to 77% in the urine and 12% in the feces in 6 days. The products excreted in the urine were maleic hydrazide (92% to 94%) and a conjugate of maleic hydrazide (6.8%). The metabolism of maleic hydrazide in tea (*Camellion sineasir*), led to succinic acid and an unidentified metabolite (508); a glycoside of maleic hydrazide from plants has also been identified as a metabolite (509).

Maleic hydrazide is frequently applied at the rate of 2.25 lb/acre to control adventitious buds when tobacco is topped. Following this treatment, the green commercial leaves and the green sucker leaves contained 37 and 482 ppm of maleic hydrazide, respectively (510); cigarettes made from tobacco so treated contained 10 to 30 ppm of maleic hydrazide (511). Information on the fate of maleic hydrazide in cigarette smoke is sparse. Experimental cigarettes containing 10 to 30 ppm of MH gave 0 and ≤ 2 ppm of unchanged MH in the smoke, respectively (511). Using ^{14}C-labeled maleic hydrazide, 23.4%

of the added radioactivity was found in CO_2, CO, and "tars" of the smoke, and 31% was calculated or found in the butt and ash. The remainder was assumed to have been lost in the sidestream smoke (512).

The chemistry and mode of action (513, 514), carcinogenicity (515–518), cytochemical effects on cultured mammalian cells (519), mutagenicity in *Drosophila* (520, 521), mitotic inhibition and chromosomal aberrations in *Vicia faba* (357, 358, 522–527), *Allium* (528), barley (529), maize (530), mitotic tissues of oats, corn and soybeans (531), and dry seeds of *Crepiscapillaris* (532, 533) of maleic hydrazide have been described.

Analysis of maleic hydrazide has been performed by conductiometric titration (534), polarography (535), ultraviolet (536, 537), infrared (538), and nmr spectroscopy (539), colorimetry (537–541), and paper (542, 543) and gas chromatography (544, 545).

16. *Captan*

Captan [*N*-(trichloromethylthio)-4-cyclohexene-1,2-dicarboximide] is produced by the reaction of perchloromethyl mercaptan on tetrahydrophthalimide via the sequence:

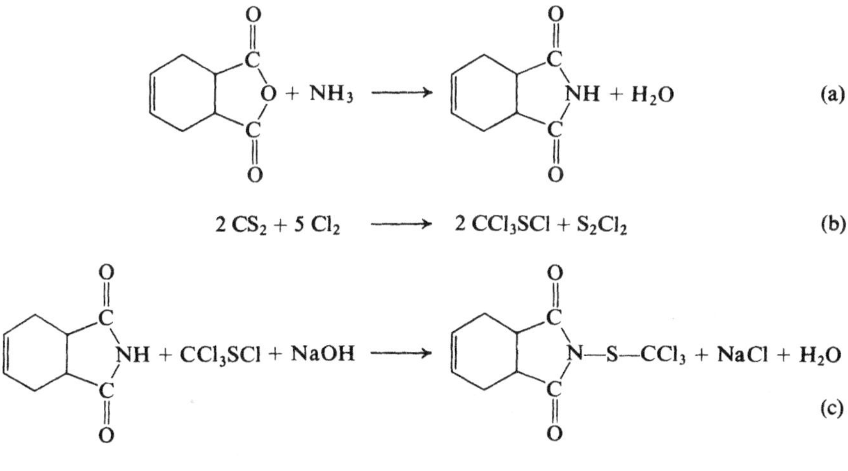

It is a general fungicide used for the treatment of folia and soil and against seedborne diseases including apple scab, grape mildews, corn seed infections, and many fruit, vegetable, and ornamental plant diseases. It is stable when dry but decomposes at or near its melting point (178°C) to yield a variety of products. The rate of hydrolytic attack increases with increasing pH and temperature.

The reaction of captan with thiols has been extensively investigated. For example, the decomposition of captan by thiols is known to produce thiocarbonyl chloride which can react in turn with cell thiols (546).

Thiocarbonyl chloride is also rapidly hydrolyzed by water to carbonyl sulfide (547) so that the extent of the thiocarbonyl chloride–SH reaction depends on this rate of hydrolysis.

A scheme for the reaction of captan with glutathione (GSH) has been proposed by several investigators (546, 548) as follows:

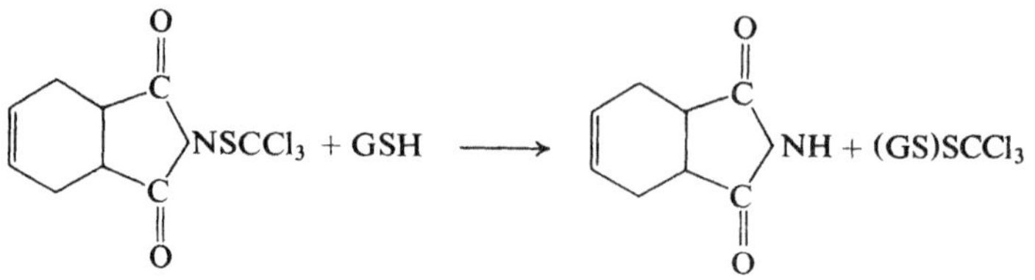

(a)

$$(GS)SCCl_3 + GSH \longrightarrow (GS)SG + HSCCl_3 \qquad \text{(b)}$$

$$HSCCl_3 \longrightarrow HCl + SCCl_2 \qquad \text{(c)}$$

It has been shown in the reaction between captan and the intercellular thiols of *Neurospora crassa* conida (549) that thiocarbonyl chloride ($SCCl_2$) can react further with GSH to form the thiazolidine derivative (I), hence interfering with the —SH oxidation-reduction cycle.

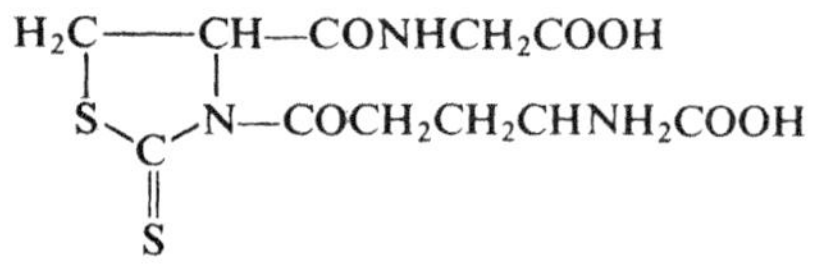

(I)

Reactions with *Saccharomyces pastorianus* (546) suggested that thiophosgene, formed in the reaction of captan with —SH groups is the ultimate toxicant, being free to combine not only with additional —SH groups, but also with —OH, —NH_2—, and —COOH.

Owens and Novotny (550) regard the fungitoxicity of captan to be due to the intact molecule and not to its decomposition products. Figure 9.2 illustrates the mode of action of captan and depicts the initial condensation of captan with thiol followed by spontaneous decomposition to products of the type characterized by Lukens and Sisler (546).

The toxicity (551, 552), teratogenicity to the chick embryo (553, 554), and mutagenicity in *E. coli* and cells cultured *in vitro* (554) have been demonstrated for captan.

Analysis of captan has been achieved via colorimetry (555–557), paper (558), thin-layer (559, 560), and gas chromatography (561–563), and bioassay (564).

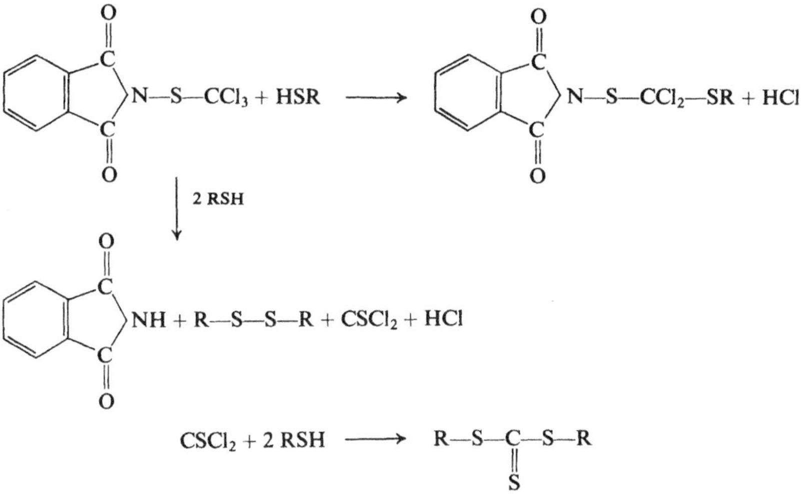

FIG. 9.2. Mode of action of captan. Owens, R. G., and Novotny, H. M., *Contrib. Boyce Thompson Inst.* **20**, 171 (1959).

17. *Hemel and Hempa*

Hemel [(2,4,6-trisdimethylamino)-1-triazine] and hempa (hexamethyl phosphoric triamide) are thermally stable nonalkylating analogs of tretamine and TEPA, respectively, developed by Chang *et al.* (565).

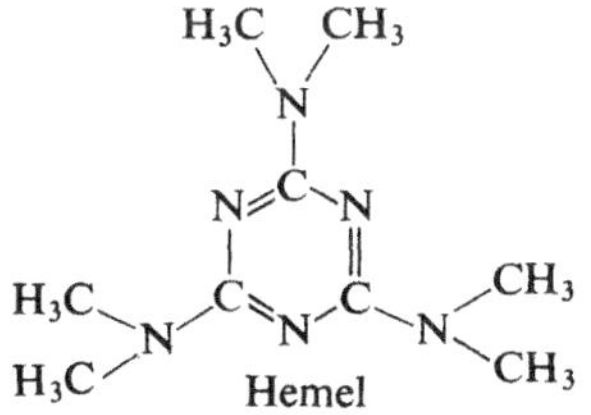

Hemel

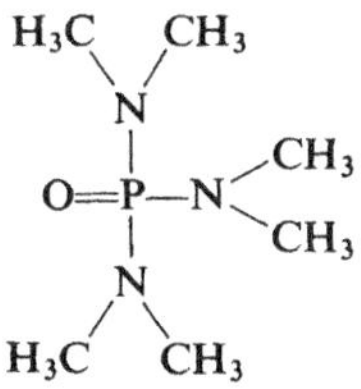

Hempa

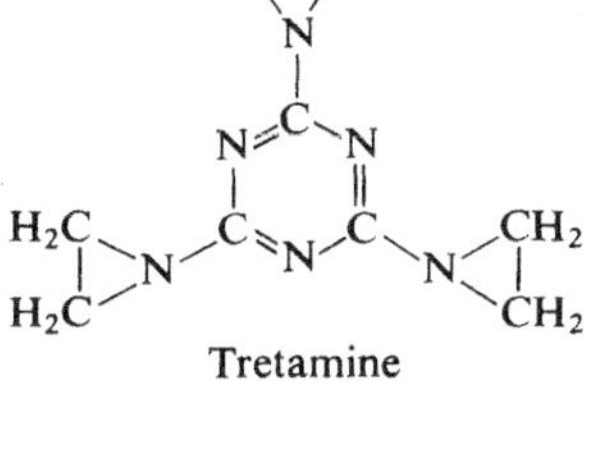

Tretamine

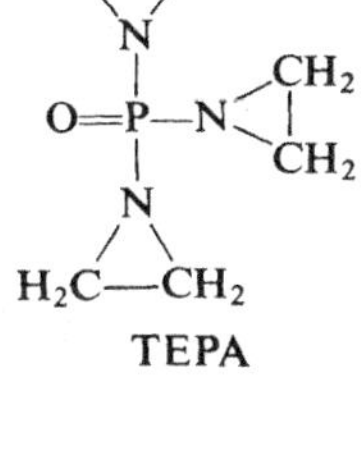

TEPA

Both hemel (566, 567) and hempa (565, 568) (prepared by the reaction of phosphorus oxychloride and triethylamine) are effective chemosterilants for houseflies (*Musca domestica*).

The major route of metabolism of hempa in houseflies *in vivo* (569) and *in vitro* (570, 571) is via *N*-demethylation to form pentamethyl phosphoramide. In the rat and mouse, hempa is metabolized to three demethylated products, viz., pentamethyl, tetramethyl, and trimethyl phosphoramide (572). Degradation of the phosphoramides by rat liver slices *in vitro* indicated that their metabolism occurs by a process of oxidative dealkylation.

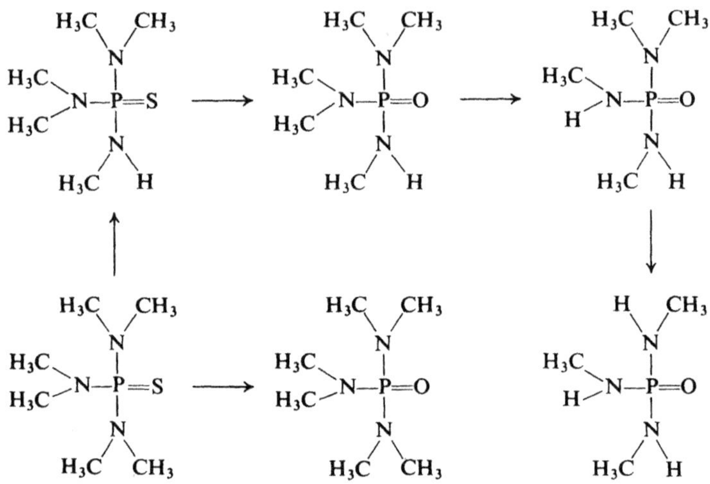

FIG. 9.3. Metabolic pathways of hempa and thiohempa in the rat and mouse.

Figure 9.3 illustrates the metabolic pathways of hempa and thiohempa in the rat and the mouse.

Although the activity of hempa in male flies is much lower than that of TEPA, both compounds produce strikingly similar physiological and cytological effects in several different organisms. Both hempa and TEPA induced a high frequency of recessive lethal mutations in the sperm of *Bracon hebetor* Say (573) (TEPA is about 100 times more efficient than hempa) and testicular atrophy in rats (574); hempa induced a marked antispermatogenic effect in rats (574, 575) and mice (574, 575). The toxicity of hempa (565, 574, 576) and its analysis by infrared spectroscopy (569), paper (577), thin-layer (569, 572, 577), and gas chromatography (569, 577, 578) have been reported.

The metabolism of ^{14}C-hemel in male *Musca domestica* was studied by Chang *et al.* (579) who isolated a number of lower methylmelamines, e.g., penta-, tetra-, and trimethylmelamines in addition to other acyclic products.

The toxicity of hemel (580, 581) as well as its mutagenicity in *Musca domestica* have also been described (582).

18. *Hydrazine and Its Derivatives*

Hydrazine is prepared from ammonia and sodium hypochlorite by two main routes:

(a)

$$NH_3 + NaOCl \rightarrow NH_2Cl + NaOH \quad \text{(i)}$$

$$NH_2Cl + NH_3 \rightarrow NH_2NH_2 + NaCl + H_2O \quad \text{(ii)}$$

(b) the Bayer process via diazocyclopropane or ketazine, viz.:

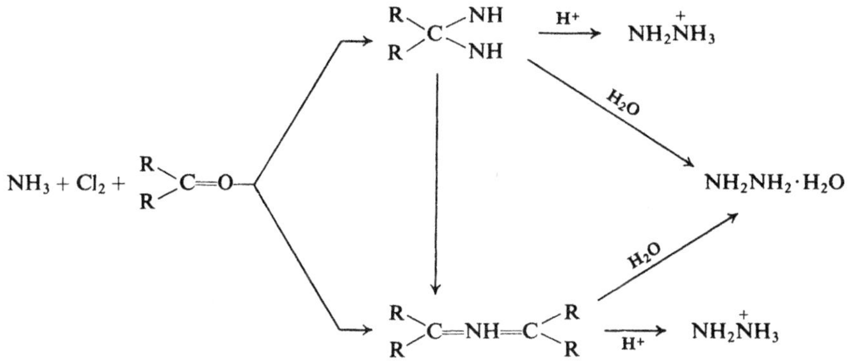

Hydrazine and its derivatives possess a broad spectrum of utility in photography, preservatives, metal processing, and the preparation of agricultural chemicals, medicinals, textile agents, explosives, fuels, and plastics. The scope of its utility and the extent of its consumption in various areas are outlined in Tables 9.2 and 9.3.

The chemistry of hydrazine (598, 599), its physiology (600), acute and chronic toxicity (601–603), hepatoxicity (604, 605), carcinogenicity (606–608), and metabolism (609–611) have been described.

A number of possible metabolic pathways for hydrazine and its derivatives have been suggested which include (1) the hydrolysis of hydrazides to hydrazine and carboxylic acid, (2) reaction of hydrazine and hydrazides with natural aldehydes and ketones, (3) acetylation, and (4) splitting of symmetrical disubstituted hydrazine to yield two amines.

a. Unsymmetrical Dimethylhydrazines. Unsymmetrical 1,1-dimethylhydrazine (UDMH) is prepared by the electrolytic reduction of dimethylnitrosamine (612), and, in addition to its primary use in rocket fuels, it has patented applications as a solvent for acetylene (613) and as a gasoline additive (614). Its toxicological and pharmacological properties (600, 615–618), metabolism (619, 620), and carcinogenicity (606) have been reported.

b. Isonicotinylhydrazide (INH). Because hydrazine is the principal metabolite of isonicotinylhydrazide (621, 622), it is germane to include a consideration of

TABLE 9.2

Utility of Hydrazine and Its Derivatives

Application	Form and References
1. As a reducing agent	
a. Corrosion inhibitor (oxygen scavenger)	Hydrazine (583)
b. Silver plating of glass and plastics	Hydrazine hydrate (584)
c. Soldering fluxes	Hydrazine hydrochloride (585)
d. Inhibitor of color and odor formation in soaps	Stearic hydrazide (586)
2. Reactive chemicals	
a. Extender for urethan polymers	Hydrazine (587)
b. Terminator of emulsion polymerization	Hydrazine with dialkyl dithiocarbamates (588)
c. Curing of epoxy resins	Dihydrazides (589)
d. Rocket fuel	Hydrazine; unsym. dimethylhydrazine (590)
e. Blowing agent	2,2′-Azoisobutyronitrile azodicarbonamide
3. Agricultural chemicals	
a. Plant growth regulators	Maleic hydrazide (591)
b. Defoliants	3-Amino-1,3,4-triazole (592)
c. Plant growth stimulator	β-Hydroxyethylhydrazine (593)
4. Medicinals	
a. Antitubercular agents	Isonicotinic acid hydrazide (594)
b. Psychic energizers	β-Phenylisopropylhydrazine (595)
c. Hypotensive agents	1-Hydrazinophthalazine (596)
d. Topical antiseptic	Nitrofurazone (597)
e. Polycythemia vera	Phenylhydrazine

INH and its properties. Its tumorigenic (623, 624) and carcinogenic behavior (625, 626) and metabolism (626–628) have been described. Its mutagenicity in *E. coli* (629–631) and the induction of chromosome breaks (632) of hydrazine and its derivatives and of mitotic recombination in *Saccharomyces cerevisiae* by *sym*-dimethylhydrazine (633) have also been described.

Antidepressant mono- and disubstituted hydrazines (isocarboxazid, nialamide, and phenelzine), as well as isonicotinylhydrazide, were reported by Freese *et al.* (634) to produce hydrogen peroxide (by interaction with oxygen) and inactivate transforming-DNA.

The analysis of hydrazine and its derivatives has been achieved by titrimetry (635), colorimetry (540, 635, 636), and paper (637) and gas chromatography (638–641).

TABLE 9.3

Estimated U.S. Yearly Consumption of Hydrazine (100% Basis), 1963 to 1964[a]

Use	Form	Million lb.
Rocket fuel	Hydrazine	11.0
	Methylhydrazine	0.1
	Unsym. dimethylhydrazine	19.0
Boiler water treatment	35% Aq. hydrazine	0.5
Blowing agents	Azodicarbonamide	0.8
	4,4′-Oxybis(benzenesulfonylhydrazide)	0.4
	Benzenesulfonylhydrazide	0.4
Agriculture	Maleic hydrazide	1.2
	3-Amino-1,2,4-triazole	0.4
Medicinals	Misc. derivatives of hydrazine	0.4
Solder flux	Hydrazine hydrobromide	0.1
Miscellaneous		
Nylon manufacture	Hydrazine	0.05
Epoxy resins	Isophthalyldihydrazide	0.03
Ignitor for explosives	Tetracene	0.03
Hydrochloric acid (food grade)	Hydrazine	0.02
Reclaiming catalysts	Hydrazine	0.1
Other	—	0.6

[a] Kirk-Othmer, "Encyclopedia of Chemical Technology," 2nd ed., Vol. 11, p. 192. Wiley (Interscience), New York, 1966.

19. *Hydroxylamine*

Hydroxylamine is prepared as its ammonium acid sulfate by the reduction of sodium nitrite with sodium bisulfite and sulfur dioxide to yield sodium hydroxylamine-*N*,*N*-disulfonate, which is then hydrolyzed to yield the product:

$$NaNO_2 + NaHSO_3 + SO_2 \rightarrow HON(SO_3Na)_2$$

$$HON(SO_3Na)_2 + 2\,H_2O \rightarrow (NH_3OH)HSO_4 + Na_2SO_4$$

Hydroxylamine is widely used in the transformation of organic compounds to derivatives which in turn may be intermediate in pharmaceutical or other industrial syntheses of complex molecules. One of the large commercial uses of hydroxylamine (estimated at over 100 million pounds) is in the synthesis of caprolactam, the raw material for nylon 6.

Hydroxylamine is also well known as a laboratory chemical used for the qualitative and quantitative determination of carbonyl compounds as oximes. Commercially, oximes are important as antishining agents in paints or for their complexing action with metallic ions.

The oxidation-reduction capabilities of hydroxylamine make it useful in many applications, e.g., as a reducing agent for many metal ions and for the termination of peroxide-catalyzed polymerizations. When functioning as a reducing agent, hydroxylamine is transformed to nitrogen or oxides of nitrogen.

The scope of commercial utility of hydroxylamine and its derivatives is illustrated by the following applications: metal salts of *N*-nitroso-*N*-arylhydroxylamines as fungicides (642), substituted nitrosohydroxylamines as stabilizers for olefins (643), use of hydroxylamine in the prevention of paper-pulp brightening and discoloration (644), in high-temperature drying of acrylic fibers (645), in photographic developers and emulsions (646), as stabilizer of adhesives from glue and lignosulfates (647), and possibly in radiation damage prevention (648).

The anticonvulsant (649, 650), antimicrobial (651, 652), and tuberculostatic activity (653) of hydroxylamine as well as its toxicity (467, 654) have been described.

The reaction of hydroxylamine with cytosine and related compounds (655) its effects and induction of mutations in transforming DNA (476, 656–659), bacteriophage [S13 and ϕX174 (490, 660), T4 (661)], *Neurospora* (662), *Schizosaccharomyces pombe* (489), and induction of chromosome aberrations in human chromosomes (663), cultured Chinese hamster cells (664), mouse embryo cells (665, 666), and in *Vicia faba* (667) have been described.

N- and *O*-derivatives of hydroxylamine, e.g., *N*-methyl- and *O*-methylhydroxylamine, have also been found to be mutagenic in transforming DNA of *B. subtillis* (656) and *Neurospora* (668), and to induce chromosome aberrations in Chinese hamster cells (665). The selective reaction of *O*-methylhydroxylamine with the cytidine nucleus has been reported by Kochetov *et al.* (669).

Hydroxylamine has been analyzed by titrimetric (670), coulometric (671), colorimetric (672–675), nmr spectroscopic (676), and paper chromatographic techniques (677, 678).

20. *N-Hydroxyurea*

N-Hydroxyurea (hydroxyurea) has been prepared by the reaction of (a) ethyl carbamate with an excess of hydroxylamine (679) and (b) potassium cyanate and hydroxylammonium chloride (680). The latter reaction has been shown to yield both isomers, i.e., *N*-hydroxyurea (m.p. 140°C) and "isohydroxyurea" (m.p. 72°C).

$$H_2N{-}\underset{\underset{O}{\|}}{C}{-}NHOH \qquad \text{and} \qquad H_2N{-}\underset{\underset{O}{\|}}{C}{-}ONH_2$$

N-Hydroxyurea — Isohydroxyurea

Hydroxyurea is of current use in cancer chemotherapy as an antineoplastic agent demonstrating activity in acute lymphoblastic leukemia, chronic myelogenous leukemia, and possibly in certain solid tumors. Its antitumor activity, clinical utility, and pharmacology have been well documented (681–684). The antiviral (685) and chemosterilant properties (686) of hydroxyurea have also been reported.

The mechanism of action of hydroxyurea in cancer chemotherapy has been studied by Fishbein and Carbone (681, 687). The presence of acetohydroxamic acid in the blood of patients suggested that the drug is hydrolyzed, yields hydroxylamine, which then cleaves thioesters, particularly acetyl coenzyme A, according to the scheme below:

$$H_2N{-}\underset{\underset{O}{\|}}{C}{-}NHOH + 2\,H_2O \rightarrow NH_2OH + NH_4^+ + HCO_3^-$$

$$NH_2OH + CH_3{-}\underset{\underset{O}{\|}}{C}{-}S{-}CoA \rightarrow CH_3{-}\underset{\underset{O}{\|}}{C}NHOH + HS{-}CoA$$

It has also been found that hydroxyurea is hydrolyzed by urease with the formation of an equimolar quantity of hydroxylamine (688, 689).

The specific inhibition of DNA synthesis by hydroxyurea (690), as well as its induction of chromosome breaks [in cultured, normal human leukocytes (666, 691)], and chromosome aberrations [in Chinese hamster cells and mouse embryo cells in culture (692) and *Vicia faba* (693)] and its teratogenicity, [in hamsters (694, 695), rats (696–698), and chick embryos (697–699)] have been described.

Hydroxyurea has been analyzed by colorimetry (700, 701), infrared spectroscopy (702, 703), and paper (687, 704–706) and gas chromatography (707).

21. *N-Hydroxy-1- and N-Hydroxy-2-naphthylamines*

The mutagenicity of both *N*-hydroxy-1- and *N*-hydroxy-2-naphthylamines in several mutants of *E. coli* has been established by Belman *et al.* (708–710) and Perez and Radomski (711). The synthesis of *N*-hydroxy-1- and *N*-hydroxy-2-naphthylamines (712, 713) and their isolation as urinary metabolites of their respective precursor naphthylamines in rat (708) and man and dog (713) have been described.

In addition to glucuronide and phosphate conjugates of *N*-hydroxy-2-naphthylamine, the existence of other metabolites such as 2-acetamido-6-naphthol and 2-amino-1-naphthylmercapturic acid has also been demonstrated in the urine of dogs and rats (714). The two suggested routes for the formation of the mercapturic acid conjugate from *N*-hydroxy-2-naphthylamine are as follows (714):

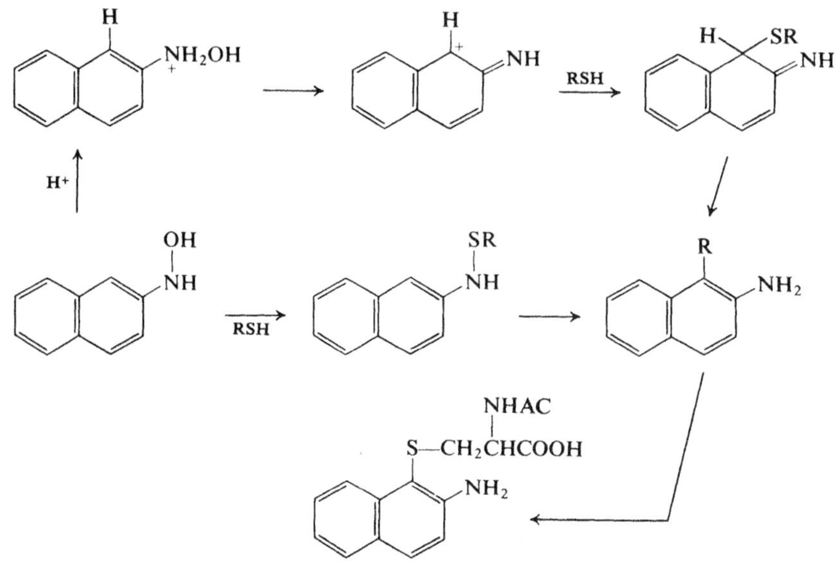

RSH = glutathione

The carcinogenicity of both isomeric *N*-hydroxynaphthylamines has also been reported (715–717). Because of the importance of both precursor naphthylamines from an environmental standpoint, a consideration of their utility, occurrence, and biological properties is relevant to this presentation.

1-Naphthylamine is prepared by reducing 1-nitronaphthalene with iron and hydrochloric acid. The commercial grade product contains up to 10% of 2-naphthylamine, and is used in the preparation of azo dyes (718) and growth regulators (719), it is also used as a corrosion inhibitor (720), antioxidant for hydrocarbons (721), and with epichlorohydrin in the synthesis of epoxy resins (722).

2-Naphthylamine is made by heating 2-naphthol with ammonium sulfite and ammonium hydroxide at 150°C, and is used as an antioxidant for hydrocarbons (721) and polyacetaldehyde polymers (723) in the softening of textile materials (724), dyeing of vinylon yarn (725), and, as phenyl-2-naphthylamine, as a corrosion inhibitor in antifreeze (726) and as an antiaging agent for rubber (727).

Formaldehyde condensation products of 1- and 2-naphthylamines which are used in the rubber industry as antioxidants contain a small proportion (2.5%) of uncombined naphthylamines.

The carcinogenicity of 2-naphthylamine is well established (715–717); it is the most commonly recognized cause of occupational bladder cancer (727). Recently both naphthylamines were found to be formed on pyrolysis of amino acids (728). Pauli *et al.* (729) found 1-naphthylamine, an aminofluorene, and an aminostilbene in cigarette smoke, and Hoffman *et al.* (730, 731) have described the identification and quantitative determination of both naphthylamines (by gas-liquid chromatography and mass spectroscopy) in cigarette smoke. The isolation of 2-naphthylamine represents the first successful identification of a bladder carcinogen in a nonoccupational respiratory environment and of possible significance because of epidemiological data which suggest an association between cigarette smoking and urinary bladder cancer in men (732).

The analysis of the isomeric *N*-hydroxynaphthylamines is achieved by the colorimetric procedure of Boyland and Nery (733) and of the naphthylamines by paper (734), gas chromatography (735, 736), and colorimetry (737).

22. *Urethan and N-Hydroxyurethan*

Urethan (ethyl carbamate) is synthesized by a variety of reactions: (a) ethyl chloroformate and ammonia (the predominant commercial process), (b) ethyl alcohol with carbamoyl chloride, cyanic acid, or urea at elevated pressure, (c) urea nitrate, ethanol, and sodium nitrite, and (d) by the ammonolysis of diethyl carbonate.

Urethan has been used medically as a mild hypnotic, sedative, and antispasmodic in the treatment of chronic myeloid leukemia and related blood diseases (738) and as an antidote for strychnine, resorcinol, and picrotoxin poisoning.

Derivatives of carbamic acid are widely used in the plastics industry, as monomers, co-monomers, plasticizers, and fiber and molding resins, in textile finishing, in agricultural chemicals as herbicides (739), in insecticides, and in insect repellants (740), fungicides (741) and molluskicides, in pharmaceutical chemicals as psychotropic drugs, hypnotics, and sedatives, anticonvulsants, miotics, anesthetics, and antiseptics. Urethans are also used as surface-active agents, selective solvents, dye intermediates, and corrosion inhibitors. *N*-Nitrocarbamates have been suggested as diesel fuel additives (742) and urethan–aldehyde condensation products have been used as latent flavor-developing substances in packaged foods.

Urethan is a pulmonary carcinogen in mice (743–745) and in rats (746), induces carcinomata of the forestomach in hamsters, and is teratogenic in mice (747), hamsters (695), fish (748), and amphibia (749).

Urethan has been reported to induce mutations in *Drosophila* (750, 751), bacteria (17, 752, 753), and plants (754, 755), to be nonmutagenic in mice (756) and *Neurospora* (757), and to induce chromosome aberrations in *Oenothera* (758). Congeners of urethan, e.g., methyl-, propyl-, and butyl carbamate are also active mutagens in bacteria (753).

Urethan is metabolized by mammals to *N*-hydroxyurethan and *N*-acetyl-*N*-hydroxyurethan (759, 760), and according to Boyland and co-workers, the carcinogenic and antileukemic effects attributed to urethan are probably caused by the hydroxyurethan metabolites which act as alkylating agents toward mercaptoamino acids and react with cytosine residues of RNA. The mechanism is similar to, but distinct from, that of the action of alkylating agents which react mainly with guanine of nucleic acid; however, in both cases, the same base pairs, guanine-cytosine, are modified.

This view has been challenged by Kaye and Trainin (761) who suggested that the carcinogenic actions of *N*-hydroxyurethan is attributed to its rapid conversion back into urethan. It has also been reported that *N*-hydroxyurethan and urethan are almost equal in their carcinogenic activity (762) as well as in their ability to produce chromosome abnormalities in cells of the Walker rat carcinoma (760).

The antitumor activity of *N*-hydroxyurethan in various mouse tumors (763) as well as its teratogenicity in the rat (696) have also been described. The induction of chromosome aberrations by *N*-hydroxyurethan in *Vicia faba* (760), mammalian cells in culture (665, 666), as well as its inactivation of transforming DNA (764, 765) have also been reported.

Freese *et al.* (634) suggested that even *N*-hydroxyurethan does not itself react with DNA but reacts with oxygen yielding peroxy radicals and other derivatives, some of which are apparently the active reagents.

The dichotomy of action of *N*-hydroxyurethan in bacteria and animals was described by Schmidt *et al.* (766) who indicated that while DNA synthesis is the primary target of *N*-hydroxyurethan in mammalian cells, in bacteria this agent primarily blocks protein production.

Urethan and *N*-hydroxyurethan have been analyzed by colorimetry (759, 767) and paper (706, 768), thin-layer (759, 769–772), and gas chromatography (773).

23. *4-Nitroquinoline-1-oxide and 4-Hydroxyaminoquinoline-1-oxide*

4-Nitroquinoline-1-oxide (4-NQO) was first synthesized by Ochiai *et al.* in 1943 (774). Its fungicidal (775), carcinostatic (776, 777), bactericidal, and virucidal (778) properties and utility in trichomonas infestation therapy were described (779). The photodynamic activity of 4-NQO (780), its reaction with thiol groups (781, 782), binding to DNA (783–785), and interference with interferon synthesis (786) have been reported.

The metabolism of 4-NQO *in vitro* (787) and *in vivo* (788) has indicated that it is reduced to 4-aminoquinoline-1-oxide in the cell via the intermediate 4-hydroxyaminoquinoline-1-oxide. It has also been shown that 4-NQO is metabolized at the site of injection to 4-hydroxyaminoquinoline-1-oxide and that this reductive metabolite might be the proximate carcinogen (789, 790). 4-Hydroxyaminoquinoline-1-oxide has been previously reported to be carcinogenic (791, 792), carcinostatic (792), mutagenic (793), and to have prophage-inducing ability (794) as well as the ability to produce intranuclear inclusion bodies in cultured Chang's liver cells (792).

Salient properties of 4-nitroquinoline-1-oxide include carcinogenicity (781, 795–798), mutagenicity in *E. coli* (157, 799), *A. niger* (800), *Streptomyces griseofulvus* (801), *Penicillium chrysogenum, Sarcina lutea* (802), and bacteriophage T2 (803), and chromosomal aberrations in mammalian cells in tissue culture (804–806).

The analysis of 4-NQO has been achieved by ultraviolet, infrared, and nmr spectroscopy (807), and paper chromatography (808).

24. *Hydrogen Peroxide*

The manufacture of hydrogen peroxide is based on the electrolysis of sulfuric acid solutions and subsequent hydrolysis of the formed peroxydisulfuric acid to yield hydrogen peroxide and sulfuric acid. The United States production of hydrogen peroxide in 1965 was estimated at 50,000 tons.

Hydrogen peroxide undergoes a variety of reactions:

Decomposition	$2\,H_2O_2 \rightarrow 2\,H_2O + O_2$	(1)
Molecular addition	$H_2O_2 + Y \rightarrow Y \cdot H_2O_2$	(2)
Substitution	$H_2O_2 + RX \rightarrow ROOH + HX$	(3)
or	$H_2O_2 + 2\,RX \rightarrow ROOR + 2\,HX$	(4)
H_2O_2 as oxidizing agent	$H_2O_2 + Z \rightarrow ZO + H_2O$	(5)
H_2O_2 as reducing agent	$H_2O_2 + W \rightarrow WH_2 + O_2$	(6)

In entering into these reactions, hydrogen peroxide may either react as a molecule or it may first ionize or be dissociated into free radicals. The largest commercial use for H_2O_2 is in the bleaching of cotton textiles and wood and chemical (Kraft and sulfite) pulps. (The rate of bleaching appears directly related to alkalinity, with the active species assumed to be the perhydroxyl anion OOH^-.) The next largest use of H_2O_2 is in the oxidation of a variety of important organic compounds. For example, soybean oil, linseed oil, and related unsaturated esters are converted to the epoxides for use as plasticizers and stabilizers for polyvinyl chloride. Other important commercial processes include the hydroxylation of olefinic compounds, the synthesis of glycerol from propylene, the conversion of tertiary amines to corresponding amine oxides, and the conversion of thiols to disulfides and sulfides to sulfoxides and sulfones. In the textile field, H_2O_2 is used to oxidize vat and sulfur dye.

In addition, many organic peroxides are made from hydrogen peroxide. These include peroxy acids such as peroxyacetic acid; hydroperoxides such as *tert*-butyl hydroperoxides; diacyl peroxides such as benzoyl peroxide; ketone derivatives such as methylethylketone peroxide. These derivatives are used as oxidants and polymerization catalysts for cross-linking agents. Other uses of hydrogen peroxide include its use as a blowing agent for the preparation of foam rubber, plastics, and elastomers, bleaching, conditioning, or sterilization of starch, flour, tobacco, paper, and fabric (809), and as a component in hypergolic fuels (810). Solutions of 3% to 6% H_2O_2 are employed for germicidal and cosmetic (bleaching) use, although concentrations as high as 30% H_2O_2 have been used in dentistry (811). Concentrations of 35% and 50% H_2O_2 are used for most industrial applications.

The biological significance of hydrogen peroxide is as yet not fully understood. Of all peroxides, it is the most efficiently destroyed, particularly by the action of catalase and formic acid oxidase.

Warburg (812) has suggested that hydrogen peroxide is the active intracellular product as a consequence of ionizing radiation and has stated that the direct action of x-rays on cell metabolism (cancer cells) was due to the hydrogen peroxide produced rather than to organic peroxides, intermediate hydrogen atoms, or hydroxyl radicals.

The question of mutagenic activity of hydrogen peroxide has been discussed, especially in connection with radiation-induced mutagenesis (813–817). The production of ions (H^+, OH^-) and inorganic radicals ($\cdot OH$, $HO_2\cdot$) with hydrogen peroxide formation in aqueous systems in response to ionizing radiation is well known.

Hydrogen peroxide has been found mutagenic in *Staphylococcus aureus* (818–820), *E. coli* (17, 159), and *Neurospora* (757, 821, 822) [including mixtures of hydrogen peroxide and acetone, and hydrogen peroxide and formaldehyde (821)]. The inactivation of transforming-DNA by hydrogen peroxide (765, 823–825) as well as by peroxide-producing agents (765) (compounds which contain a free N—OH group such as hydroxylamine, *N*-methylhydroxylamine, hydroxyurea, hydroxyurethan, and hydrazines on exposure to oxygen) has been described. The nonmutagenic effect of hydrogen peroxide in ultraviolet-light exposed and photoreactivated bacteria has been described by Doudney

(826). Hydrogen peroxide has induced chromosomal aberrations in strains of ascites tumors in mice (827) and in *Vicia faba* (828).

The determination of hydrogen peroxide is made by a variety of techniques including titration with potassium permanganate, polarography, or by procedures involving the oxidation of iodide ion to iodine, the reduction of ceric sulfate to cerous sulfate, or the titanium sulfate method (829) followed by spectrophotometric measurement of the resulting complex (830). The spectrophotometric determination of hydrogen peroxide in biological systems as shown has been most recently reported (831).

$$\underset{}{2\,H_2O_2} + \underset{(O\text{-dianisidine})}{\text{acceptor}} \xrightarrow{\text{peroxidase}} 2\,H_2O + \text{oxidized acceptor}$$

25. *Organic Peroxides*

The utility of a variety of organic peroxides in a broad spectrum of commercial and laboratory polymerization reactions is well established. Their formation and mechanisms of reaction have been reviewed (832–835). Organic peroxides find 90% of their market in the polymer industry (rubbers, elastomers, plastics, resins, etc.). The 1964 consumption of the seventeen major commercial organic peroxides in the United States amounted to 8.7 million pounds. The commercial organic peroxides are sources of free radicals: $RO{:}OR \rightarrow RO\cdot + \cdot OR$. In the polymer industry organic peroxides are used as (a) initiators for the free-radical polymerizations and/or copolymerizations of vinyl and diene monomers, (b) curing agents for resins and elastomers, and (c) cross-linking agents for polyolefins.

Polyester thermosets are made by curing an unsaturated polyester resin and a polymerizable monomer with an organic peroxide. These resins are normally cured with reinforcing agents such as glass fibers for a variety of uses, e.g., furniture and electrical parts. Many of these applications require room-temperature curing; this is achieved by the use of activators or promoters such as dimethylaniline which cause the peroxide, such as dibenzoyl peroxide, to decompose and initiate curing at room temperature. Organic peroxides have been used in the bleaching of various materials such as flour, gums, waxes, fats, and oils. They have also been used as cosmetic and pharmaceutical additives and intermediates, and as a free-radical source in many organic syntheses. The miscellaneous uses of organic peroxides in the polymer industry include vulcanization of natural and butadiene rubbers, curing polyurethans and adhesives, preparing graft copolymers, flame retardant synergist for polystyrene, solidification of soils with calcium acrylate, cross-linking polyethylene and ethylene-containing copolymers.

a. tert-Butyl Hydroperoxide [$(CH_3)_3COOH$]. The commercial form consists of a 90% liquid and 60% to 75% solutions containing di-*tert*-butyl peroxide. It is used as an initiator for vinyl monomer polymerizations and copolymerizations with styrene, vinyl acetate, acrylics, and as a curing agent for thermoset polyesters. The 1964 consumption of *tert*-butyl hydroperoxide was 70,000 pounds.

The mutagenic effect of *tert*-butyl hydroperoxide in *Drosophila* (836, 837), *E. coli* (838), and *Neurospora* (821), and its induction of chromosome aberrations in *Vicia faba* (754, 839, 840) and *Oenothera* (754) has been described.

b. Cumene Hydroperoxide [$C_6H_5C(CH_3)_2OOH$]. The commercial form consists of 70% liquid containing cumene, acetophenone and cumyl alcohol. It is employed as an initiator for vinyl monomer polymerizations and copolymerizations with styrene, acrylics, butadiene-styrene, and as a curing agent for thermoset polyesters and styrenated alkyds and oils. The 1964 consumption of cumene hydroperoxide was 640,000 pounds. Cumene hydroperoxide has been found mutagenic in *E. coli* (838) and *Neurospora* (757, 825).

c. Succinic Acid Peroxide and Disuccinyl Peroxide. Succinic acid peroxide

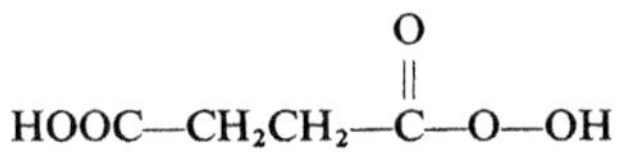

is commercially available as a 95% powder and is used an an initiator for vinyl monomer polymerizations and copolymerizations with ethylene and fluoroolefins. It has been found to inactivate T2 phage (841), transforming-DNA of *H. influenzae* (823), and to be mutagenic in *E. coli* (838, 842).

Disuccinyl peroxide has utility analogous to succinic acid peroxide. It has been found to inactivate transforming-DNA (765, 823, 825, 841). Luzzati *et al.* (823) found that the hydrolysis of disuccinyl peroxide resulted in the formation of both succinic acid and succinic acid peroxide, viz.:

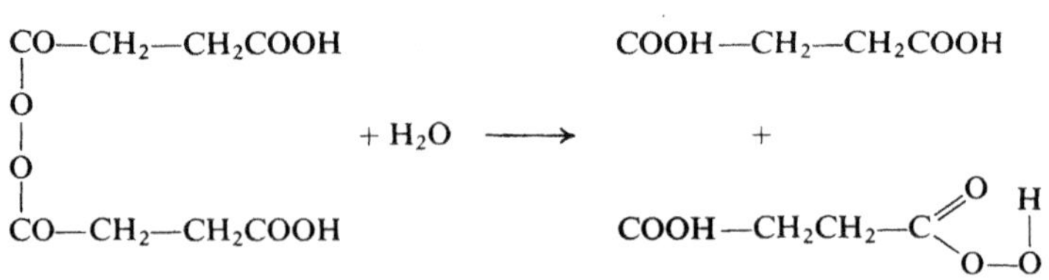

Aqueous solutions of succinic peroxide are also unstable, yielding oxygen and succinic acid. Thus succinic peroxide acts as an oxygen "donor."

d. Dihydroxydimethyl Peroxide (*HO—CH₂—O—O—CH₂—OH*). Dihydroxydimethyl peroxide is prepared via the reaction of formaldehyde and hydrogen peroxide in ether (843). Its formation in polluted air has been postulated by Kotin and Falk (844) as depicted in Fig. 9.4.

Mutations in *Drosophila* have been produced by direct application of dihydroxydimethyl peroxide (845, 846). The peroxide like formaldehyde induces mutations in mature sperm, supporting the hypothesis that formaldehyde exerts at least part of its mutagenic effects via the formation of an organic peroxide. In addition to their action on mature sperm, both dihydroxydimethyl

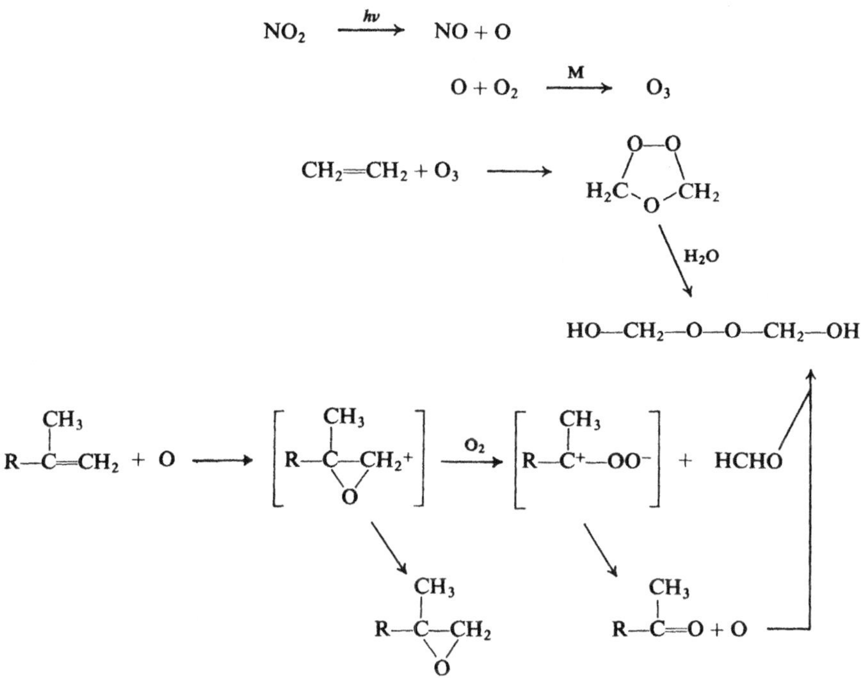

FIG. 9.4. Anticipated reactions leading to the formation of dihydroxydimethyl peroxide in polluted air. Kotin, P., and Falk, H. L., *Radiation Res. Suppl.* 3, 193 (1963).

peroxide and formaldehyde (847) produce mutations in earlier stages of spermatogenesis. This further suggests that the formation of a peroxide may also be involved in the mutagenic effects of formaldehyde on earlier stages of sperm development.

It has also been shown that a mixture of formaldehyde and hydrogen peroxide is a much stronger mutagen for *Neurospora cassida* (757) than either substance alone; this is attributed (841) to the formation of dihydroxymethyl peroxide. However, in *Drosophila* a mixture of formaldehyde and hydrogen peroxide did not produce more mutations than formaldehyde alone (847).

Sobels (847) suggested that this may be due to the high catalase content of the body fluids of the fly.

The radiomimetic effect of dihydroxydimethyl peroxide on roots of *Vicia faba* has been described by Loveless (840). Freese *et al.* (765) found that dihydroxydimethyl peroxide (as well as formaldehyde and its oxidation products, formic and performic acids) did not inactivate or mutate transforming-DNA at a significant rate.

e. Di-tert-butyl Peroxide [$(CH_3)_3COOC(CH_3)_3$]. Di-*tert*-butyl peroxide is commercially available as a 99% liquid and has an annual consumption in the United States of approximately 500,000 pounds. It is used as an initiator for vinyl monomer polymerizations and copolymerizations of ethylene, styrene, vinyl acetate, and acrylics and as a curing agent for thermoset polyesters, styrenated alkyds and oils, and silicone rubbers. It has also found use as an ignition accelerator for diesel fuels and is used extensively in organic synthesis as a free radical catalyst and as a source of reactive methyl radicals.

The decomposition of di-*tert*-butyl peroxide in the dilute gas phase has been shown to proceed by the following mechanism (848):

$$(CH_3)_3COOC(CH_3)_3 \rightarrow (CH_3)_3CO\cdot$$
$$(CH_3)_3\cdot \rightarrow (CH_3)_2C{=}O + CH_3\cdot$$

The products are acetone and ethane plus small amounts of other compounds. The *tert*-butoxy radical is reactive and can be utilized for the initiation of polymerizations and antioxidations. When di-*tert*-butyl peroxide as the neat liquid is decomposed at 100°C, isobutylene oxide is formed as a major product. This arises from a radical attack on the peroxide molecule, viz.:

$$\left.\begin{array}{l}CH_3\cdot\\(CH_3)_3CO\cdot\end{array}\right\} + (CH_3)_3COOC(CH_3)_3 \rightarrow (CH_3)_3CO{-}O{-}\overset{\displaystyle \cdot CH_2}{\overset{|}{C}}(CH_3)_2 + \left\{\begin{array}{l}CH_4\\(CH_3)_3COH\end{array}\right.$$

The isobutylene is formed by the following step which is an intramolecular radical displacement (849):

Di-*tert*-butyl peroxide is mutagenic in *Neurospora* (757) but has been reported to be inactive toward transforming-DNA (825).

The analysis of organic peroxides has been reviewed by Martin *et al.* (850), Siggia (851), and Cosija (852). Since organic peroxides are oxidizing agents, the most common chemical methods of analysis for this class of compounds involves the reduction of the peroxide group followed by the determination of the excess reducing agent or of the oxidized form of the reducing agent, e.g.:

$$\mathrm{R{-}\underset{\underset{O}{\|}}{C}{-}O{-}O{-}\underset{\underset{O}{\|}}{C}{-}R} + 2\,I^- \rightarrow 2\,\mathrm{R{-}\underset{\underset{O}{\|}}{C}{-}O^-} + I_2$$

The liberated iodine can then be titrated with standard thiosulfate solution or in trace analysis, determined by spectrophotometric methods (853, 854). Organic peroxides generally absorb in the 800 to 900 cm^{-1} infrared region which has been used for analysis, particularly for kinetic studies (855). Gas chromatography has been used for the estimation of some thermally stable hydroperoxides, peroxyesters, and dialkyl peroxides (856) (e.g., di-*tert*-butyl peroxide) (857). Other techniques that have been used for the analysis of organic peroxides include polarography (858) and paper (859) [*tert*-butyl-, and cumenehydroperoxide (860)] and thin-layer chromatography (861, 862).

26. *Aflatoxin*

Aflatoxins are toxic mold metabolites which are produced by a limited number of strains of a few fungi, e.g., *Aspergillus flavus*, *A. parasiticus*, and *Penicillium puberulum*. Although the collective term "aflatoxin" was formerly used to describe the toxic products of these fungi, it is now well recognized that there are eight compounds of related molecular configuration (i.e. aflatoxins, B_1, B_2, B_{2a}, G_1, G_2, G_{2a}, M_1, and M_2) as shown in Fig. 9.5.

The chemical, physical, and biological properties and isolation of the aflatoxins have been reviewed by Wogan (863–865), the chromatography by Fishbein and Falk (866), the occurrence in feeds and foods by Borker *et al.* (867), and their role in carcinogenesis by Kraybill and Shimkin (868), Wogan (869), and Newberne and Butler (870).

The high order of carcinogenic potency of the aflatoxins has caused considerable speculation as to whether they might be involved in the etiology of human liver disease, including primary carcinoma. It has been suggested by Wogan (871) that epidemiological patterns of primary liver cancer incidence, together with what is known about the risks of aflatoxin contamination of foodstuffs and the potency of the compounds in animals, provide suggestive circumstantial evidence that aflatoxins (or other mold toxins) play a role in the etiology of the disease (869).

The metabolism of aflatoxin B_1 has been studied in rats (870–873), sheep (874, 875), and cows (876). The facility of aflatoxin B_1 for binding to DNA (877–880), the effect on DNA polymerase of *E. coli* (881), the effect on physical profile and RNA synthesis in rat liver (882), the modes of action of aflatoxin on the transcription process (883) have been studied. The biological effects of aflatoxin in cell culture (884), its effect on human leukocytes (885), its suppression of mitosis in cultured human diploid and heteroploid embryonic lung cells (886, 887), and its inhibition of DNA synthesis and giant cell forma-

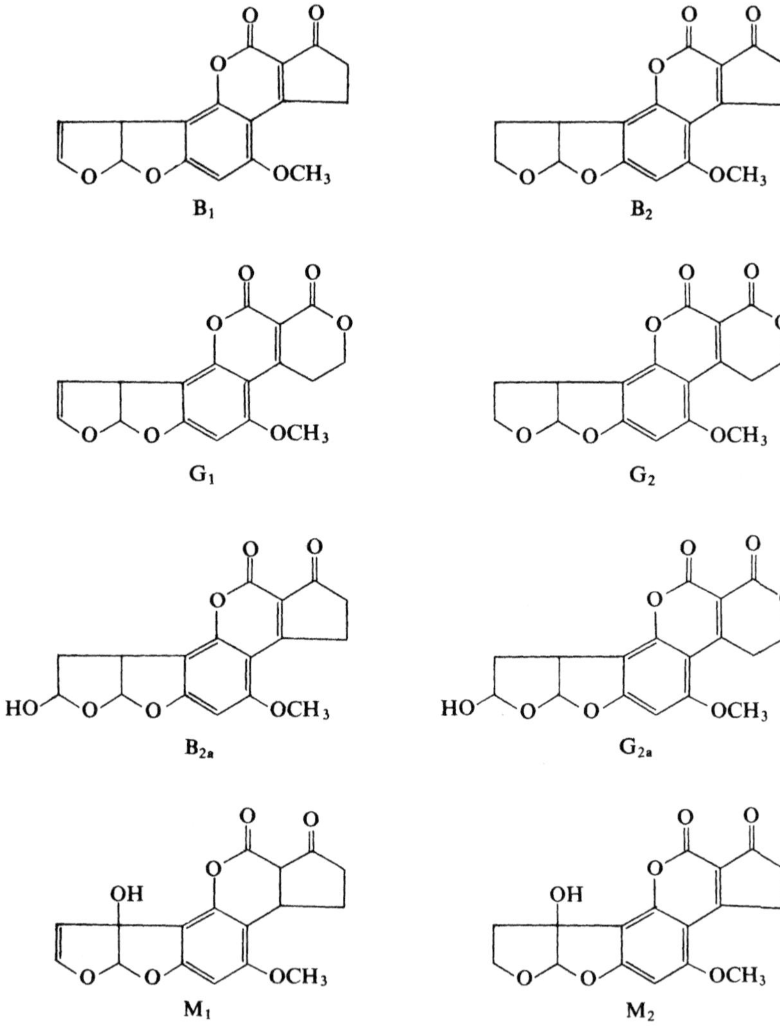

FIG. 9.5. Structures of the aflatoxins.

tions in tissue culture in a manner similar to some of the alkylating agents (888) have been reported. The mutagenicity of aflatoxin B_1 in the dominant lethal test in mice has been most recently reported by Epstein and Shafner (889).

27. *Polynuclear Aromatic Hydrocarbons*

The general aspects of polynuclear aromatic hydrocarbons (PAH) that have been reviewed and are germane for our consideration include the following: the occurrence, isolation, and identification in the human environment [air, water, soil (890–893), tobacco and cigarette smoke (894), foods (895, 896)], analysis and bioassay (897–899) [in foodstuffs (899–902), atmosphere (903, 904)], carcinogenic aspects and mode of action (905–908), reaction with nucleic acid (909–911), and metabolism (912–915).

TABLE 9.4

Occurrence of 3,4-Benzpyrene in Environmental Sources

Source	References
Cigar, cigarette, and tobacco smoke condensate	915–917
Cigarette additive (DL-menthol) pyrolysis	918
Polluted air	919, 920
Rubber-tire dust	921
Gasoline	922
Exhaust gas from gasoline engines	923, 924
Refined oils (mineral oil)	925
Roasting coffee and coffee substitutes	926
Baking bread and biscuits	927
Margarine and mayonnaise	928
Smoked foods	895, 929, 930
Liquid smoke and curing smoke	930–932
Oranges	933
Commercial wax	934
Marine fauna and flora	935
Freshly mined asbestos	936

a. 3,4-Benzpyrene. 3,4-Benzpyrene is the most ubiquitous and potent known carcinogenic agent in polluted air. Its presence has been reported for decades. The primary source of benzpyrene is the incomplete combustion of solid, liquid, and gaseous fuels. Table 9.4 depicts a list of the occurrence of benzpyrene in a variety of environmental sources.

Badger *et al.* (937, 938) have proposed the formation of benzpyrene during pyrolysis to proceed as depicted in Fig. 9.6. In this scheme possible intermediate compounds would be ethylene or acetylene for (I), butadiene or vinylacetylene for (II), styrene or ethylbenzene for (III), and phenylbutadiene, *n*-butylbenzene, or Tetralin for (IV) and (V).

The biochemistry and metabolism of benzpyrene have been described by Weigert (939), Harper (940), Kotin and Falk *et al.* (941, 942), and Heidelberger *et al.* (943). Figure 9.7 illustrates the anatomic and biochemical pathways for the metabolism of benzpyrene as postulated by Falk *et al.* (942).

The interaction of 3,4-benzpyrene with DNA (944) and the role of peroxide-induced binding of PAH compounds (e.g., 3,4-benzpyrene, 3-methylcholanthrene, and 9,10-dimethyl-1,2-benzanthracene) to DNA (911) have been described.

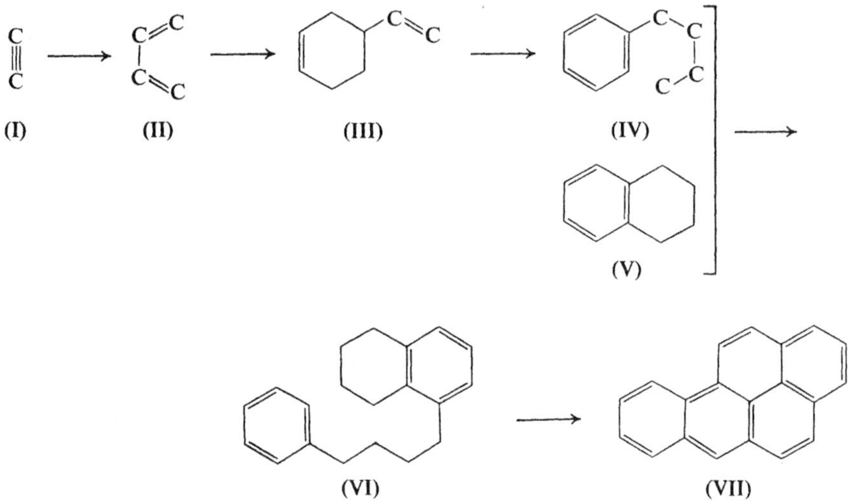

FIG. 9.6. Hypothesis of benzpyrene formation during pyrolysis. Badger, G. M., Kimber, R. W. C., and Novotny, J., *Australian J. Chem.* **15**, 616 (1962).

3,4-Benzpyrene has been found to be mutagenic in *Drosophila* (945), *E. coli* (946), and in mice (dominant lethal test) (889).

b. 1,2,5,6-Dibenzanthracene (Dibenz[a,h]anthracene). The isolation of 1,2,5,6-dibenzanthracene from cigarette smoke (947), mineral oils (925), "dairy wax" (948), smoked meat (949) and fish (950), and oranges (933) has been reported. The carcinogenicity of dibenzanthracene was first described by Kennaway in 1930 (951), Boyland and Burrows (952), and Steiner and Falk (953), and its metabolism by Heidelberger and West (954) and Boyland and Sims (955). Figure 9.8 depicts the metabolism of dibenzanthracene in mouse and rabbit as proposed by Boyland. The binding of dibenzanthracene by body constituents (nucleic acids, proteins, carbohydrates) has been described (956).

Dibenzanthracene is mutagenic in *Drosophila* (945, 957, 958), *Neurospora* (959, 960), *E. coli* (946) and mice (961); its nonmutagenicity in mice (962)

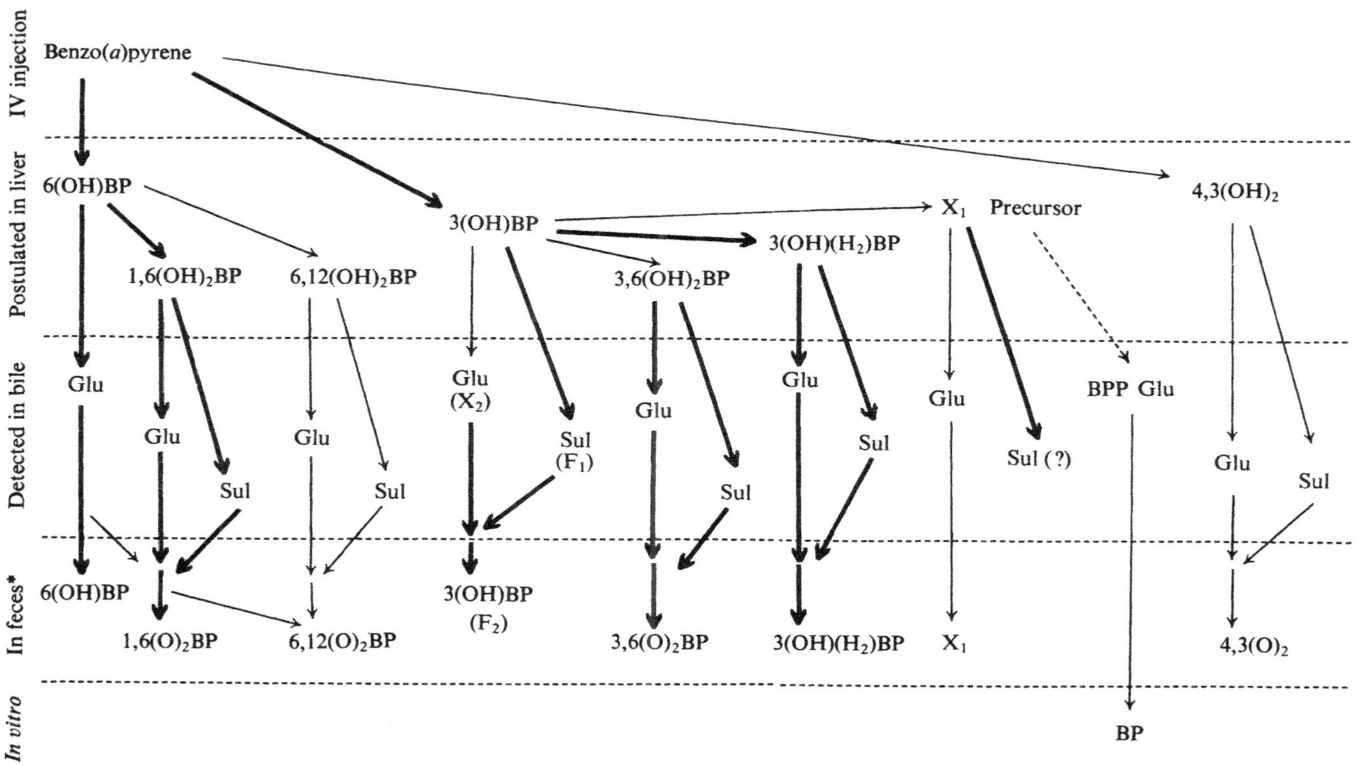

FIG. 9.7. Anatomic and biochemical pathways for the metabolism of benzpyrene. BPP = BP precursor; BP = benzo(*a*)pyrene; Sul = sulfate; Glu = glucuronide. Sulfate identification is only tentative. Asterisk indicates that metabolites anticipated in feces were detected after β-glucuronidase incubation of bile fractions. Falk, H. L., Kotin, P., Lee, S. K., and Nathan, A., *J. Natl. Cancer Inst.* **28**, 699 (1962).

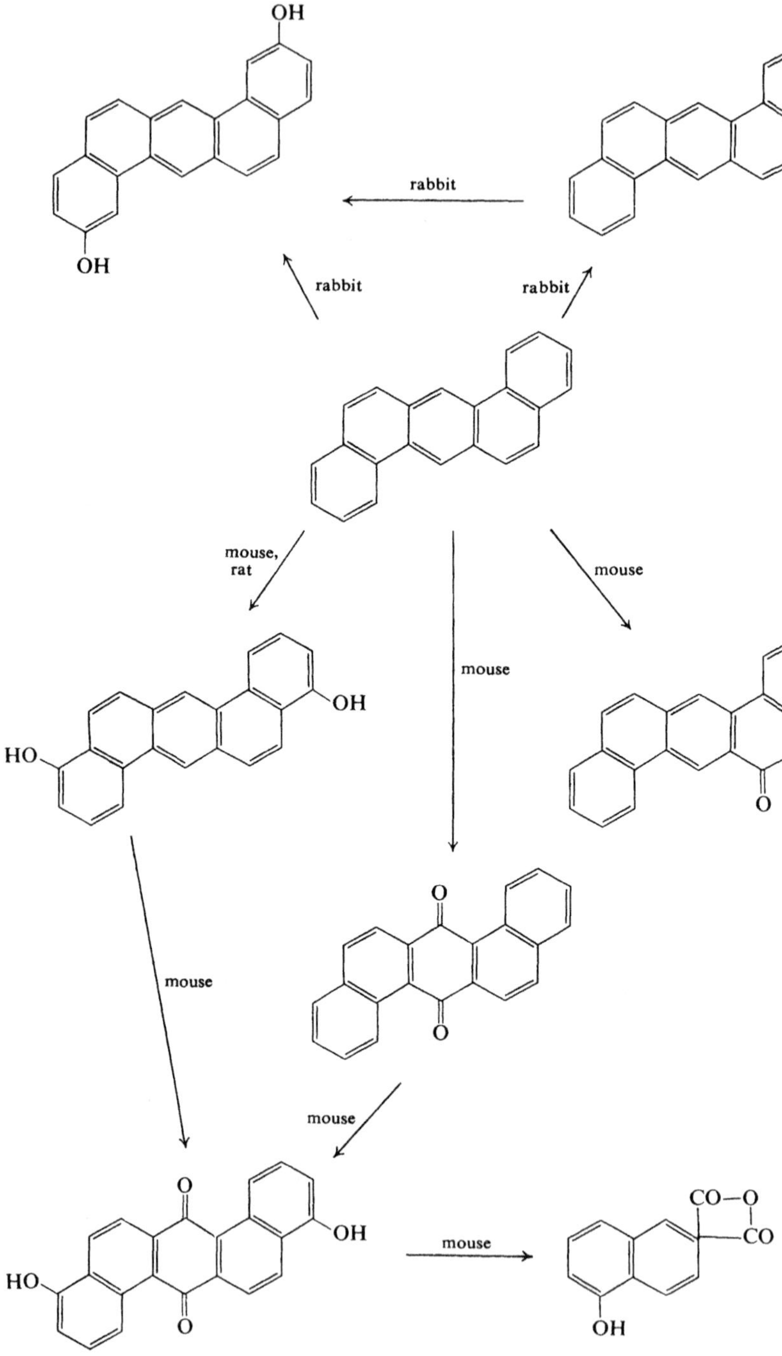

FIG. 9.8. Metabolism of 1,2,5,6-dibenzanthracene.

has been described. Dibenzanthracene (as well as 3,4-benzpyrene and 1,2-benzanthracene) have been reported to be unable to induce mitotic gene conversion in *Saccharomyces cerevisiae* (963).

The inactivity of these compounds was ascribed either to their inability to react directly with DNA or else to the lack of activating enzymes in the yeast cells used.

The mutagenic action of other polynuclear aromatic hydrocarbons has been reported, e.g., for 20-methylcholanthrene: in *Drosophila* (945, 957, 964), *E. coli* (946), *Neurospora* (959, 960), and mice (965–967).

The nonmutagenicity of 20-methylcholanthrene in *Drosophila* (968, 969) and *E. coli* (970) has also been reported.

9,10-Dimethyl-1,2-dibenzanthracene is mutagenic in *Neurospora* (959, 960) and induces mitotic gene conversion in *Saccharomyces cerevisiae* (963).

28. *Metallic Salts*

a. Manganous Chloride. Manganese is found in a large variety of minerals, the most important of which is the oxide MnO_2. The principal uses of manganese are (1) in steel manufacturing (to reduce oxygen and sulfur), (2) in dry cell batteries as MnO_2, and (3) as an oxidizing agent in chemical industry for the production of $KMnO_4$ and other manganese chemicals. It is also used in antiknock compounds (as manganese cyclopentadienyl tricarbonyl), pigments, paint, and varnish dryers, welding rods, drugs, soaps, rubber and wood preservatives, fungicides, and fertilizers, and as a chemical catalyst (in the form of manganous sulfate), as a trace element in poultry and animal feeds, and as manganates and permanganates for disinfection, bleaching, and laboratory reagents.

Manganous chloride is used specifically as a catalyst in the polymerization of butadiene (971), in urethan polymer manufacture (972), in acrylonitrile polymerization (973), in acrylonitrile polymer fiber drying (974), as an additive in fluxes for degassing aluminum alloys (975), in lubricating oils for use under extreme pressures (976), in phosphors (977), in fertilizer sprays (978), as an egg-hatching agent for the beet cyst nematode *Heterodera schachtii* (979), as well as in the purification of natural gas and in the drying of linseed oil.

In addition to industrial exposures to manganese fumes and dusts per se, manganese is found in tobacco. Cogbill and Hobbs (980) found an average of 180 ppm of manganese in tobacco but only 0.4 μg in the mainstream smoke of 100 cigarettes. Voss and Nical (981) found 135 ppm of manganese in tobacco leaf while Tso (982) reported 70 mg/100 g of leaf tobacco. Tobacco, as well as other plant tissues, contains minerals and other inorganic constituents derived from soil, fertilizers, or agricultural sprays. The fungicide maneb (manganese ethylenebis[dithiocarbamate]) and the antiozonant manganous-

1,2-naphthoquinone-2-oxime could serve as additional exogenous sources of manganese.

Manganese is an essential element for normal metabolism. Human requirements are estimated between 3 mg and 9 mg per day (983). On this daily average intake, the blood contains from 12 to 15 $\mu g\%$ manganese, the urine somewhat less than 10 μg/liter, most of the intake appearing in the feces. Small quantities are found in most tissues (liver, bone, and lymph nodes containing the highest concentrations).

The distribution, pharmacology, and health hazards of manganese have been reviewed by von Oettingen (984), Fairhall and Neal (985), and Cotzias (986). Acute (987, 988) and chronic (989) animal and human (990, 991) toxicity as well as the metabolism of manganous chloride (983, 984, 992) and its potentiation of lymphosarcomas in the mouse (993) have been described. The toxicity of manganese in plants has also been reported (994, 995).

Manganous chloride has been reported to be mutagenic in *E. coli* (17, 996–999) and *Serratia marscescens* (1000). Demerec and Hanson (996) found the mutagenicity of $MnCl_2$ enhanced in *E. coli* by washing the bacteria in a number of hypertonic solutions before treatment and diminished by similar washing with water or hypotonic solutions. It was suggested (996) that $MnCl_2$ alters metabolism of bacterial cells so as to make the gene systems more unstable than usual. Manganous acetate, nitrate, and sulfate salts have also been found to be mutagenic in *E. coli* (17).

Manganese has been determined in biological materials by spectrophotometry (1001) and in water by polarography (1002), colorimetry (1003, 1004), and atomic absorption spectroscopy (1005).

b. Miscellaneous Inorganic Salts. Ferrous chloride has also been found mutagenic in *E. coli* (17), while negative results were obtained for nickelous sulfate and nitrate, stannic chloride, potassium ferrocyanide and ferricyanide, cerous chloride, and nitrate and mercurous chloride (1000). Several other compounds including ferric chloride and ferric sulfate were slightly mutagenic in *E. coli* (17).

Muro and Goyer (1006) reported that chromosomes from leukocyte cultures from mice fed a diet containing 1% lead acetate exhibit an increased number of gap-break type aberrations (the observed chromosome abnormalities involve only single chromatids). It was hypothesized that chromosome damage may result from activation of lysosomal enzymes, particularly DNase, or that lead impairs adenosine triphosphate and protein synthesis required for repair. The authors suggested (1006) that chromosomal damage induced by administration of lead salts may contribute to reduction in fertility and the formation of congenital malformations. Excessive fetal loss and malformed offsprings resulting from lead intoxication in both experimental animals and humans have been reported (1007, 1008). The induction of chromosome breaks in root tips of *Vicia faba* has been reported for aluminum chloride (754, 755, 1009) and cadmium nitrate (754).

29. *Phosphoric Acid Esters*

a. Trimethyl Phosphate. Trimethyl phosphate (TMP), the simplest trialkyl ester of phosphoric acid, is a methylating agent largely employed as a low-cost gasoline additive and in the preparation of organophosphorus insecticides. The chemosterilant action of TMP in rodents has been reported by Jackson and Jones (1010). Using ^{32}P-TMP, the sole phosphorus-containing metabolite is dimethyl phosphate which has no antifertility activity. With ^{14}C-trimethyl phosphate, *S*-methyl cysteine was identified as a urinary metabolite indicating that trimethyl phosphate is involved as an alkylating agent (at least in its detoxification process). Jackson and Jones postulated that the antifertility action of TMP is probably related to methyl alkylation bearing an analogy with the methyl ester of methane sulfonic acid which also produces the "functional" type of sterility in rats and mice (1011). Like methyl methane sulfonate (MMS), trimethyl phosphate in substerilizing doses induces dominant lethal mutations in mice (1012).

TMP is mutagenic in *Neurospora* and most recently has been reported by Epstein (1013) to be mutagenic in the dominant lethal test.

b. 2,2-Dichlorovinyl Dimethyl Phosphate. 2,2-Dichlorovinyl dimethyl phosphate (Dichlorovos, DDVP) is prepared by the reaction of trimethyl phosphate and chloral and is employed as an insecticide and fumigant and in veterinary medicine as a helminthic. It is widely used as the active compound in household vaporizing resin strips. (Dog collars of resin strips impregnated with DDVP are also used.) Using the resin strips as recommended, the air concentration of DDVP will be between 0.5 and 1 mg/m^3 during the first week and above 0.1 mg/m^3 during the first 5 to 6 weeks of the total usage period (1013, 1014).

Saturated aqueous solutions of DDVP at room temperature convert to dimethyl phosphoric acid and dichloroacetaldehyde at a rate of about 3% per day; enhanced hydrolysis occurs in alkali media.

Löfroth *et al.* (1015) recently described the alkylating and biological properties of DDVP. In tests with secondary roots of *Vicia faba*, DDVP concentrations down to 1 m*M* were mostly lethal. In root tips surviving this and lower concentrations, chromatid breaks and gaps were observed. Also, in preliminary tests with a streptomycin-dependent Sd-4 mutant of *E. coli* strain B, concentrations of 1 to 3 m*M* of DDVP caused no significant lethality, but increased mutation rate was observed measured as revertants to streptomycin independency. Sax and Sax (380) previously showed that DDVP causes chromosome aberrations in onion root tip cells.

Since many organophosphorus insecticides are of the triester type and are likely to be alkylating, investigation of their mutagenicity in appropriate test systems has been advocated (1016).

REFERENCES

1. A. Albert, "The Acridines." Arnold, London, 1951.
2. A. R. Peacock, *Chem. & Ind. London* p. 642 (1969).
3. S. Brenner, L. Barnett, F. H. C. Crick, and A. Orgel, *J. Mol. Biol.* **3**, 121 (1961).
4. L. E. Orgel, *Advan. Enzymol.* **27**, 289 (1965).
5. L. S. Lerman, *Proc. Natl. Acad. Sci. U.S.* **49**, 94 (1963).
6. B. L. Van Duuren, B. M. Goldschmidt, and H. H. Seltzman, *Ann. N.Y. Acad. Sci.* **153**, 744 (1969).
7. C. H. Browning, *Exptl. Chemotherapy* **2**, 1 (1964).
8. D. Ligat, *Brit. Med. J.* **II**, 78 (1917).
9. C. H. Browning, R. Gulbransen, and L. H. D. Thornton, *Brit. Med. J.* **II**, 70 (1917).
10. R. I. DeMars, *Nature* **172**, 964 (1953).
11. D. A. Ritchie, *Genet. Res.* **5**, 168 (1964).
12. D. A. Ritchie, *Genet. Res.* **6**, 474 (1965).
13. C. M. Calberg-Bacq, M. Delmelle, and J. Duchesne, *Mutation Res.* **6**, 15 (1968).
14. A. Orgel and S. Brenner, *J. Mol. Biol.* **3**, 762 (1961).
15. W. Ostertag and W. Kersten, *Exptl. Cell Res.* **39**, 29 (1965).
16. E. M. Witkin, *Cold Spring Harbor Symp. Quant. Biol.* **12**, 256 (1947).
17. M. Demerec, G. Bertani, and J. Flint, *Am. Naturalist* **85**, 119 (1951).
18. C. J. Avers and C. D. Dryfuss, *Nature* **206**, 850 (1965).
19. C. J. Avers, C. R. Pfeiffer, and M. W. Rancourt, *J. Bacteriol.* **90**, 481 (1965).
20. W. Lotz, R. W. Kaplan, and H. D. Mennigmann, *Mutation Res.* **6**, 329 (1968).
21. E. Witkin, *J. Cellular Comp. Physiol.* **58**, Suppl. 1, 135 (1961).
22. N. Trainin, A. M. Kaye, and I. Berenblum, *Biochem. Pharmacol.* **13**, 263 (1964).
23. A. V. Karyakin and L. A. Chmutina, *Biofizika* **9**, 558 (1964).
24. R. E. Ballard, A. J. McCaffery, and S. F. Mason, *Biopolymers* **4**, 97 (1966).
25. L. Stryer and E. R. Blout, *J. Am. Chem. Soc.* **83**, 1411 (1961).
26. D. Freifelder, P. F. Davison, and E. P. Geiduschek, *Biochem. J.* **1**, 389 (1961).
27. A. F. Harris, A. Saifer, and S. K. Weintraub, *Arch. Biochem.* **95**, 106 (1961).
28. D. F. Bradley and M. K. Wold, *Proc. Natl. Acad. Sci. U.S.* **45**, 944 (1959).
29. R. B. Webb and H. E. Kubitschek, *Biochem. Biophys. Res. Commun.* **13**, 90 (1963).
30. F. D'Amato, *Caryologia* Suppl. 6, 831 (1950).
31. T. Alderson and A. H. Khan, *Mutation Res.* **5**, 147 (1968).
32. P. P. Puglesi, *Mutation Res.* **4**, 289 (1967).
33. G. E. Magni, R. C. von Borstel, and S. Sora, *Mutation Res.* **1**, 227 (1964).
34. R. M. Peck, E. R. Breuninger, J. A. Miller, and H. J. Creech, *J. Med. Chem.* **7**, 480 (1964).
35. R. M. Peck, R. K. Preston, and H. J. Creech, *J. Am. Chem. Soc.* **81**, 3984 (1959).
36. H. J. Creech, E. Breuninger, R. F. Hankwitz, Jr., G. Polsky, and M. L. Wilson, *Cancer Res.* **20**, 471 (1960).

37. I. I. Oster and E. Pooley, *Genetics* **45**, 1004 (1960).
38. L. A. Snyder and I. I. Oster, *Mutation Res.* **1**, 437 (1964).
39. E. A. Carlson and I. I. Oster, *Genetics* **47**, $61 (1962).
40. J. L. Southin, *Mutation Res.* **3**, 54 (1966).
41. E. A. Carlson and I. I. Oster, *Genetics* **46**, 856 (1961).
42. B. N. Ames and H. J. Whitfield, Jr., *Cold Spring Harbor Symp. Quant. Biol.* **31**, 217 (1966).
43. H. E. Brockman and W. Goben, *Science* **147**, 750 (1965).
44. H. V. Malling, *Mutation Res.* **4**, 265 (1967).
45. H. V. Malling and F. J. deSerres, *Mutation Res.* **4**, 425 (1967).
46. S. Kumar, U. Aggarwal, and M. S. Swaminathan, *Mutation Res.* **4**, 155 (1967).
47. V. S. Gupta, S. C. Kraft, and J. S. Samuelson, *J. Chromatog.* **26**, 158 (1967).
48. J. E. Gill, *J. Chromatog.* **26**, 315 (1967).
49. T. Hata, Y. Sano, R. Sugawara, A. Matsumae, K. Kanamori, T. Shima, and T. Hoshi, *J. Antibiot.* (*Tokyo*) **A9**, 141 (1956).
50. S. K. Carter, *Cancer Chemotherapy Rept.* **1**, No. 1, Part 3, 99 (1968).
51. D. A. Karnofsky and B. D. Clarkson, *Ann. Rev. Pharmacol.* **3**, 357 (1963).
52. C. L. Stevens *et al.*, *J. Med. Chem.* **8**, 1 (1964).
53. S. Wakaki, H. Marumo, K. Tomioka, G. Shimizu, E. Kao, H. Kamada, S. Kudo, and Y. Fugimoto, *Antibiot. Chemotherapy* **8**, 228 (1958).
54. F. S. Philips, H. S. Schwartz, and S. S. Sternberg, *Cancer Res.* **20**, 1354 (1960).
55. L. H. Manheimer and J. Vital, *Cancer* **19**, 207 (1966).
56. I. H. Goldberg, *Am. J. Med.* **39**, 722 (1965).
57. V. N. Iyer and W. Szybalski, *Proc. Natl. Acad. Sci. U.S.* **50**, 355 (1963).
58. M. J. Waring, *Nature* **219**, 1320 (1968).
59. V. N. Iyer and W. Szybalski, *Science* **145**, 55 (1964).
60. S. Shika, A. Tirawaki, T. Taguchi, and J. Kawamata, *Nature* **183**, 1056 (1959).
61. R. Mukherjee, *Genetics* **51**, 947 (1965).
62. D. T. Suzuki, *Genetics* **51**, 635 (1965).
63. R. H. Smith, *Mutation Res.* **7**, 231 (1969).
64. M. W. Shaw and M. M. Cohen, *Genetics* **51**, 181 (1965).
65. M. M. Cohen and M. W. Shaw, *J. Cell Biol.* **23**, 386 (1964).
66. P. C. Nowell, *Exptl. Cell Res.* **33**, 445 (1964).
67. T. Merz, *Science* **133**, 329 (1961).
68. K. V. Rao and W. P. Cullen, *Antibiot. Ann.* **7**, 950 (1960).
69. K. V. Rao, K. Biemann, and R. B. Woodward, *J. Am. Chem. Soc.* **85**, 2532 (1963).
70. H. C. Reilly and K. Suguira, *Antibiot. Chemotherapy* **11**, 174 (1961).
71. W. I. Wilson, C. Labra, and E. Barrist, *Antibiot. Chemotherapy* **11**, 147 (1961).
72. J. J. Oleson, L. A. Calderella, K. J. Mjos, A. R. Reith, R. S. Thil, and I. Toplin, *Antibiot. Chemotherapy* **11**, 158 (1961).
73. P. F. Nora, J. C. Kukral, T. Soper, and F. W. Preston, *Cancer Chemotherapy* **48**, 41 (1965).
74. J. D. Hurley, *Proc. Am. Assoc. Cancer Res.* **5**, 29 (1964).
75. M. N. Harris, N. J. Medrek, F. M. Golumb, S. L. Gumport, A. H. Postel, and J. C. Wright, *Cancer* **18**, 49 (1965).
76. R. D. Sullivan, E. Miller, T. Chryssochoos, and E. Watkins, Jr., *Cancer Chemotherapy* **16**, 499 (1962).
77. M. Levine and M. Borthwick, *Virology* **21**, 568 (1963).
78. C. M. Radding, *Proc. 11th Intern. Congr. Genet., The Hague, 1963* p. 22. Pergamon Press, Oxford, 1965.

79. H. L. White and J. R. White, *Biochim. Biophys. Acta* **123**, 648 (1966).
80. B. A. Kihlman, *Mutation Res.* **1**, 54 (1964).
81. G. Zetterberg and B. A. Kihlman, *Mutation Res.* **2**, 470 (1965).
82. M. M. Cohen, M. W. Shaw, and A. P. Craig, *Proc. Natl. Acad. Sci. U.S.* **50**, 16 (1963).
83. T. T. Puck, *Science* **144**, 565 (1964).
84. B. Kihlman and G. Odmark, *Mutation Res.* **2**, 494 (1965).
85. R. R. Herr, H. Jahnke, and A. Argoudelis, *J. Am. Chem. Soc.* **89**, 4808 (1967).
86. R. R. Herr, T. E. Eble, M. E. Bergy, and H. K. Jahnke, *Antibiot. Ann.* p. 236 (1960).
87. L. Lewis and A. R. Barbiers, *Antibiot. Ann.* p. 247 (1960).
88. W. T. Sokolski, J. J. Vavra, and L. J. Hanka, *Antibiot. Ann.* p. 241 (1960).
89. L. J. Hanka and W. T. Sokolski, *Antibiot. Ann.* p. 255 (1960).
90. B. Heinmann and A. J. Howard, *Appl. Microbiol.* **12**, 234 (1964).
91. K. E. Price, R. L. Buck, and J. Lein, *Antimicrobial Agents Chemotherapy* p. 505 (1964).
92. J. S. Evans, G. C. Gerritsen, K. M. Mann, and S. P. Owen, *Cancer Chemotherapy Rept.* **48**, 1 (1965).
93. N. Rakieten, M. L. Rakieten, and M. V. Nadkarni, *Cancer Chemotherapy Rept.* **29**, 91 (1963).
94. S. M. Kolbye and M. S. Legator, *Mutation Res.* **6**, 387 (1968).
95. M. G. Gabridge, A. Denunzio, and M. S. Legator, *Nature* **221**, 68 (1968).
96. M. G. Gabridge, E. J. Oswald, and M. S. Legator, *Mutation Res.* **7**, 117 (1969).
97. E. R. Garrett, *J. Am. Pharm. Assoc., Sci. Ed.* **49**, 767 (1960).
98. A. A. Forist, *Anal. Chem.* **36**, 1338 (1964).
99. J. J. Vavra, C. DeBoer, A. Dietz, L. J. Hanka, and W. T. Sokolski, *Antibiot. Ann.* p. 230 (1960).
100. S. A. Waksman, E. S. Horning, and E. L. Spenser, *Science* **96**, 202 (1944).
101. J. Kent and N. G. Heatley, *Nature* **156**, 295 (1945).
102. W. H. Wilkens and G. G. M. Harris, *Brit. J. Exptl. Pathol.* **23**, 166 (1942).
103. F. A. Norstadt and T. M. McCalla, *Science* **140**, 410 (1963).
104. R. B. Woodward and G. Singh, *J. Am. Chem. Soc.* **72**, 1428 (1950).
105. S. Tannenbaum and E. Bassett, *J. Biol. Chem.* **234**, 1861 (1959).
106. R. R. Graves and C. W. Hesseltine, *Mycopathol. Mycol. Appl.* **29**, 277 (1966).
107. H. H. Kuehn, *Mycologia* **50**, 417 (1958).
108. E. O. Karow and J. W. Foster, *Science* **99**, 265 (1944).
109. E. P. Abraham and H. W. Florey, *in* "Antibiotics" (N. W. Florey *et al.*, eds.), Vol. I, p. 273. Oxford Univ. Press, London and New York, 1949.
110. O. M. Efimenko and P. A. Yakimov, *Tru. Leningr. Khim.-Farmatseut. Inst.* p. 88 (1960); *Chem. Abstr.* **55**, 21470 (1961).
111. W. P. Brian, G. W. Elson, and D. Lowe, *Nature* **178**, 263 (1956).
112. T. Ukai, Y. Yamamoto, and T. Yamamoto, *J. Pharm. Soc. Japan* **74**, 450 (1954).
113. H. F. Kraybill and M. B. Shimkin, *Advan. Cancer Res.* **8**, 191 (1964).
114. V. W. Mayer and M. S. Legator, *J. Agr. Food Chem.* **17**, 454 (1969).
115. P. M. Scott and E. Somers, *J. Agr. Food Chem.* **16**, 483 (1968).
116. E. G. Jefferys, *J. Gen. Microbiol.* **7**, 295 (1952).
117. F. Dickens and J. Cooke, *Brit. J. Cancer* **15**, 85 (1961).
118. W. B. Geiger and J. E. Conn, *J. Am. Chem. Soc.* **67**, 112 (1945).
119. F. Dickens and H. E. H. Jones, *Brit. J. Cancer* **15**, 85 (1961).
120. B. Babudieri, *Rend. Ist. Super. Sanita* **11**, 577 (1968); *Chem. Abstr.* **46**, 7178 (1952).
121. F. H. Wang, *Botan. Bull. Acad. Sinica* **2**, 265 (1948).
122. H. Keilova-Rodova, *Experientia* **5**, 242 (1949).
123. P. Sentein, *Compt. Rend. Soc. Biol.* **149**, 1621 (1955).

124. R. F. J. Withers, *Symp. Mutational Process, Mech. Mutation Inducing Factors, Prague, 1965* p. 359.
125. K. Maeda, H. Kosaka, K. Yagishita, and H. A. Umezawa, *J. Antibiot. (Tokyo)* **A9**, 82 (1956).
126. W. T. Bradner and M. H. Pindell, *Nature* **196**, 682 (1962).
127. M. Ishizuka, H. Takayama, T. Takeuchi, and H. Umezawa, *J. Antibiot. (Tokyo)* **A19**, 260 (1966).
128. H. Umezawa, M. Hori, M. Ishizuka, and T. Takeuchi, *J. Antibiot. (Tokyo)* **A19**, 200 (1966).
129. H. Umezawa, *J. Antibiot. (Tokyo)* **A19**, 256 (1966).
130. N. Rakieten, M. V. Nadkarni, M. L. Rakieten, and B. S. Gordon, *Toxicol. Appl. Pharmacol.* **14**, 590 (1969).
131. B. C. Smale, M. D. Montgillion, and T. G. Pridham, *Plant Disease Reptr.* **45**, 244 (1961).
132. N. Tanaka, H. Yamaguchi, and H. Umezawa, *J. Antibiot. (Tokyo)* **A16**, 86 (1963).
133. A. Falaschi and A. Kornberg, *Federation Proc.* **23**, 940 (1964).
134. P. Pietsch and H. Garrett, *Nature* **219**, 488 (1968).
135. P. Pietsch and H. Garrett, *Cytobios.* **1A**, 7 (1969).
136. N. Tanaka, H. Yamaguchi, and H. Umezawa, *Biochem. Biophys. Res. Commun.* **10**, 171 (1963).
137. N. F. Jacobs, R. L. Neu, and L. I. Gardner, *Mutation Res.* **7**, 251 (1969).
138. G. M. Jagiello, *Mutation Res.* **6**, 289 (1968).
139. K. Kajiwara, J. H. Kim, and G. C. Mueller, *Cancer Res.* **26**, 233 (1966).
140. B. Djordjevic and J. H. Kim, *Cancer Res.* **27**, 2255 (1967).
141. E. Mattingly, *Mutation Res.* **4**, 51 (1967).
142. B. A. Kihlman, G. Odmark, and B. Hartley, *Mutation Res.* **4**, 783 (1967).
143. Q. R. Bartz, T. H. Haskell, C. C. Elder, D. W. Johannessen, R. P. Frohardt, A. Ryder, and S. A. Fusari, *Nature* **173**, 72 (1954).
144. J. A. Moore, J. R. Dice, E. D. Nicolaides, R. D. Westland, and E. L. Wittle, *J. Am. Chem. Soc.* **76**, 2884 (1954).
145. S. S. Sternberg and F. S. Philips, *Cancer* **10**, 809 (1957).
146. C. C. Stock, H. C. Reilly, S. M. Buckley, D. A. Clarke, and C. P. Rhoads, *Acta, Unio Intern. Contra Cancrum* **11**, 186 (1955).
147. H. C. Reilly, *Bacterial Proc. (Soc. Am. Bacteriologists)* **54**, 79 (1954); *Chem. Abstr.* **51**, 15012 (1957).
148. J. F. Henderson, G. A. LePage, and F. A. McIver, *Cancer Res.* **17**, 609 (1957).
149. C. P. Dagg and D. A. Karnofsky, *J. Exptl. Zool.* **130**, 555 (1955).
150. M. L. Murphy and D. A. Karnofsky, *Cancer* **9**, 955 (1956).
151. R. J. Blattner, A. P. Williamson, and L. Simonsen, *Proc. Soc. Exptl. Biol. Med.* **97**, 500 (1958).
152. M.Nakamura, *Nature* **178**, 1119 (1956).
153. L. Kaplan, H. C. Reilly, and C. C. Stock, *J. Bacteriol.* **78**, 511 (1959).
154. D. L. Kohberger, H. C. Reilly, G. L. Coffey, A. B. Hillegas, and J. Ehrlich, *Antibiot. Chemotherapy* **5**, 59 (1955).
155. A. J. Tomisek, H. J. Kelley, and H. E. Skipper, *Abstr. 128th Meeting Am. Chem. Soc., Minneapolis* p. 5C (1955).
156. J. M. Buchanan, *Ciba Found. Symp., Amnio Acids Peptides Antimetab. Activity* p. 75 (1959).
157. W. Szybalski, *Ann. N.Y. Acad. Sci.* **76**, 475 (1958).
158. A. Zampieri and J. Greenberg, *Genetics* **57**, 41 (1967).

159. V. N. Iyer and W. Szybalski, *Appl. Microbiol.* **6**, 23 (1958).
160. S. Nakamura, S. Omura, M. Hamada, T. Nishimura, H. Yamaki, N. Tanaka, Y. Okami, and H. Umezawa, *J. Antibiot.* (*Tokyo*) **A20**, 217 (1967); *Chem. Abstr.* **67**, 97818 (1967).
161. C. A. Amman and R. S. Safferman, *Antibiot. Chemotherapy* **8**, 1 (1958).
162. R. Rieger and A. Michaelis, *Kulturpflanze* **10**, 212 (1962).
163. N. Tanaka and A. Sugimura, *Proc. Intern. Genet. Symp., Tokyo Kyoto, 1956* p. 189. Sci. Council Japan, Veno Park, Tokyo, 1957.
164. A. S. Perry and S. Miller, *J. Insect Physiol.* **11**, 1277 (1965).
165. M. DuBost, O. Ganter, R. Maral, L. Minet, S. Pinnert, J. Preudhomme, and G. H. Werner, *Compt. Rend.* **257**, 1813 (1963).
166. E. Calendi, A. D. Marco, M. Reggiani, B. Scarpinato, and L. Valenti, *Biochim. Biophys. Acta* **103**, 25 (1965).
167. G. Cassinelli and P. Orezzi, *Giorn. Microbiol.* **11**, 167 (1963).
168. A. DiMarco, M. Gaetani, L. Dorigotti, M. Soldati, and O. Bellini, *Tumori* **43**, 203 (1963).
169. F. Arcamone, G. Franceschi, P. Orezzi, G. Cassinelli, W. Barbieri, and R. Mondelli, *J. Am. Chem. Soc.* **86**, 5334 (1964).
170. F. Arcamone, G. Cassinelli, P. Orezzi, G. Franceschi, and R. Mondelli, *J. Am. Chem. Soc.* **86**, 5335 (1964).
171. C. Tan and H. Tasaka, *Proc. 9th Intern. Cancer Congr. Tokyo, 1966* p. 442.
172. C. Jacquillat, M. Boiron, M. Well, J. Tanzer, Y. Majean, and J. Bernard, *Lancet* **II**, 27 (1966).
173. C. Tan, H. Tasaka, K. P. Yu, M. L. Murphy, and D. A. Karnofsky, *Cancer* **20**, 333 (1967).
174. A. DiMarco, M. Gaetani, P. Orezzi, B. M. Scarpinato, R. Silvestrin, M. Soldati, T. Dasdia, and L. Valenti, *Nature* **201**, 706 (1964).
175. A. C. Grein, C. Spalla, A. DiMarco, and G. Canevazzi, *Giorn. Microbiol.* **11**, 109 (1963).
176. A. DiMarco, M. Soldati, A. Fionette, and T. Dasdia, *Tumori* **49**, 235 (1963).
177. G. Costa and G. Astaldi, *Tumori* **50**, 477 (1964).
178. A. DiMarco, T. Silvestrini, S. DiMarco, and T. Dasdia, *J. Cell. Biol.* **27**, 545 (1965).
179. A. Theologides, J. W. Yarbro, and J. Kennedy, *Cancer* **21**, 16 (1968).
180. B. K. Vig, S. B. Kontras, and A. M. Aubele, *Mutation Res.* **7**, 91 (1969).
181. B. K. Vig, S. B. Kontras, E. F. Paddock, and L. D. Samuels, *Mutation Res.* **5**, 279 (1968).
182. B. K. Vig, S. B. Kontras, and L. D. Samuels, *Experientia* **24**, 271 (1968).
183. T. I. Watkins, *J. Chem. Soc.* p. 3059 (1952).
184. R. Tomchick and H. G. Mantel, *J. Gen. Microbiol.* **36**, 225 (1964).
185. B. A. Newton, *in* "Metabolic Inhibitors" (R. M. Hochster and J. H. Quastel, eds.), Vol. 2, p. 285. Academic Press, New York, 1963.
186. W. H. Elliott, *Biochem. J.* **86**, p. 562 (1963).
187. M. J. Waring, *Biochim. Biophys. Acta* **87**, 358 (1964).
188. M. J. Waring, *Mol. Pharmacol.* **1**, 1 (1965).
189. M. J. Waring, *J. Mol. Biol.* **13**, 269 (1965).
190. P. P. Slonimski, G. Perrodin, and J. H. Croft, *Biochem. Biophys. Res. Commun.* **30**, 232 (1968).
191. J. B. Adams, *J. Pharm. Pharmacol.* **10**, 507 (1958).
192. H. G. Shirk, *Natl. Acad. Sci.—Natl. Res. Council, Publ.* **514**, 23 (1956).
193. A. B. Durkee, *J. Agr. Food Chem.* **6**, 194 (1958).

194. W. R. Jarvis, *Plant Pathol.* **9**, 150 (1960).
195. G. J. M. A. Gorter and W. Krüger, *S. African J. Agr. Sci.* **1**, 305 (1958).
196. T. Swarbrick, U.S. Patent 2,765,255 (1966); *Chem. Abstr.* **51**, 9998i (1957).
197. G. Eisenschiml, *Offic. Dig., Federation Paint & Varnish Prod. Clubs* **30**, 398 (1958); *Chem. Abstr.* **53**, 20834 (1959).
198. L. Marcou and G. Guillemaille, French Patent 1,099,606 (1955); *Chem. Abstr.* **53**, 22713 (1959).
199. W. A. Miller, *Poultry Sci.* **36**, 579 (1957).
200. R. G. W. Hollingshead, "Oxine and its Derivatives," Vol. II, Part 1, Derivatives of Oxine; Vol. IV, Part 2, Derivatives of Oxine. Lewis, London, 1956.
201. D. B. Clayson, J. W. Jull, and G. M. Bonser, *Brit. J. Cancer* **12**, 222 (1958).
202. E. Mattea, *Z. Urol.* **53**, 149 (1960); *Chem. Abstr.* **55**, 5734 (1961).
203. K. Yamagata and M. Oda, *Hakko Kogaku Zasshi* **35**, 67 (1957); *Chem. Abstr.* **51**, 16687b (1957).
204. K. Yamagata, M. Oda, and T. Ando, *J. Ferment. Technol.* **34**, 378 (1956); *Chem. Abstr.* **51**, 12231 (1957).
205. M. R. Hanna, *Can. J. Botany* **39**, 757 (1961).
206. E. Gebhart, *Mutation Res.* **6**, 309 (1968).
207. C. H. Hill, H. Tai, A. L. Underwood, and R. A. Day, Jr., *Anal. Chem.* **28**,1688 (1956).
208. B. Breyer, *Australian J. Sci.* **23**, 225 (1961).
209. C. Duval, *Anal. Chim. Acta* **16**, 545 (1957).
210. G. Schoffa, O. Ristau, and B. E. Wahler, *Z. Physik. Chem.* (*Leipzig*) **215**, 203 (1960).
211. L. Reio, *J. Chromatog.* **4**, 76 (1960).
212. C. B. Coulson and W. C. Evans, *J. Chromatog.* **1**, 374 (1958).
213. A. Waksmundzki and E. Soczewinski, *Roczniki Chem.* **33**, 1423 (1959).
214. K. Zahn and C. R. Koch, *Chem. Ber.* **71B**, 172 (1938).
215. H. Ippen, *Arch. Klin. Exptl. Dermatol.* **227**, 202 (1966).
216. H. Ippen and D. Montag, *Arzneimittel-Forsch.* **8**, 778 (1958).
217. A. Krebs and H. Schaltegger, *Experientia* **21**, 128 (1965).
218. G. Swanbeck, *Biochim. Biophys. Acta* **123**, 630 (1966).
219. G. Swanbeck and J. Liden, *Acta Dermato-Venereol.* **46**, 628 (1966).
220. G. Swanbeck and N. Thyresson, *Acta Dermato-Venereol.* **45**, 344 (1965).
221. H W. Fox and F. G. Bock, *J. Natl. Cancer Inst.* **38**, 789 (1967).
222. B. O. Gillberg, G. Zetterberg, and G. Swanbeck, *Nature* **214**, 415 (1967).
223. C. Auerbach, *J. Am. Pharm. Assoc., Sci. Ed.* **34**, 310 (1945).
224. P. M. Parikh, D. J. Vadodaria, and S. P. Mukherji, *J. Pharm. Pharmacol.* **11**, 314 (1959).
225. W. Kikuth, R. Gönnert, and H. Mauss, *Naturwissenschaften* **33**, 253 (1946).
226. T. H. Sharp, *J. Chem. Soc.* p. 2961 (1951).
227. E. J. Blanz, Jr. and F. A. French, *J. Med. Chem.* **6**, 185 (1963).
228. E. Hirschberg, A. Gellhorn, M. R. Murray, and E. Elsager, *J. Natl. Cancer Inst.* **22**, 567 (1959).
229. D. M. Blair, *Bull. World Health Organ.* **18**, 989 (1958).
230. I. B. Weinstein, R. Carchman, E. Marner, and E. Hirschberg, *Biochim. Biophys. Acta* **142**, 440 (1967).
231. E. Hirschberg, I. B. Weinstein, and R. Carchman, *Proc. 9th Intern. Cancer Congr., Tokyo, 1966* p. 347.
232. H. Lüers, R. Gönnert, and H. Mauss, *Z. Vererbungslehre* **87**, 93 (1955).
233. G. Obe, *Mol. Gen. Genet.* **103**, 326 (1969).
234. M. M. Janot, F. Laine, and R. Goutarel, *Ann. Pharm. Franc.* **18**, 673 (1960).

235. F. K. Huu-Laine and W. Pinto-Scognamiglio, *Arch. Intern. Pharmacodyn.* **147**, 209, (1964).
236. M. M. Janot, F. Laine, Q. Khuong-Huu, and R. Goutarel, *Bull. Soc. Chim. France* p. 111 (1962).
237. H. R. Mahler and G. Dutton, *J. Mol. Biol.* **10**, 157 (1964).
238. H. R. Mahler, R. Goutarel, Q. Khuong-Huu, and M. T. Ho, *Biochemistry* **5**, 2177 (1966).
239. H. R. Mahler and M. B. Baylor, *Proc. Natl. Acad. Sci. U.S.* **58**, 256 (1967).
240. T. Matsuda, *J. Ferment. Technol.* **44**, 495 (1966).
241. J. A. Buzard, *Giorn. Ital. Chemioterap.* **5**, 1 (1962).
242. Eaton Laboratories, "The Nitrofurans." Norwich, New York, 1958.
243. C. E. Friedgood and L. B. Ripstein, *Cancer Res.* **11**, 248 (1951).
244. M. W. Green and C. E. Friedgood, *Proc. Soc. Exptl. Biol. Med.* **69**, 603 (1948).
245. R. J. Stein, D. Yost, F. Petroliunas, and A. von Esch, *Federation Proc.* **25**, 291 (1966).
246. D. R. McCalla, *Can. J. Biochem.* **42**, 1245 (1964).
247. P. Woody-Karrer and J. Greenberg, *J. Bacteriol.* **85**, 1208 (1963).
248. D. R. McCalla, *Can. J. Microbiol.* **11**, 185 (1965).
249. A. Zampieri and J. Greenberg, *Biochem. Biophys. Res. Commun.* **14**, 172 (1964).
250. A. Stoll and A. Hoffmann, *Helv. Chim. Acta* **26**, 944 (1943).
251. A. Hoffmann, "Die Mutterkorn Alkaloide," p. 190. Enke, Stuttgart, 1964.
252. E. S. Boyd, E. Rothlin, J. F. Bonner, I. H. Slater, and H. C. Hodge, *J. Pharmacol. Exptl. Therap.* **113**, 6 (1955).
253. A. Stoll, E. Rothlin, J. Rutschmann, and W. R. Schalch, *Experientia* **11**, 396 (1955).
254. U. A. Lanz, A. Cerletti, and E. Rothlin, *Helv. Physiol. Pharmacol. Acta* **13**, 207 (1955).
255. J. Axelrod, R. O. Brady, B. Witkop, and E. V. Evarts, *Ann. N.Y. Acad. Sci.* **66**, 435 (1967).
256. J. Axelrod, S. Udenfriend, and B. B. Brodie, *J. Pharmacol. Exptl. Therap.* **111**, 17 (1954).
257. J. Axelrod, R. O. Brady, B. Witkop, and E. V. Evarts, *Nature* **178**, 143 (1956).
258. L. Sokoloff, S. Perlin, C. Kornetsky, and S. S. Kety, *Federation Proc.* **15**, 174 (1956).
259. A. Laitha, *Psychopharmacology* **3**, 126 (1958).
260. E. S. Boyd, *Arch. Intern. Pharmacodyn.* **120**, 292 (1959).
261. R. Auerbach and J. A. Rugowski, *Science* **157**, 1325 (1967).
262. J. K. Hanaway, *Science* **164**, 574 (1969).
263. G. J. Alexander, B. E. Miles, G. M. Gold, and R. B. Alexander, *Science* **157**, 459 (1967).
264. W. F. Geber, *Science* **158**, 265 (1967).
265. S. Fabro and S. M. Sieber, *Lancet* **I**, 639 (1968).
266. S. Irwin and J. Egozcue, *Science* **157**, 313 (1967).
267. J. E. Idanpaan-Heikkila and J. C. Schoolar, *Science* **164**, 1295 (1969).
268. G. Carakushansky, R. L. Neu, and L. I. Gardner, *Lancet* **I**, 150 (1969).
269. H. Zellweger, J. S. McDonald, and G. Abbo, *Lancet* **II**, 1066 (1967).
270. F. Hecht, R. K. Beals, M. H. Lees, H. Jolly, and P. Roberts, *Lancet* **II**, 1087 (1968).
271. M. M. Cohen, M. J. Marinello, and N. Beck, *Science* **155**, 1417 (1967).
272. M. M. Cohen, K. Hirschhorn, and W. A. Forsch, *New Engl. J. Med.* **277**, 1043 (1967).
273. L. F. Oarvik and T. Lato, *Lancet* **I**, 250 (1968).
274. J. Egozcue, S. Irwin, and C. A. Maruffo, *J. Am. Med. Assoc.* **204**, 214 (1968).
275. N. E. Skakkebaek, J. Phillip, and O. J. Rafaelson, *Science* **160**, 1246 (1968).
276. M. M. Cohen and A. B. Mukherjee, *Nature* **219**, 1072 (1968).
277. W. D. Loughman, T. W. Sargeant, and D. M. Irsaelstam, *Science* **158**, 508 (1967).

278. R. S. Sparkes, J. Melynk, and L. P. Bozzetti, *Science* **160**, 1343 (1968).
279. L. Bender and S. Sankar, *Science* **159**, 749 (1968).
280. S. Sturelid and B. A. Kihlman, *Hereditas* **62**, 259 (1969).
281. L. S. Browning, *Science* **161**, 1022 (1968).
282. E. Vann, *Nature* **223**, 95 (1969).
283. D. Grace, E. A. Carlson, and P. Goodman, *Science* **161**, 694 (1968).
284. G. Zetterberg, *Hereditas* **62**, 262 (1969).
285. A. Stoll, A. Hoffmann, and F. Trokler, *Helv. Chim. Acta* **32**, 506 (1949).
286. A. Stoll and A. Hoffmann, *Helv. Chim. Acta* **38**, 421 (1955)
287. A. Laitha, *in* "Psychopharmacology" (H. H. Pennes, ed.), pp. 126–151. Harper (Hoeber), New York, 1958.
288. E. Jacobsen, *Clin. Pharmacol. Therap.* **4**, 480 (1963).
289. T. M. Itil, A. Keskiner, and J. M. C. Holden, *Gwan* Suppl. 30, 93 (1969).
290. D. Bente, T. Itil and E. E. Schmid, *Psychiat. Neurol.* **135**, 273 (1958).
291. I. Isbell, *Psychopharmacologia* **1**, 29 (1959).
292. R. C. DeBold and R. C. Leaf, "LSD, Man and Society, A Symposium." Wesleyan Univ. Press, Middleton, Connecticut, 1967.
293. O. M. Garson and M. K. Robson, *Brit. Med. J.* **1**, 800 (1969).
294. A. Macieira-Coelho, I. J. Hiu, and E. Garcia-Giralt, *Nature* **222**, 1170 (1969).
295. L. Genest and C. G. Farmilo, *J. Pharm. Pharmacol.* **16**, 250 (1964).
296. G. K. Aghajanian and O. H. L. Bing, *Clin. Pharmacol. Therap.* **5**, 611 (1964).
297. H. Hellberg, *Acta Chem. Scand.* **11**, 219 (1957).
298. R. J. Martin and T. G. Alexander, *J. Assoc. Offic. Anal. Chemists* **51**, 159 (1968).
299. S. W. Bellman, *J. Assoc. Offic. Anal. Chemists* **51**, 164 (1968).
300. J. Look, *J. Assoc. Offic. Anal. Chemists* **51**, 1318 (1968).
301. E. G. Clarke, *J. Forensic Sci. Soc.* **7**, 46 (1967).
302. S. Agurell, *Acta Pharm. Suecica* **2**, 362 (1956).
303. W. N. French and A. Wehrli, *J. Pharm. Sci.* **54**, 1515 (1965).
304. C. Radecka and I. C. Nigan, *J. Pharm. Sci.* **55**, 861 (1966).
305. M. A. Katz, G. Tadjer, and W. A. Aufricht, *J. Chromatog.* **31**, 545 (1967).
306. R. P. Pioch, U.S. Patent 2,736,728 (1956); *Chem. Abstr.* **50**, 10803c (1956).
307. J. Arbit, *J. Appl. Physiol.* **10**, 317 (1957).
308. L. S. Goodman and A. Gilman, eds., "The Pharmacological Basis of Therapeutics," 3rd ed., p. 354. Macmillan, New York, 1967.
309. J. Axelrod and J. Reichenthal, *J. Pharmacol. Exptl. Therap.* **107**, 519 (1953).
310. R. N. Warren, *J. Chromatog.* **40**, 468 (1969).
311. H. H. Cornish and A. A. Christman, *J. Biol. Chem.* **228**, 315 (1957).
312. G. Schmidt and R. Schoyerer, *Deut. Z. Ges. Gerichtl. Med.* **57**, 402 (1966).
313. L. E. Andrew, *Am. Naturalist* **93**, 135 (1959).
314. T. Alderson and A. H. Khan, *Nature* **215**, 1080 (1967).
315. W. Ostertag and J. Haake, *Z. Vererbungslehre* **98**, 299 (1966).
316. S. Mittler, J. E. Mittler, and S. L. Owens, *Nature* **214**, 424 (1967).
317. A. M. Clark and E. G. Clark, *Mutation Res.* **6**, 227 (1968).
318. L. E. Andrew, *Am. Naturalist* **4**, 708 (1967).
319. M. Demerec, B. Wallace, and E. M. Witkin, *Carnegie Inst. Wash.; Year Book* **47**, 169 (1948).
320. K. Gezelius and N. Fries, *Hereditas* **38**, 112 (1952).
321. A. Novick, *Brookhaven Symp. Biol.* **8**, 101 (1956).
322. E. A. Gläss and A. Novick, *J. Bacteriol.* **77**, 10 (1959).
323. A. Novick and L. Szilard, *Cold Spring Harbor Symp. Quant. Biol.* **16**, 337 (1951).

324. H. E. Kubitscheck and H. E. Bendigkeit, *Genetics* **46**, 105 (1961).
325. N. Fries and B. Kihlman, *Nature* **162**, 573 (1948).
326. G. Zetterberg, *Hereditas* **46**, 229 (1960).
327. N. Fries, *Hereditas* **36**, 134 (1950).
328. G. Zetterberg, *Hereditas* **46**, 279 (1959).
329. W. Ostertag, *Mutation Res.* **3**, 249 (1966).
330. W. Ostertag, E. Duisberg, and M. Stürmann, *Mutation Res.* **2**, 293 (1965).
331. B. A. Kihlman and A. Levan, *Hereditas* **35**, 109 (1949).
332. H. E. Kubitschek and H. E. Bendigkeit, *Mutation Res.* **1**, 113 (1964).
333. W. Kuhlman, H. G. Fromme, E. M. Heege, and W. Ostertag, *Cancer Res.* **28**, 2375 (1968).
334. M. F. Lyon, J. S. R. Phillips, and A. G. Searle, *Z. Vererbungslehre* **93**, 7 (1962).
335. B. M. Cattanach, *Z. Vererbungslehre* **93**, 215 (1962).
336. H. Nishimura and K. Nakai, *Proc. Soc. Exptl. Biol. Med.* **104**, 140 (1960).
337. S. Fabro and S. M. Sieber, *Nature* **223**, 410 (1969).
338. A. Goldstein and R. Warren, *Biochem. Pharmacol.* **11**, 166 (1962).
339. H. Raber, *Sci. Pharm.* **34**, 202 (1966).
340. K. A. Connors, *in* "Pharmaceutical Analysis" (T. Higuchi and E. Brochmann-Hanssen, eds.), p. 240. Wiley (Interscience), New York, 1961.
341. U. Tanaka and Y. Ohkubo, *J. Coll. Agr., Tokyo Imp. Univ.* **14**, 153 (1937).
342. H. Nash, "Alcohol and Caffeine." Thomas, Springfield, Illinois, 1962.
343. R. Smith and D. Rees, *Analyst* **88**, 310 (1963).
344. D. Abelson and D. Borcherds, *Nature* **179**, 1135 (1957).
345. R. Munier and M. Macheboeuf, *Bull. Soc. Chim. Biol.* **32**, 192 (1950).
346. U. M. Senanayake and R. O. B. Wijesekera, *J. Chromatog.* **32**, 175 (1968).
347. B. Davidow, N. L. Petri, and B. Quame, *Am. J. Clin. Pathol.* **38**, 714 (1968).
348. F. L. Grab and J. A. Reinstein, *J. Pharm. Sci.* **57**, 1703 (1968).
349. A. Monard, *J. Pharm. Belg.* **(N.S.) 23**, 323 (1968).
350. R. C. Huston and W. F. Allen, *J. Am. Chem. Soc.* **56**, 1356 (1934).
351. P. O. P. Ts'o, G. K. Helmkamp, and G. Sander, *Proc. Natl. Acad. Sci. U.S.* **48**, 686 (1962).
352. B. A. Kihlman, *Symbolae Botan. Upsalienses* **11**, 1 (1951).
353. B. A. Kihlman, *Advan. Genet.* **10**, 1 (1961).
354. B. A. Kihlman, "Actions of Chemicals on Dividing Cells." Prentice-Hall, Englewood Cliffs, New Jersey, 1966.
355. J. F. Jackson, *J. Cell Biol.* **22**, 291 (1964).
356. J. MacLeish, *Heredity* **6**, 385 (1953).
357. B. A. Kihlman, *J. Biophys. Biochem. Cytol.* **2**, 543 (1956).
358. B. A. Kihlman, *Exptl. Cell Res.* **8**, 345 (1955).
359. The Ad Hoc Committee on Non-Nutritive Sweeteners, Food Protection Committee. Natl. Acad. Sci.—Natl. Res. Council, Washington, D.C., 1968.
360. J. D. Taylor, R. K. Richards, R. G. Wiegand, and M. S. Weinberg, *Food Cosmet. Toxicol.* **6**, 313 (1968).
361. P. H. Derse and R. J. Daun, *J. Assoc. Offic. Anal. Chemists* **49**, 1090 (1966).
362. S. Kojima and H. Ichibagase, *Chem. & Pharm. Bull. (Tokyo)* **14**, 971 (1966).
363. J. S. Leahy, M. Wakefield, and T. Taylor, *Food Cosmet. Toxicol.* **5**, 447 (1967).
364. J. S. Leahy, T. Taylor, and C. J. Rudd, *Food Cosmet. Toxicol.* **5**, 595 (1967).
365. R. C. Sonders and R. G. Wiegand, *Toxicol. Appl. Pharmacol.* **11**, 13 (1968).
366. "Food Additives and Contaminants Committee Second Report on Cyclamates." Min. Agr. Fisheries & Food, 1967.

367. L. Goldberg, C. Parekh, A. Patti, and K. Soike, *Toxicol. Appl. Pharmacol.* **14**, 654 (1969).
368. B. L. Oser, S. Carson, E. E. Vogin, and R. C. Sonders, *Science* **220**, 178 (1968).
369. R. Higuchi *et al.*, *J. Food Hyg. Soc. Japan* **6**, 448 (1966).
370. M. L. Richardson and P. E. Luton, *Analyst* **91**, 520 (1966).
371. K. Maruyama and K. Kawanabe, *J. Food Hyg. Soc. Japan* **4**, 265 (1963).
372. T. E. Furia, "Handbook of Food Additives," p. 502. Chem. Rubber Publ. Co., Cleveland, Ohio, 1968.
373. K. M. Beck, *Food Eng.* **26**, 87 (1954).
374. K. M. Beck, *Food Technol.* **11**, 156 (1957).
375. L. F. Dalderup and W. Visser, *Nature* **221**, 91 (1969).
376. O. G. Fitzhugh, A. A. Nelson, and J. P. Frawley, *J. Am. Pharm. Assoc., Sci. Ed.* **11**, 583 (1951).
377. P. O. Nees and P. H. Derse, *Nature* **81**, 81 (1965).
378. P. O. Nees and P. H. Derse, "Calcium Cyclamate Feeding Study." Wisconsin Alumni Res. Found., Madison, Wisconsin, 1965.
379. E. Bajusz, *Nature* **223**, 407 (1969).
380. K. Sax and H. J. Sax, *Japan. J. Genet.* **43**, 89 (1968).
381. D. Stone, E. Lamson, Y. S. Chang, and K. Pickering, *Abstr. 19th Ann. Meeting Tissue Culture Assoc., Puerto Rico* p. 60 (1968).
382. D. Stone, E. Lamson, Y. S. Chang, and K. W. Pickering, *Science* **164**, 568 (1969).
383. M. L. Richardson, *Talanta* **14**, 385 (1967).
384. D. E. Johnson and H. B. Nunn, *J. Assoc. Offic. Anal. Chemists* **51**, 1274 (1968).
385. E. Bradford and R. E. Weston, *Analyst* **94**, 68 (1969).
386. D. I. Rees, *Analyst* **90**, 568 (1965).
387. A. Castiglioni, *Z. Anal. Chem.* **145**, 188 (1955).
388. T. Korbelak and J. N. Bartlett, *J. Chromatog.* **41**, 124 (1969).
389. W. Kamp, *Pharm. Weekblad* **101**, 57 (1966).
390. G. J. Dickes, *J. Assoc. Public Analysts* **3**, 118 (1965).
391. R. C. Sonders, R. G. Wiegand, and J. C. Netwal, *J. Assoc. Offic. Anal. Chemists* **51**, 136 (1968).
392. M. Legator, *Med. World News* **9**, 25 (1968).
393. M. Legator, K. A. Palmer, S. Green, and K. W. Petersen, *Chem. Eng. News* **48**, 37 (1968).
394. I. P. Lee and R. L. Dixon, *J. Pharm. Sci.* **57**, 1132 (1968).
395. E. E. Swanson and K. K. Chen, *J. Pharmacol. Exptl. Therap.* **88**, 10 (1946).
396. E. E. Swanson and K. K. Chen, *J. Pharmacol. Exptl. Therap.* **93**, 423 (1948).
397. I. P. Lee and R. L. Dixon, *Toxicol. Appl. Pharmacol.* **14**, 654 (1969)
398. W. L. Sutton, *Ind. Hyg. Toxicol.* **2**, 2058 (1967).
399. F. S. Mallette and E. von Haam, *A.M.A. Arch. Ind. Hyg. Occupational Med.* **5**, 311 (1952).
400. G. V. Lomonova, *Federation Proc.* **24**, T96 (1965).
401. M. Rubin, S. Gignac, S. P. Bessman, and E. L. Belknap, *Science* **117**, 659 (1953).
402. A. Guarino and S. Biondi, *Folia Med. (Naples)* **40**, 111 (1957); *Chem. Abstr.* **51**, 9927 (1957).
403. J. F. Fried, A. Lindenbaum, and J. Schubert, *Proc. Soc. Exptl. Biol. Med.* **100**, 570 (1959).
404. H. Foreman, *J. Am. Pharm. Assoc., Sci. Ed.* **42**, 629 (1953).
405. F. Proescher, *Proc. Soc. Exptl. Biol. Med.* **76**, 619 (1951).
406. H. Spencer, V. Vankinscott, I. Lewin, and D. Lazlo, *J. Clin. Invest.* **31**, 1023 (1952).

407. T. J. Suen and R. L. Webb, U.S. Patent 2,917,477 (1959); *Chem. Abstr.* **54**, 16011e (1960).
408. S. Zaima and T. Shiraishi, Japanese Patent 3194 (1961); *Chem. Abstr.* **55**, 21495b (1961).
409. K. Lindner, German Patent 1,069,926 (1959); *Chem. Abstr.* **55**, 11752i (1961).
410. J. A. Kelley and J. L. Ridgeway, U.S. Patent 2,931,716 (1960); *Chem. Abstr.* **54**, 13521a (1960).
411. C. H. Fawcett, R. L. Wain, and F. Wightman, *Nature* **178**, 972 (1956).
412. R. A. Reed, British Patent 781,491 (1958); *Chem. Abstr.* **52**, 626g (1958).
413. J. Farrar and R. W. Sprague, U.S. Patent 3,283,507 (1966); *Chem. Abstr.* **66**, 20699a (1967).
414. S. N. H. Stothart and G. C. Beecroft, British Patent 858,030 (1961); *Chem. Abstr.* **55**, 12783d (1961).
415. C. H. Davis and C. G. Grand, U.S. Patent 2,904,468 (1959); *Chem. Abstr.* **54**, 2670c (1960).
416. J. Bailar, "The Chemistry of the Coordination Compounds." Reinhold, New York, 1956.
417. S. Chaberek and A. E. Martell, "Organic Sequestering Agents." Wiley, New York, 1959.
418. L. S. Goodman and A. Gilman, eds. "The Pharmacological Basis of Therapeutics," 3rd ed., p. 934. Macmillan, New York, 1967.
419. S. Shibata, *Nippon Yakurigaku Zasshi* **52**, 113 (1956).
420. M. J. Seven, "Metal Binding in Medicine," Chapter 10. Lippincott, Philadelphia, Pennsylvania, 1960.
421. H. Foreman and T. T. Trujillo, *J. Lab. Clin. Med.* **43**, 566 (1954).
422. H. Foreman, M. Vier, and M. Magee, *J. Biol. Chem.* **203**, 1045 (1953).
423. L. S. Tsarapin, *Zashch. Vosstanov. Luchevykh Povrezhdeniyakh, Akad. Nauk SSSR* p. 142 (1966); *Chem. Abstr.* **67**, 18400y (1967).
424. N. L. Delone, *Biofizika* **3**, 717 (1958); *Chem. Abstr.* **53**, 4448 (1959).
425. N. L. Delone, *Dokl. Akad. Nauk SSSR* **119**, 800 (1958); *Chem. Abstr.* **52**, 15660 (1958).
426. M. R. McDonald and B. P. Kaufman, *Exptl. Cell Res.* **12**, 415 (1957).
427. B. P. Kaufman and M. R. McDonald, *Proc. Natl. Acad. Sci. U.S.* **43**, 262 and 255–261 (1957).
428. N. B. Khristol Yubova, *Dokl. Akad. Nauk SSSR* **138**, 681 (1961); *Chem. Abstr.* **55**. 21396 (1961).
429. R. Wakonig and T. J. Arnason, *Proc. Benet. Soc. Can.* **3**, 37 (1958).
430. R. Rieger, H. Nicoloff, and A. Michaelis, *Biol. Zentr.* **82**, 393 (1963).
431. T. S. West and A. S. Sykes, "Analytical Applications of Diamino-Ethane-Tetra Acetic Acid," 2nd ed. British Drug Houses, Poole-Dorset, England, 1960.
432. G. Balica, M. Margina, and L. Lipcovski, *Rev. Chim. (Bucharest)* **17**, 435 (1966).
433. C. L. Atkin, E. P. Parry, and D. H. Hern, *Anal. Chem.* **39**, 672 (1967).
434. L. Erdey and I. Buzas, *Anal. Chim. Acta* **22**, 524 (1960).
435. W. Hoyle, I. P. Sanderson, and T. S. West, *J. Electroanal. Chem.* **2**, 166 (1961).
436. A. Iguchi, Y. Yoshino, and M. Kojima, *Bunseki Kagaku* **8**, 123 (1959); *Chem. Abstr.* **55**, 25602f (1961).
437. P. F. Cox and E. O. Morgan, *J. Am. Chem. Soc.* **81**, 6409 (1959).
438. J. Sykora and V. Eybl, *Collection Czech. Chem. Commun.* **32**, 89051a (1967).
439. D. Bacikova, P. Nemec, L. Drobnica, K. Antos, P. Kristian, and A. Hulka, *J. Antibiot. (Tokyo)* **A18**, 162 (1965).
440. T. Zsolnai, *Arzneimittel-Forsch.* **16**, 870 (1966).

441. L. Drobnica, M. Zemanova, P. Nemec, K. Antos, P. Kristian, A. Stullerova, V. Knoppova, and D. Nemec, Jr., *Appl. Microbiol.* **15**, 701 (1967).
442. G. B. Dubrova, *Tr. Leningr. Tekhnol. Inst. Kholodil'n. Prom.* **10**, 74 (1956); *Chem. Abstr.* **53**, 22582 (1959).
443. C. Izard and C. Papaioannou, *Compt. Rend.* **249**, 300 (1959).
444. E. C. Hagan, W. H. Hansen, O. G. Fitzhugh, P. M. Jenner, W. I. Jones, J. M. Taylor, E. L. Long, A. A. Nelson, and J. B. Brouwer, *Food Cosmet. Toxicol.* **5**, 141 (1967).
445. K. Horakova, *Naturwissenschaften* **53**, 383 (1966).
446. K. Horakova, L. Drobnica, P. Nemec, K. Antos, and P. Kristian, *Neoplasma* **15**, 160 (1968).
447. C. Auerbach and J. M. Robson, *Nature* **154**, 81 (1944).
448. C. Auerbach and J. M. Robson, *Proc. Roy. Soc. Edinburgh* **B62**, 284 (1947).
449. C. Auerbach, M. Y. Ansari, and J. M. Robson, *Rept. Min. Supply* p. Y18171 (1943).
450. C. Auerbach, *Biol. Rev. Cambridge Phil. Soc.* **24**, 355 (1949).
451. N. Fries, *Physiol. Plantarum* **1**, 330 (1948).
452. K. Sharma and A. Sharma, *Nucleus (Calcutta)* **5**, 127 (1962).
453. H. G. Maier and W. Diemaier, *Z. Anal. Chem.* **227**, 187 (1967).
454. L. Rosenthaler, *Pharm. Acta Helv.* **33**, 269 (1958).
455. I. Nishioka, R.Matsuo, Y. Ohkura, and T. Momose, *Yakugaku Zasshi* **88**, 1281 (1968).
456. J. Murad, *Tribuna Farm. (Brazil)* **23**, 137 (1955).
457. H. Wagner, L. Horhammer, and H. Nufer, *Arzneimittel-Forsch.* **15**, 453 (1965).
458. H. Binder, *J. Chromatog.* **41**, 448 (1969).
459. A. Mathias, *Tetrahedron* **21**, 1073 (1965).
460. H. L. A. Tarr and P. A. Sunderland, *Mod. Refrig.* **43**, 41 (1940).
461. A. G. Castellani and C. F. Niven, Jr., *Appl. Microbiol.* **3**, 154 (1955).
462. C. Simon, *Lancet* **I**, 872 (1966).
463. C. Simon, *Arch. Franc. Pediat.* **23**, 231 (1966).
464. P. Hölscher and J. Natzschka, *Ger. Med. Monthly* **9**, 325 (1964).
465. P. K. McIlwain and I. A. Schipper, *J. Am. Vet. Med. Assoc.* **142**, 502 (1963).
466. E. E. Hatfield *et al.*, *Proc. Soc. Animal Prod. Western Sect.* **12**, 41 (1961).
467. R. P. Smith and W. R. Layne, *J. Pharmacol. Exptl. Therap.* **165**, 30 (1969).
468. J. L. Sell and W. K. Roberts, *J. Nutr.* **79**, 171 (1963).
469. D. Keilin and E. F. Hartree, *Nature* **157**, 210 (1946).
470. F. Vella, *Experientia* **15**, 433 (1959).
471. E. Beutler and G. M. Kelly, *Experientia* **19**, 96 (1963).
472. A. T. Austin, *Nature* **185**, 1086 (1960).
473. H. B. Strack, E. B. Freese, and E. Freese, *Mutation Res.* **1**, 10 (1964).
474. E. E. Horn and R. M. Herriot, *Proc. Natl. Acad. Sci. U.S.* **48**, 1409 (1962).
475. R. Litman and H. Ephrussi-Taylor, *Compt. Rend.* **249**, 838 (1959).
476. S. E. Bresler, V. L. Kalinin, and D. A. Perumov, *Mutation Res.* **5**, 329 (1968).
477. F. Kaudewitz, *Nature* **183**, 1829 (1959).
478. W. G. Verly, H. Barbason, J. Dusart, and A. Petispas-Dewandre, *Biochim. Biophys. Acta* **145**, 752 (1967).
479. A. Reisenstart and J. L. Rosner, *Genetics* **49**, 343 (1964).
480. R. Rudner, *Z. Vererbungslehre* **92**, 336 (1961).
481. F. J. deSerres, H. E. Brockman, W. E. Barnett, and H. G. Kølmark, *Mutation Res.* **4**, 415 (1967).
482. H. V. Malling, *Mutation Res.* **2**, 320 (1965).
483. F. K. Zimmermann and R. Schwaier, *Mol. Gen. Genet.* **100**, 63 (1967).
484. F. K. Zimmermann, R. Schwaier, and U. von Laer, *Z. Vererbungslehre* **98**, 230 (1966).

485. R. Schwaier, N. Nashed, and F. K. Zimmermann, *Mol. Gen. Genet.* **102**, 290 (1968).
486. N. Nashed and G. Jabbur, *Z. Vererbungslehre* **98**, 106 (1966).
487. F. K. Zimmermann, R. Schwaier, and U. von Laer, *Z. Vererbungslehre* **97**, 68 (1965).
488. A. Nasim and C. H. Clarke, *Mutation Res.* **2**, 395 (1965).
489. N. Loprieno, R. Guglielminetti, S. Bonatti, and A. Abbondandalo, *Mutation Res.* **8**, 65 (1969).
490. I. Tessman, R. K. Poddar, and S. Kumar, *J. Mol. Biol.* **9**, 352 (1964).
491. I. Tessman, *Virology* **9**, 375 (1959).
492. S. Benzer, *Proc. Natl. Acad. Sci. U.S.* **47**, 403 (1961).
493. F. Kaudewitz, *Abhandl. Deut. Akad. Wiss. Berlin. Kl. Med.* p. 86 (1960); *Chem. Abstr.* **55**, 671 (1961).
494. W. Vielmetter and C. M. Wiedner, *Z. Naturforsch.* **14b**, 312 (1959).
495. O. H. Siddiqi, *Genet. Res.* **3**, 303 (1962).
496. R. A. Steinberg and C. Thom, *Proc. Natl. Acad. Sci. U.S.* **26**, 363 (1940).
497. K. W. Mundry and A. Gierer, *Z. Vererbungslehre* **89**, 614 (1958).
498. H. Schuster, A. Gierer, and K. W. Mundry, *Abhandl. Deut. Akad. Wiss Berlin, Kl. Med.* p. 76 (1960); *Chem. Abstr.* **55**, 670 (1961).
499. O. P. Sehgal, *Proc. 12th Intern. Congr. Genet. 1964* Vol. 1, p. 81.
500. O. P. Sehgal and G. F. Krause, *J. Virol.* **2**, 966 (1968).
501. R. Grau and A. Mirna, *Z. Anal. Chim.* **158**, 182 (1957).
502. J. P. Griess, *Chem. Ber.* **12**, 426 (1879).
503. L. Kamm, D. F. Bray, and D. E. Coffin, *J. Assoc. Offic. Anal. Chemists* **51**, 140 (1968).
504. D. Daiber and R. Preussmann, *Z. Anal. Chem.* **206**, 344 (1964).
505. C. Hughes and S. P. Spragg, *Biochem. J.* **70**, 205 (1958).
506. W. A. Andrae and S. R. Andrae, *Can. J. Botany* **31**, 426 (1953).
507. D L. Mays, G. S. Born, J. E. Christian, and B. J. Liska, *J. Agr. Food Chem.* **16**, 356 (1968).
508. B. K. Biswas, O. Hall, and P. D. Mayberry, *Physiol. Plantarum* **20**, 819 (1967).
509. B. H. N. Towers, A. Hutchinson, and W. A. Andrae, *Nature* **181**, 1535 (1958).
510. I. Hoffman and E. V. Parups, *J. Agr. Food Chem.* **10**, 453 (1962).
511. G. J. Stone, *in* "A Literature Summary on Maleic Hydrazide, 1949–1957" (J. W. Zukel, ed.). U.S. Rubber Co., Naugatuck Chem. Div., Naugatuck, Connecticut, 1957.
512. C. Anglin and J. H. Mahon, *J. Assoc. Offic. Agr. Chemists* **41**, 177 (1958).
513. A. S. Crafts, "The Chemistry and Mode of Action of Herbicides," Chapter 15. Wiley (Interscience), New York, 1961.
514. I. Hoffman and E. V. Parups, *Residue Rev.* **7**, 96 (1964).
515. F. Dickens and H. E. H. Jones, *Brit. J. Cancer* **19**, 392 (1965).
516. S. S. Epstein, J. Andrea, H. Jaffee, H. Falk, and N. Mantel, *Nature* **215**, 1388 (1967).
517. S. S. Epstein and N. Mantel, *Intern. J. Cancer* **3**, 325 (1968).
518. K. Fujii and S. S. Epstein, *Toxicol. Appl. Pharmacol.* **14**, 613 (1969). (abstr.).
519. R. E. McCarthy and S. S. Epstein, *Life Sci.* **7**, 1 (1968).
520. C. E. Nasrat, *Nature* **207**, 439 (1965).
521. J. H. Northrup, *J. Gen. Physiol.* **46**, 971 (1963).
522. C. D. Darlington and J. MacLeish, *Nature* **167**, 407 (1951).
523. J. MacLeish, *Heredity* **6**, Suppl., 125 (1952).
524. J. MacLeish, *Heredity* **8**, 385 (1954).
525. A. Michaelis and R. Rieger, *Nature* **199**, 1014 (1963).
526. D. Scott, *Mutation Res.* **5**, 65 (1968).
527. H. G. Evans and D. Scott, *Genetics* **49**, 17 (1964).
528. M. A. McManus, *Nature* **185**, 44 (1960).

529. K. D. Wuu and W. F. Grant, *Botan. Bull. Acad. Sinica* **8**, 191 (1967).
530. G. E. Graf, *J. Heredity* **48**, 155 (1957).
531. J. M. Carlson, *Iowa State Coll. J. Sci.* **29**, 105 (1954).
532. V. V. Shevchenko, *Genetika* **1**, 86 (1965).
533. L. G. Dubinina and N. P. Dubinin, *Genetika* **4**, 5 (1968).
534. S. Shimomura, *J. Pharm. Soc. Japan* **6**, 589 (1958).
535. T. Takeuchi, N. Yokouchi, and K. Onoda, *Japan. Analyst* **5**, 399 (1956).
536. D. M. Miller, *Can. J. Chem.* **34**, 1510 (1956).
537. I. Hoffman, *J. Assoc. Offic. Agr. Chemists* **44**, 723 (1961).
538. M. Mashima, *J. Chem. Soc. Japan, Pure Chem. Sect.* **83**, 981 (1962).
539. O. Ohashi, M. Mashima, and M. Kubo, *Can. J. Chem.* **42**, 970 (1964).
540. M. Pesez and A. Petit, *Bull. Soc. Chim. France* p. 122 (1947).
541. G. R. Lane, D. K. Gullstrom, and J. E. Newell, *J. Agr. Food Chem.* **6**, 671 (1958).
542. W. A. Andrae, *Can. J. Biochem. Physiol.* **36**, 71 (1958).
543. V. A. Gruelach and J. G. Haesloop, *Anal. Chem.* **33**, 1446 (1961).
544. L. Fishbein and W. L. Zielinski, Jr., *J. Chromatog.* **18**, 581 (1965).
545. W. L. Zielinski, Jr. and L. Fishbein, *J. Chromatog.* **20**, 140 (1965).
546. R. J. Lukens and H. D. Sisler, *Phytopathology* **48**, 235 (1958).
547. E. Somers, D. V. Richmond, and J. A. Pickard, *Nature* **215**, 214 (1967).
548. R. G. Owens and G. Blaak, *Contrib. Boyce Thompson Inst.* **20**, 475 (1960).
549. D. V. Richmond and E. Somers, *Ann. Appl. Biol.* **62**, 35 (1968).
550. R. G. Owens and H. M. Novotny, *Contrib. Boyce Thompson Inst.* **20**, 171 (1959).
551. E. M. Boyd and C. J. Krijnen, *J. Clin. Pharmacol.* **8**, 225 (1968).
552. J. N. Ospenson, D. E. Pack, G. K. Kohn, H. P. Burchfield, and E. E. Storrs, *in* "Analytical Methods for Pesticides, Plant Growth Regulators, and Food Additives" (G. Zweig, ed.) Vol. 2, p. 7. Academic Press, New York, 1964.
553. J. McLaughlin, Jr., E. F. Reynaldo, J. K. Lamar, and J. P. Marliac, *Toxicol. Appl. Pharmacol.* **14**, 641 (1969).
554. M. Legator and J. Verrett, *in* "Conference on Biological Effects of Pesticides in Mammalian Systems," Abstr. No. 6, p. 17. N.Y. Acad. Sci., New York, 1967.
555. A. R. Kittleson, *Anal. Chem.* **24**, 1173 (1952).
556. J. Wagner, V. Wallace, and J. M. Lawrence, *J. Agr. Food Chem.* **4**, 1035 (1956).
557. H. P. Burchfield and P. H. Schuldt, *J. Agr. Food Chem.* **6**, 106 (1958).
558. L. C. Mitchell, *J. Assoc. Offic. Agr. Chemists* **41**, 781 (1958).
559. L. Fishbein, J. Fawkes, and P. Jones, *J. Chromatog.* **23**, 529 (1968).
560. R. Engst and W. Schnaak, *Nahrung* **11**, 95 (1967).
561. A. Bevenue and J. N. Ogata, *J. Chromatog.* **36**, 529 (1968).
562. W. W. Kilgore and E. R. White, *J. Agr. Food Chem.* **15**, 1118 (1967).
563. I. H. Pomerantz and R. Ross, *J. Assoc. Offic. Anal. Chemists* **51**, 1058 (1968).
564. Z. H. Abedi and W. P. McKinley, *Nature* **216**, 1321 (1967).
565. S. C. Chang, P. H. Terry, and A. B. Borkovec, *Science* **144**, 57 (1964).
566. R. L. Fye, H. K. Gouck, and G. C. LaBrecque, *J. Econ. Entomol.* **58**, 446 (1965).
567. J. T. Chang and Y. C. Chiang, *Acta Entomol. Sinica* **13**, 679 (1964).
568. A. B. Borkovec, *Advan. Pest Control Res.* **7**, 45 (1966).
569. S. C. Chang, P. H. Terry, C. W. Woods, and A. B. Borkovec, *J. Econ. Entomol.* **60**, 1623 (1967).
570. S. Akov and A. B. Borkovec, *Life Sci.* **7**, 1215 (1968).
571. S. Akov, J. E. Oliver, and A. B. Borkovec, *Life Sci.* **7**, 1207 (1968).
572. A. R. Jones and H. Jackson, *Biochem. Pharmacol.* **17**, 2247 (1968).
573. J. Palmquist and L. E. LaChance, *Science* **154**, 915 (1966).

574. R. M. Kimbrough and T. B. Gaines, *Nature* **211**, 146 (1966).
575. H. Jackson and A. W. Craig, *Proc. 5th I.P.P.F. Conf., 1967* p. 49.
576. T. B. Gaines and R. Kimbrough, *Bull. World Health Organ.* **13**, 737 (1964).
577. A. C. Terranova and C. H. Schmidt, *J. Econ. Entomol.* **60**, 1659 (1967).
578. M. C. Bowman and M. Beroza, *J. Assoc. Offic. Anal. Chemists* **49**, 1046 (1966).
579. S. C. Chang, A. B. DeMilo, C. W. Woods, and A. B. Borkovec, *J. Econ. Entomol.* **61**, 1357 (1968).
580. R. L. Jasper, E. L. Silvers, and H. O. Williamson, *Federation Proc.* **24**, 641 (1965).
581. F. S. Philips and J. B. Thiersch, *J. Pharmacol. Exptl. Therap.* **100**, 398 (1950).
582. D. T. North, *Mutation Res.* **4**, 225 (1967).
583. F. W. Schremp, J. F. Chittum, and T. S. Arczynski, *J. Petrol. Technol.* **13**, 703 (1961).
584. W. Peacock, British Patent 524,753 (1940); *Chem. Abstr.* **35**, 6755 (1941).
585. H. H. Willard and W. S. Gale, U.S. Patent 2,612,459 (1952); *Chem. Abstr.* **49**, 476e (1953).
586. R. C. Harshman and V. C. Fusco, U.S. Patent 2,963,438 (1960); *Chem. Abstr.* **55**, 5997b (1961).
587. H. Oertel, H. Rinkle, and W. Thomas, German Patent 1,123,467 (1962); *Chem. Abstr.* **57**, 1012 (1962).
588. G. R. Himes and H. O. Spang, U.S. Patent 3,091,602 (1963); *Chem. Abstr.* **59**, 5374g (1963).
589. R. L. Wear, U.S. Patent 2,847,395 (1958); *Chem. Abstr.* **54**, 1924b (1958).
590. E. M. Wilson and A. F. Graefe, U.S. Patent 2,940,843 (1960); *Chem. Abstr.* **54**, 18962g (1960).
591. D. L. Schoene and O. L. Hoffmann, *Science* **109**, 588 (1949).
592. W. W. Allen, U.S. Patent 2,670,282 (1954); *Chem. Abstr.* **48**, 6642d (1954).
593. R. W. Leeper and V. C. Fusco, U.S. Patent 2,876,090 (1959); *Chem. Abstr.* **54**, 1795d (1960).
594. H. W. Fox, *J. Org. Chem.* **17**, 542 (1952).
595. J. H. Biel, *in* "Molecular Modification in Drug Design" (F. W. Schueler, ed.), Chapter 10, p. 114. Am. Chem. Soc., Washington, D.C., 1964.
596. J. Druey and B. H. Ringier, *Helv. Chim. Acta* **34**, 195 (1951).
597. W. B. Stillman and A. B. Scott, U.S. Patent 2,416,234 (1947); *Chem. Abstr.* **41**, 3488i (1947).
598. L. F. Audrieth and B. A. Off, "The Chemistry of Hydrazine." Wiley, New York, 1951.
599. G. D. Byrkit and G. A. Michalek, *Ind. Eng. Chem.* **42**, 1862 (1950).
600. L. B. Witkin, *A.M.A. Arch. Ind. Health* **13**, 34 (1956).
601. P. McGrath, C. C. Comstock, and F. W. Oberst, *Federation Proc.* **11**, 374 (1952).
602. C. C. Comstock and F. W. Oberst, *Federation Proc.* **11**, 333 (1952).
603. S. Krop, *A.M.A. Arch. Ind. Hyg. occupational Med.* **9**, 199 (1954).
604. A. M. Dominguez, J. S. Amenta, C. S. Hill, and T. T. Domanski, *Aerospace Med.* **33**, 1094 (1962).
605. C. Biancifiori, E. Bucciarelli, D. B. Clayson, and F. E. Santilli, *Brit. J. Cancer* **18**, 543 (1964).
606. F. J. C. Roe, G. A. Grant, and S. M. Millican, *Nature* **216**, 375 (1967).
607. H. Druckrey, *Z. Krebsforsch.* **67**, 31 (1965).
608. B. Toth, *J. Natl. Cancer Inst.* **42**, 469 (1969).
609. H. McKennis, Jr., A. S. Yard, J. H. Weatherby, and J. A. Hagy, *J. Pharmacol. Exptl. Therap.* **126**, 109 (1959).
610. H. McKennis, Jr., A. S. Yard, and E. V. Dahnelas, *Am. Rev. Tuberc. Pulmonary Diseases* **73**, 956 (1956).

611. T. Dambrauskus and H. C. Cornish, *Toxicol. Appl. Pharmacol.* **6**, 653 (1964).
612. D. Horvitz and E. Cerwonka, U.S. Patent 2,916,426 (1959); *Chem. Abstr.* **54**, 6370 (1960).
613. W. T. Nelson, U.S. Patent 2,926,750 (1960); *Chem. Abstr.* **54**, 11444c (1960).
614. Olin Mathieson Chem. Corp., British Patent 799,923 (1958); *Chem. Abstr.* **53**, 2598 (1959).
615. E. H. Jenny and C. C. Pfeifer, *J. Pharmacol. Exptl. Therap.* **122**, 110 (1958).
616. K. H. Jacobson, J. H. Clem, H. J. Wheelwright, W. E. Rinehart, and N. Mayes, *A.M.A. Arch. Ind. Health* **12**, 609 (1955).
617. W. E. Rinehart, E. Donati, and E. A. Greene, *Am. Ind. Hyg. Assoc. J.* **21**, 207 (1960).
618. H. C. Cornish and R. Hartung, *Toxicol. Appl. Pharmacol.* **15**, 62 (1969).
619. K. C. Back, M. K. Pinkerton, A. B. Cooper, and A. A. Thomas, *Toxicol. Appl. Pharmacol.* **5**, 401 (1963).
620. F. N. Dost, D. J. Reed, and C. W. Wang, *Biochem. Pharmacol.* **15**, 1325 (1966).
621. G. Porcellati and P. Preziosi, *Enzymologia* **17**, 47 (1954).
622. I. Toida, *Am. Rev. Respitat. Diseases* **85**, 720 (1962).
623. J. Juhász, J. Baló, and G. Kendrey, *Tuberkulozis* **3–4**, 49 (1957).
624. A. Peacock and P. R. Peacock, *Brit. J. Cancer* **20**, 307 (1966).
625. J. Juhász, J. Baló, and B. Szende, *Z. Krebsforsch.* **65**, 434 (1963).
626. L. Severi and C. Biancifiori, *J. Natl. Cancer Inst.* **41**, 381 (1968).
627. C. Cattaneo, *Ann. Ist "Carlo Forlanini"* **27**, 78 (1967).
628. H. B. Hughes, *J. Pharmacol. Exptl. Therap.* **109**, 444 (1953).
629. E. Freese, E. Bautz, and E. B. Freese, *Proc. Natl. Acad. Sci. U.S.* **47**, 845 (1961).
630. H. K. Jain and R. N. Raut, *Nature* **211**, 652 (1966).
631. F. Lingens, *Z. Naturforsch.* **19b**, 151 (1964).
632. A. Rutishauser and W. Bollag, *Experientia* **19**, 131 (1963).
633. F. K. Zimmermann and R. Schwaier, *Naturwissenschaften* **54**, 251 (1967).
634. E. Freese, S. Sklarow, and E. B. Freese, *Mutation Res.* **5**, 343 (1968).
635. L. Feinsilver, J. A. Perregrino, and C. J. Smith, *Am. Ind. Hyg. Assoc. J.* **20**, 26 (1959).
636. R. Preussmann, H. Hengy, and A. von Hodenberg, *Anal. Chim. Acta* **42**, 95 (1968).
637. R. C. R. Barreto and S. O. Sabino, *J. Chromatog.* **13**, 435 (1964).
638. C. Bighi and G. Saglietto, *J. Chromatog.* **18**, 297 (1965).
639. G. Neurath and W. Lüttich, *J. Chromatog.* **34**, 257 (1968).
640. G. Bighi and G. Saglietto, *J. Gas Chromatog.* 303 (1966).
641. N. A. Kirschen and G. H. Olsen, *Anal. Chem.* **40**, 1341 (1968).
642. G. J. Leitner, U.S. Patent 2,951,008 (1960); *Chem. Abstr.* **55**, 25144g (1961).
643. H. Wilhelm, H. Metzger, and K. Dachs, British Patent 846,239 (1960); *Chem. Abstr.* **55**, 7288e (1961).
644. D. H. Andrews, Canadian Patent 611,510 (1960); *Chem. Abstr.* **55**, 10888h (1961).
645. J. Khachoyan and J. P. Niederhauser, *J. Soc. Dyers Colourists* **74**, 133 (1958).
646. L. F. A. Mason, German Patent 1,057,875 (1959); *Chem. Abstr.* **55**, 8137c (1961).
647. K. A. E. Blackmore, U.S. Patent 3,314,807 (1967); *Chem. Abstr.* **67**, 3924v (1967).
648. B. E. Hietbrink, A. B. Raymond, G. R. Zins, and K. P. DuBois, *Toxicol. Appl. Pharmacol.* **3**, 266 (1961).
649. E. Roberts, U.S. Patent 3,179,563 (1965); *Chem. Abstr.* **57**, 8670a (1962).
650. J. D. Gabourel, *Biochem. Pharmacol.* **5**, 283 (1961).
651. S. A. Price, P. Mamalis, D. McHale, and J. Green, *Brit. J. Pharmacol.* **15**, 243 (1960).
652. T. Elsässer and E. Lange, *Pharmazie* **11**, 391 (1956).
653. E. Jeney and T. Zsolnai, *Zentr. Bakteriol., Parasitenk., Abt. I. Orig.* **167**, 55 (1956); *Chem. Abstr.* **51**, 7579 (1957).

654. H. Riemann, *Acta Pharmacol. Toxicol.* **6**, 285 (1950).
655. D. M. Brown and P. Schell, *J. Mol. Biol.* **3**, 709 (1961).
656. E. B. Freese and E. Freese, *Proc. Natl. Acad. Sci. U.S.* **52**, 1289 (1964).
657. E. Freese and H. B. Strack, *Proc. Natl. Acad. Sci. U.S.* **48**, 1796 (1962).
658. R. T. Kapadia and M. Srogl, *Folia Microbiol. (Prague)* **14**, 51 (1968).
659. E. Freese and E. B. Freese, *Radiation Res.* Suppl. 6, 97 (1966).
660. I. Tessman, H. Ishiwa, and S. Kumar, *Science* **148**, 507 (1965).
661. E. Freese, E. B. Freese, and E. Bautz, *J. Mol. Biol.* **3**, 133 (1961).
662. H. V. Malling, *Mutation Res.* **3**, 470 (1966).
663. W. Engel, W. Krone, and U. Wold, *Mutation Res.* **4**, 353 (1967).
664. C. F. Somers and T. C. Hsu, *Proc. Natl. Acad. Sci. U.S.* **48**, 937 (1962).
665. E. Borenfreund, M. Krim, and A. Bendich, *J. Natl. Cancer Inst.* **32**, 667 (1964).
666. A. Bendich, E. Borenfreund, G. C. Korngold, and M. Krim, *Federation Proc.* **22**, 582 (1963).
667. A. T. Natarajan and M. D. Upadhya, *Chromosoma* **15**, 156 (1964).
668. H. V. Malling, *Mutation Res.* **4**, 559 (1967).
669. N. K. Kochetkov, E. I. Budowsky, and R. P. Shibaeva, *Biochim. Biophys. Acta* **68**, 493 (1964).
670. B. R. Sant, *Anal. Chim. Acta* **20**, 371 (1959).
671. T. Takahashi and H. Sakurai, *Rept. Inst. Ind. Sci., Univ. Tokyo* **13**, 1 (1963); *Chem. Abstr.* **67**, 7748s (1967).
672. W. N. Fishbein, *Anal. Chim. Acta* **37**, 484 (1967).
673. S. Moore and W. H. Stein, *J. Biol. Chem.* **176**, 367 (1948).
674. J. Blom, *Chem. Ber.* **59**, 121 (1926).
675. T. Z. Czaky, *Acta Chem. Scand.* **2**, 450 (1948).
676. B. M. Schmidt, L. C. Brown, and D. Williams, *J. Mol. Spectry.* **3**, 30 (1959).
677. H. M. Stevens, *Anal. Chim. Acta* **21**, 456 (1959).
678. S. Ohkuma, *Yakugaku Zasshi* **79**, 1582 (1959); *Chem. Abstr.* **54**, 9591i (1960).
679. C. Runti and R. Deghengi, *Ann. Triest.* **22–23**, 185 (1953); *Chem. Abstr.* **49**, 1568 (1955).
680. H. Kofod, *Acta Chem. Scand.* **7**, 938 (1953).
681. W. N. Fishbein, P. P. Carbone, E. J. Friereich, J. Misra, and E. Frei, *Clin. Pharmacol. Therap.* 574 (1964).
682. R. H. Adamson, S. T. Yancey, M. Ben, T. L. Loo, and D. P. Rall, *Arch. Intern. Pharmacodyn.* **2**, 153 (1965).
683. B. J. Kennedy and J. W. Yarbro, *J. Am. Med. Assoc.* **195**, 1038 (1960).
684. I. H. Krakoff, M. L. Murphy, and H. Savel, *Proc. Am. Assoc. Cancer Res.* **4**, 35 (1963).
685. H. S. Rosenkranz, H. M. Rose, C. Morgan, and K. C. Hsu, *Virology* **28**, 510 (1966).
686. J. B. Kissam, J. A. Wilson, and S. B. Hays, *J. Econ. Entomol.* **60**, 1130 (1967).
687. W. N. Fishbein and P. P. Carbone, *Science* **142**, 1069 (1963).
688. W. N. Fishbein and T. S. Winter, *Federation Proc.* **23**, 385 (1964).
689. W. N. Fishbein, T. S. Winter, and J. D. Davidson, *J. Biol. Chem.* **240**, 2402 (1965).
690. H. S. Rosenkranz and J. A. Levy, *Biochim. Biophys. Acta* **95**, 181 (1965).
691. J. J. Oppenheim and W. N. Fishbein, *Cancer Res.* **25**, 980 (1965).
692. W. K. Sinclair, *Science* **24**, 1729 (1965).
693. B. A. Kihlman, T. Eriksson, and G. Odmark, *Hereditas* **55**, 386 (1966).
694. V. H. Ferm, *Lancet* **I**, 1338 (1965).
695. V. H. Ferm, *Arch. Pathol.* **81**, 174 (1966).
696. S. Chaube and M. L. Murphy, *Cancer Res.* **26**, 1448 (1966).
697. M. L. Murphy and S. Chaube, *Cancer Chemotherapy Rept.* **40**, 1 (1964).
698. S. Chaube, E. Simmel, and C. Lacon, *Proc. Am. Assoc. Cancer Res.* **4**, 10 (1963).

699. F. Bodit, R. Stoll, and R. Maraud, *Compt. Rend. Soc. Biol.* **160**, 960 (1966); *Chem. Abstr.* **66**, 26889j (1967).
700. J. D. Davidson and T. S. Winter, *Cancer Chemotherapy Rept.* **27**, 97 (1963).
701. B. H. Bolton, L. A. Woods, D. T. Kaung, and R. L. Lawton, *Cancer Chemotherapy Rept.* **46**, 1 (1965).
702. H. Kofod, *Acta Chem. Scand.* **11**, 1236 (1957).
703. H. Kofod, *Acta Chem. Scand.* **13**, 461 (1959).
704. R. H. Adamson, S. L. Ague, S. M. Hess, and J. D. Davidson, *J. Pharmacol. Exptl. Therap.* **150**, 322 (1965).
705. H. Kofod, *Acta Chem. Scand.* **9**, 1575 (1955).
706. L. Fishbein and M. A. Cavanaugh, *J. Chromatog.* **20**, 283 (1965).
707. W. L. Zielinski, Jr. and L. Fishbein, *J. Chromatog.* **25**, 475 (1966).
708. S. Bellman, W. Troll, G. Teebor, and F. Muakaki, *Cancer Res.* **28**, 535 (1968).
709. F. Mukai, S. Belman, W. Troll, and I. Hawryluk, *Proc. Am. Assoc. Cancer Res.* **8**, 49 (1967).
710. S. Belman, W. Troll, G. Teebor, R. Reinhold, B. Fishbein, and F. Mukai, *Proc. Am. Assoc. Cancer Res.* **7**, 6 (1966).
711. G. Perez and J. L. Radomski, *Ind. Med. Surg.* **34**, 714 (1965).
712. R. Willstätter and H. Kubli, *Chem. Ber.* **41**, 1936 (1908).
713. W. Troll and N. Nelson, *Federation Proc.* **20**, 41 (1961).
714. E. Boyland, D. Manson, and R. Nery, *Biochem. J.* **86**, 263 (1963).
715. E. Boyland, C. E. Dukes, and P. L. Grover, *Brit. J. Cancer* **17**, 79 (1963).
716. E. Boyland, E. R. Busby, C. E. Dukes, P. L. Grover, and D. Manson, *Brit. J. Cancer* **18**, 575 (1964).
717. M. A. Walters, F. J. C. Roe, B. C. V. Mitchley, and A. Walsh, *Brit. J. Cancer* **21**, 367 (1967).
718. Société Monsavon-L'Oreal, Austrian Patent 203,458 (1959); *Chem. Abstr.* **53**, 14432e (1959).
719. S. Tanaka and K. Uemura, *Nippon Dojo-Hiryogaku Zasshi* **28**, 114 (1957); *Chem. Abstr.* **52**, 15815d (1958).
720. K. S. Rajagopalan, S. K. Gupta, and B. Sanyal, *Defence Sci. J. (New Delhi)* **9**, 212 (1959); *Chem. Abstr.* **55**, 17453f (1961).
721. K. I. Ivanov and E. D. Vilyanskaya, *Khim. i Tekhnol. Topliva i Masel* p. 11 (1957); *Chem. Abstr.* **51**, 14245h (1957).
722. V. D. Yasnopol'Skii and A. A. Medzhidov, *Vysokomolekul. Soedin.* **3**, 1 (1961); *Chem. Abstr.* **55**, 26507i (1961).
723. M. Letort and J. Petry, French Patent 1,020,456 (1953); *Chem. Abstr.* **52**, 19243 (1958).
724. J. Nüsslein and A. Hartmann, German Patent 1,042,519 (1958); *Chem. Abstr.* **54**, 20234d (1960).
725. J. H. Onions, British Patent 795,190 (1958); *Chem. Abstr.* **52**, 20784g (1958).
726. A. S. Kuz'Minskii and L. G. Angert, *Gosudarst. Nauch. Tekh. Izdatel. Khim. Lit.* p. 17 (1955); *Chem. Abstr.* **54**, 8130c (1960).
727. D. F. McDonald, *Ind. Hyg. Digest* p. 17 (1969).
728. Y. Masuda, K. Mori, and M. Kuratsune, *Intern. J. Cancer* **2**, 489 (1967).
729. M. Pauli, W. J. Hübsch, and H. Kuhn, *Fachliche Mitt. Oesterr. Tabakregie* **7**, 109 (1967).
730. D. H. Hoffmann, Y. Masuda, and E. L. Wynder, *Nature* **211**, 253 (1969).
731. Y. Masuda and D. Hoffmann, *Anal. Chem.* **41**, 650 (1969).
732. Rept. of the Advisory Comm. to the Surgeon General on Smoking and Health, *U.S. Public Health Serv., Publ.* **1103** (1964); The Health Conferences of Smoking, *ibid.* **1696** (1967); Suppl. (1968).

733. E. Boyland and R. Nery, *Analyst* **89**, 95 (1964).
734. J. Latinak, *Collection Czech. Chem. Commun.* **24**, 2939 (1959).
735. F. Fujiwara and C. Sugimoto, *Kagaku To Kogyo* (*Osaka*) **35**, 258 (1961).
736. Q. Quick, R. F. Layton, H. R. Harless, and O. R. Haynes, *J. Gas Chromatog.* **6**, 46 (1968).
737. L. T. Butt and N. Strafford, *J. Appl. Chem.* **6**, 525 (1956).
738. A. P. Dustin, *Ciba Found. Symp., Leukaemia Res.* p. 244 (1954).
739. D. K. George, D. H. Moore, W. P. Brian, and J. A. Garman, *J. Agr. Food Chem.* **2**, 353 (1954).
740. G. R. Ferguson and C. C. Alexander, *J. Agr. Food Chem.* **1**, 888 (1953).
741. D. T. Mowry and N. R. Piesbergew, U.S. Patent 2,537,690 (1951); *Chem. Abstr.* **45**, 2142d (1951).
742. R. T. Pollock, U.S. Patent 2,438,452 (1948); *Chem. Abstr.* **42**, 4334a (1948).
743. A. Nettleship, P. S. Henshaw, and H. L. Meyer, *J. Natl. Cancer Inst.* **4**, 523 (1943).
744. P. N. Cowen, *Brit. J. Cancer* **1**, 401 (1947).
745. J. W. Orr, *Brit. J. Cancer* **1**, 311 (1947).
746. W. G. Jaffe, *Cancer Res.* **7**, 107 (1947).
747. J. G. Sinclair, *Texas Rept. Biol. Med. Med.* **8**, 623 (1950).
748. H. I. Battle and K. K. Hisaoka, *Cancer Res.* **12**, 334 (1952).
749. D. B. McMillan and H. I. Battle, *Cancer Res.* **14**, 319 (1954).
750. M. Vogt, *Experientia* **4**, 68 (1948).
751. M. Vogt, *Pubbl. Staz. Zool. Napoli* **22**, Suppl., 114 (1950).
752. R. Latarjet, N. P. Buu-Hoi, and C. A. Elias, *Pubbl. Staz. Zool. Napoli* **22**, Suppl., 76 (1950).
753. V. Bryson, *Hereditas* Suppl., p. 545 (1945).
754. F. Oehlkers, *Heredity* Suppl. 6, 95 (1953).
755. F. Oehlkers, *Z. Vererbungslehre* **81**, 313 (1943).
756. A. Bateman, *Mutation Res.* **4**, 710 (1967).
757. K. A. Jensen, I. Kirk, G. Kølmark, and M. Westergaard, *Cold Spring Harbor Symp. Quant. Biol.* **16**, 245 (1951).
758. F. Oehlkers, *Z. Induktive Abstammungs- Vererbungslehre* **81**, 313 (1943).
759. E. Boyland and R. Nery, *Biochem. J.* **94**, 198 (1965).
760. E. Boyland, R. Nery, and K. S. Peggie, *Brit. J. Cancer* **19**, 878 (1965).
761. A. M. Kaye and N. Trainin, *Cancer Res.* **26**, 2206 (1966).
762. I. Berenblum, D. Ben-Ishai, N. Haran-Ghera, A. Lapidot, E. Simon, and N. Trainin, *Biochem. Pharmacol.* **2**, 168 (1959).
763. M. A. Hahn, C. C. Botkin, and R. H. Adamson, *Nature* **211**, 984 (1966).
764. E. B. Freese, *Genetics* **51**, 953 (1965).
765. E. B. Freese, J. Gerson, H. Taber, H. J. Rhaese, and E. Freese, *Mutation Res.* **4**, 517 (1967).
766. R. M. Schmidt, R. D. Pollack, and H. S. Rosenkranz, *Biochim. Biophys. Acta* **138**, 645 (1967).
767. R. Nery, *Analyst* **91**, 388 (1966).
768. A. M. Kaye, *Cancer Res.* **20**, 44 (1964).
769. E. Boyland and R. Nery, *J. Chem. Soc., C* p. 346 (1966).
770. S. S. Mirvish, *Biochim. Biophys. Acta* **93**, 673 (1964).
771. S. S. Mirvish, *Biochim. Biophys. Acta* **117**, 1 (1966).
772. S. S. Mirvish, *Analyst* **90**, 244 (1965).
773. W. L. Zielinski, Jr. and L. Fishbein, *J. Gas Chromatog.* **3**, 142 (1965).
774. E. Ochiai, M. Ishikura, and Z. Sai, *J. Pharm. Soc. Japan* **63**, 280 (1943).

775. H. Freytag, F. Lober, and P. E. Frohberger, German Patent 1,013,113 (1957); *Chem. Abstr.* **54**, P16732g (1960).
776. F. Fukuoka, *Acta, Unio Intern. Contra Cancrum* **15**, 109 (1959).
777. S. Sakai, K. Minoda, G. Saito, S. Akagi, and F. Fukuoka, *Gann* **46**, 605 (1955).
778. I. H. Pang, *T'ai-Wan I Hsueh Hui Tsa Chih* **59**, 1105 (1960); *Chem. Abstr.* **55**, 24914f (1961).
779. K. Minoda, S. Akagi, and N. Okoti, *J. Sci. Res. Inst.* (*Tokyo*) **51**, 91 (1957); *Chem. Abstr.* **51**, 16941c (1957).
780. C. Nagata, K. Jujii, and S. S. Epstein, *Nature* **215**, 972 (1967).
781. H. Endo, *Gann* **49**, 151 (1958).
782. H. Endo, *Symp. Cell Chem.* **10**, 105 (1960).
783. C. Nagata, M. Kodama, Y. Tagashira, and A. Imamura, *Biopolymers* **4**, 409 (1966).
784. M. F. Malkin and A. C. Zahalsky, *Science* **154**, 1665 (1966).
785. T. Okano and K. Uekama, *Chem. & Pharm. Bull.* (*Tokyo*) **15**, 1812 (1967).
786. E. DeMaeyer and J. DeMayer-Guignard, *Arch. Ges. Virusforsch.* **22**, 61 (1967).
787. T. Sugimura, K. Okabe, and H. Endo, *Gann* **56**, 489 (1965).
788. T. Okabayashi and A. Yoshimoto, *Chem. & Pharm. Bull.* (*Tokyo*) **10**, 1221 (1962).
789. Y. Kawazoe, M. Tachibana, K. Aoki, and W. Nakahara, *Biochem. Pharmacol.* **16**, 631 (1967).
790. T. Sugimura, K. Okabe, and M. Nagao, *Cancer Res.* **26**, 1717 (1966).
791. H. Endo and F. Kume, *Gann* **54**, 443 (1963).
792. Y. Shirasu and A. Ohta, *Gann* **54**, 24 (1963).
793. T. Okabayashi, *Chem. & Pharm. Bull.* (*Tokyo*) **10**, 1127 (1962).
794. H. Endo, M. Ishizawa, and T. Kamiya, *Nature* **198**, 195 (1963).
795. C. Huggins, *in* "Horizons in Biochemistry" (M. Kasha and B. Pullman, eds.), p. 497. Academic Press, New York, 1962.
796. K. Mori, *Gann* **58**, 389 (1967).
797. D. J. Parish and C. E. Searle, *Brit. J. Cancer* **20**, 200 (1966).
798. G. Karreman, *Ann. N.Y. Acad. Sci.* **96**, 1092 (1962).
799. T. Okabayashi, M. Ide, A. Yoshimoto, and M. Otsubo, *Chem. & Pharm. Bull.* (*Tokyo*) **13**, 610 (1965).
800. T. Okabayashi, *Ferment. Ind.* **33**, 513 (1955).
801. S. Mashima and Y. Ikeda, *J. Appl. Microbiol.* **6**, 45 (1958).
802. I. H. Pan, *J. Formosan Med. Assoc.* **59**, 41 (1960).
803. I. H. Pan, *J Formosan Med. Assoc.* **62**, 107 (1963).
804. T. H. Yoshida, Y. Kurita, and K. Moriwaki, *Gann* **56**, 513 (1956).
805. T. Mita, R. Tikuzen, F. Fukuoka, and W. Nakahara, *Gann* **56**, 293 (1965).
806. Y. Kurita, T. H. Yoshida, and K. Moriwaki, *Japan. J. Genet.* **40**, 365 (1965).
807. Y. Kawazoe and M. Tachibana, *Chem. & Pharm. Bull.* (*Tokyo*) **15**, 1 (1967).
808. T. Okabayashi and M. Ide, *J. Chromatog.* **9**, 523 (1962).
809. J. H. Young, U.S. Patent 2,777,749 (1957); *Chem. Abstr.* **51**, 5442g (1957).
810. A. L. Ayers and C. R. Scott, U.S. Patent 2,874,535 (1959); *Chem. Abstr.* **53**, 9621b (1959).
811. R. Ludewig, *Z. Deut. Zahnaerztl.* **15**, 444 (1960); *Chem. Abstr.* **54**, 10138d (1960).
812. O. Warburg, "New Methods of Cell Physiology." Wiley (Interscience), New York, 1962.
813. F. H. Sobels, *Radiation Res.* Suppl. 3, 171 (1963).
814. R. G. Zimmer, *Abhandl. Math.-Naturw. Kl., Akad. Wiss. Literatur. Mainz* No. 3, 825 (1960).
815. R. G. Zimmer, *Acta Radiol.* **46**, 595 (1956).

816. T. Alper, *Radiation Res.* **5**, 573 (1956).
817. I. A. Butler, *Radiation Res.* **4**, 20 (1956).
818. O. Wyss, W. S. Stone, and J. B. Clark, *J. Bacteriol.* **54**, 767 (1947).
819. O. Wyss, J. B. Clark, F. Haas, and W. S. Stone, *J. Bacteriol.* **56**, 51 (1948).
820. F. L. Haas, J. B. Clark, O. Wyss, and W. S. Stone, *Am. Naturalist* **74**, 261 (1950).
821. F. H. Dickey, G. H. Cleland, and C. Lotz, *Proc. Natl. Acad. Sci. U.S.* **35**, 581 (1949).
822. R. P. Wagner, C. R. Haddox, R. Fuerst, and W. S. Stone, *Genetics* **35**, 237 (1950).
823. D. Luzzati, H. Schweitz, M. L. Bach, and M. R. Chevallier, *J. Chim. Phys.* **58**, 1021 (1961).
824. A. Zamenhof, H. E. Alexander, and G. Leidy, *J. Exptl. Med.* **98**, 373 (1953).
825. R. Latarjet, N. Rebeyrotte, and P. Demerseman, *in* "Organic Peroxides in Radiobiology" (M. Haissinsky, ed.), p. 61. Pergamon Press, Oxford, 1958.
826. C. O. Doudney, *Mutation Res.* **6**, 345 (1968).
827. J. Schöneich, *Mutation Res.* **4**, 385 (1967).
828. L. J. Lilly, *Nature* **177**, 338 (1956).
829. E. Freese and E. B. Freese, *Biochemistry* **4**, 2419 (1965).
830. G. M. Eisenberg, *Ind. Eng. Chem., Anal. Ed.* **15**, 327 (1943).
831. M. Kminkova, M. Gottwaldova, and J. Hanus, *Chem. & Ind. London* p. 519 (1969).
832. M. C. R. Symons, *in* "Peroxide Reaction Mechanisms" (J. O. Edwards, ed.), p. 137. Wiley (Interscience), New York, 1962.
833. J. G. Wallace, "Hydrogen Peroxide in Organic Chemistry." E. I. duPont de Nemours & Co., Inc., Wilmington, Delaware, 1962.
834. C. A. Bunton, *in* "Peroxide Reaction Mechanisms" (J. O. Edwards, ed.), pp. 11–29. Wiley (Interscience), New York, 1962.
835. W. Machu, "Das Wasserstoffperoxyd und Die Perverbindungen." Springer, Vienna, 1951.
836. L. S. Altenberg, *Proc. Natl. Acad. Sci. U.S.* **40**, 1037 (1940).
837. L. S. Altenberg, *Genetics* **43**, 662 (1958).
838. M. R. Chevallier and D. Luzatti, *Compt. Rend.* **250**, 1572 (1960).
839. S. H. Revell, *Heredity* **6**, Suppl., 107 (1953).
840. A. Loveless, *Nature* **167**, 338 (1951).
841. R. Latarjet, *Ciba Found. Symp., Ionizing Radiations Cell Metab.* p. 275 (1957).
842. D. Luzzati and M. R. Chevallier, *Ann. Inst. Pasteur* **93**, 366 (1957).
843. H. Wieland and A. Wingler, *Ann. Chem.* **431**, 301 (1923).
844. P. Kotin and H. L. Falk, *Radiation Res.* Suppl. 3, 193 (1963).
845. F. H. Sobels, *Drosophila Inform. Serv.* **28**, 150 (1954).
846. F. H. Sobels, *Nature* **177**, 979 (1956).
847. F. H. Sobels, *Am. Naturalist* **88**, 109 (1954).
848. C. Walling, *in* "Free Radicals in Solution" p. 469. Wiley, New York, 1957.
849. W. A. Pryor, "Free Radicals," p. 99. McGraw-Hill, New York, 1966.
850. A. J. Martin, *in* "Organic Analysis" (J. Mitchell, Sr. *et al.*, eds.), Vol. 4, pp. 1–64. Wiley (Interscience), New York, 1960.
851. S. Siggia, "Quantitative Organic Analysis via Functional Groups," 3rd ed., pp. 255–295. Wiley, New York, 1963.
852. A. H. M. Cosija, *Chem. Weekblad* **60**, 585 (1964).
853. D. K. Banerjee and C. C. Budke, *Anal. Chem.* **36**, 2367 (1964).
854. C. D. Wagner, R. H. Smith, and E. D. Peters, *Anal. Chem.* **19**, 976 (1947).
855. Y. N. Anisimov and S. S. Ivanchev, *Zh. Anal. Khim.* **21**, 113 (1966).
856. S. W. Bukata, L. L. Zabrocki, and M. F. McLaughlin, *Anal. Chem.* **35**, 885 (1963).
857. D. B. Adams, *Analyst* **91**, 397 (1966).

858. L. Dulog, *Z. Anal. Chem.* **202**, 258 (1964).
859. N. A. Milas and I. Belic, *J. Am. Chem. Soc.* **81**, 3358 (1959).
860. J. Cartlidge and C. F. H. Tipper, *Anal. Chim. Acta* **12**, 106 (1960).
861. S. Hayano and T. Ota, *Japan. Analyst* **15**, 365 (1966).
862. G. Dobson and G. Hughes, *J. Chromatog.* **16**, 416 (1964).
863. G. N. Wogan, *Bacteriol. Rev.* **30**, 460 (1966).
864. G. N. Wogan, "Mycotoxins in Foodstuffs." M.I.T. Press, Cambridge, Massachusetts, 1968.
865. G. N. Wogan, *Trout Hepatoma Res. Conf. Papers, 1967* Res. Rept. No. 70, p. 121. U.S. Govt. Printing Office, Washington, D.C., 1968.
866. L. Fishbein and H. L. Falk, *Chromatog. Rev.* **12**, 42 (1970).
867. E. Borker, N. F. Insalata, C. P. Levi, and J. S. Witzeman, *Advan. Appl. Microbiol.* **8**, 315 (1966).
868. H. F. Kraybill and M. B. Shimkin, *Advan. Cancer Res.* **8**, 208 (1964).
869. G. N. Wogan, *Progr. Exptl. Tumor Res.* **11**, 134 (1969).
870. P. M. Newberne and W. H. Butler, *Cancer Res.* **29**, 236 (1969).
871. G. N. Wogan, *Federation Proc.* **27**, 932 (1968).
872. G. N. Wogan, G. S. Edwards, and R. C. Shank, *Cancer Res.* **27**, 1729 (1967).
873. O. Bassir and F. Osiyemi, *Nature* **215**, 882 (1967).
874. R. Allcroft, H. Roger, G. Lewis, J. Nabney, and P. E. Best, *Nature* **209**, 154 (1966).
875. C. W. Holzapfel, P. S. Steyn, and I. F. H. Purchase, *Tetrahedron Letters* **25**, 2799 (1966).
876. M. S. Masri, D. E. Lundin, J. R. Page, and V. C. Garcia, *Nature* **215**, 753 (1967).
877. M. B. Sporn, C. W. Dingman, H. L. Phelps, and G. N. Wogan, *Science* **151**, 1539 (1966).
878. J. I. Clifford and K. R. Rees, *Nature* **209**, 312 (1966).
879. H. S. Black and B. Kirgensons, *Plant Physiol.* **42**, 731 (1967).
880. A. M. DeRecondo, C. Frayssinet, C. LaFarge, and E. LeBreton, *Biochim. Biophys. Acta* **119**, 332 (1960).
881. J. B. Wragg, V. C. Ross, and M. S. Legator, *Proc. Soc. Exptl. Biol. Med.* **125**, 1052 (1967).
882. A. K. Roy, *Biochim. Biophys. Acta* **169**, 106 (1968).
883. Y. Moule and C. Frayssinet, *Nature* **218**, 93 (1968).
884. M. S. Legator, *Bacteriol. Rev.* **30**, 471 (1966).
885. D. A. Dolimpo, C. Jacobson, and M. S. Legator, *Proc. Soc. Exptl. Biol. Med.* **127**, 559 (1968).
886. M. S. Legator and A. Withrow, *J. Assoc. Offic. Agr. Chemists* **47**, 1007 (1964).
887. M. S. Legator, S. M. Zuffante, and A. R. Harp, *Nature* **208**, 345 (1965).
888. J. Gabliks, W. Schaeffer, L. Friedman, and G. Wogan, *J. Bacteriol.* **90**, 720 (1965).
889. S. S. Epstein and H. Shafner, *Nature* **219**, 385 (1968).
890. F. A. Gunther and F. Buzzetti, *Residue Rev.* **9**, 90 (1965).
891. E. Sawicki, *Chemist-Analyst* **53**, 24 (1964).
892. E. Sawicki, T. W. Stanley, W. C. Elbert, J. Meeker, and S. McPherson, *Atmospheric Environ.* **1**, 131 (1967).
893. H. Kunte, *Arch. Hyg. Bakteriol.* **151**, 13 (1967).
894. E. L. Wynder and D. Hoffmann, "Tobacco and Tobacco Smoke," p. 330. Academic Press, New York, 1967.
895. J. W. Howard and T. Fazio, *J. Agr. Food Chem.* **17**, 527 (1969).
896. D. J. Tilgner, *Food Manuf.* **43**, 37 (1968).
897. A. C. Stern, ed., "Air Pollution," 2nd ed., Vol. 2. Academic Press, New York, 1968.
898. E. Sawicki, *Chemist-Analyst* **53**, 56 and 88 (1964).

899. E. O. Haenni, *Residue Rev.* **24**, 42 (1968).
900. G. Grimmer and A. Hildebrant, *J. Chromatog.* **20**, 89 (1965).
901. J. W. Howard, R. T. Teague, Jr., R. H. White, and B. E. Fry, Jr., *J. Assoc. Offic. Anal. Chemists* **49**, 595 (1966).
902. F. A. Gunther, F. Buzzetti, and W. E. Westlake, *Residue Rev.* **17**, 81 (1967).
903. E. Sawicki *et al.*, *Am. Ind. Hyg. Assoc. J.* **22**, 137 (1962).
904. E. Sawicki, *Arch. Environ. Health* **14**, 46 (1967).
905. H. L. Falk, P. Kotin, and A. Miller, *Intern. J. Air Pollution* **2**, 201 (1960).
906. H. L. Falk, P. Kotin, and A. Miller, *Arch. Environ. Health* **8**, 721 (1964).
907. A. Haddow, *Brit. Med. Bull.* **4**, 331 (1947).
908. W. C. Hueper and W. D. Conway, "Chemical Carcinogenesis and Cancers," p. 197. Thomas, Springfield, Illinois, 1964.
909. P. Brookes and P. D. Lawley, *J. Cellular Comp. Physiol.* **64**, Suppl., 111 (1964).
910. E. Boyland, *Brit. Med. Bull.* **20**, 121 (1964).
911. C. E. Morreac, T. L. Dao, K. Eskins, C. L. King, and J. Dienstag, *Biochim. Biophys. Acta* **169**, 224 (1968).
912. E. Boyland and F. Weigert, *Brit. Med. Bull.* **4**, 354 (1947).
913. C. Heidelberger and H. B. Jones, *Cancer* **1**, 252 (1948).
914. P. M. Daniel, O. E. Pratt, and M. M. L. Prichard, *Nature* **215**, 1142 (1967).
915. B. L. Van Duuren, *J. Natl. Cancer Inst.* **21**, 1 (1958).
916. B. L. Van Duuren and N. Nelson, *Proc. Am. Assoc. Cancer Res.* **2**, 353 (1958).
917. A. J. Lindsey, *Brit. J. Cancer* **13**, 195 (1959).
918. I. Schmelz and W. S. Schlotzhauer, *Nature* **219**, 370 (1968).
919. M. Katz and J. L. Monkman, *Occupational Health Rev.* **16**, No. 1, (1964).
920. E. Sawicki, F. T. Fox, W. Elbert, T. R. Hauser, and J. Meeker, *Am. Ind. Hyg. Assoc. J.* **23**, 482 (1962).
921. H. L. Falk, P. E. Steiner, S. Goldfein, A. Breslow, and R. Hykes, *Cancer Res.* **11**, 318 (1951).
922. C. R. Begeman and J. M. Colucci, *Science* **161**, 271 (1968).
923. C. R. Begeman and J. M. Colucci, *Natl. Cancer Inst. Monograph* **9**, 17 (1962).
924. P. Kotin, H. Falk, and M. Thomas, *A.M.A. Arch. Ind. Health* **9**, 164 (1954).
925. E. O. Haenni and M. A. Hall, *J. Assoc. Offic. Agr. Chemists* **43**, 92 (1960).
926. W. Fritz, *Nahrung* **12**, 79 (1968).
927. W. Fritz, *Nahrung* **12**, 805 (1968).
928. W. Fritz, *Nahrung* **12**, 495 (1968).
929. W. Lijinsky and P. Shubik, *Toxicol. Appl. Pharmacol.* **7**, 337 (1965).
930. W. Lijinsky and P. Shubik, *Science* **145**, 53 (1964).
931. H. N. Draudt, *Food Technol.* **17**, 85 (1963).
932. Z. E. Sikorski and D. J. Tilgner, *Z. Lebensm.-Untersuch. -Forsch.* **124**, 274 (1964).
933. F. A. Gunther and W. E. Westlake, *Residue Rev.* **17**, 81 (1967).
934. J. W. Howard and E. O. Haenni, *J. Assoc. Offic. Agr. Chemists* **46**, 933 (1963).
935. L. Mallet, *Compt. Rend.* **253**, 168 (1961).
936. J. S. Harrington, *Nature* **193**, 43 (1962).
937. G. M. Badger, R. W. C. Kimber, and J. Novotny, *Australian J. Chem.* **15**, 616 (1962).
938. G. M. Badger, R. W. C. Kimber, and T. M. Spotswood, *Nature* **187**, 663 (1960).
939. F. Weigert, *Cancer Res.* **6**, 109 (1946).
940. K. H. Harper, *Brit. J. Cancer* **12**, 645 (1958).
941. P. Kotin, H. L. Falk, and R. Busser, *J. Natl. Cancer Inst.* **23**, 541 (1959).
942. H. L. Falk, P. Kotin, S. S. Lee, and A. Nathan, *J. Natl. Cancer Inst.* **28**, 699 (1962).
943. C. Heidelberger, H. S. Rieke, and M. Weiss, *Cancer Res.* **11**, 255 (1951).

944. S. A. Lesko, Jr., P. O. P. Ts'o, and R. S. Umans, *Biochemistry* **8**, 2291 (1969).
945. M. Demerec, *Brit. J. Cancer* **2**, 114 (1948).
946. G. H. Scherr, M. Fishman, and R. H. Weaver, *Genetics* **39**, 141 (1954).
947. E. L. Wynder and D. Hoffmann, *Deut. Med. Wochschr.* **88**, 623 (1963).
948. H. L. Falk, P. Kotin, and A. Miller, *Nature* **183**, 1184 (1959).
949. D. J. Tilgner, *Fleishwirtschaft* **10**, 649 (1958).
950. P. P. Dikun, *Rybn. Khozy.* **41**, 60 (1965); *Chem. Abstr.* **63**, 7574 (1965).
951. E. L. Kennaway, *Biochem. J.* **24**, 497 (1930).
952. E. Boyland and H. Burrows, *J. Pathol. Bacteriol.* **41**, 231 (1935).
953. P. E. Steiner and H. L. Falk, *Cancer Res.* **11**, 56 (1951).
954. C. Heidelberger and W. G. West, *Cancer Res.* **11**, 255 (1951).
955. E. Boyland and P. Sims, *Biochem. J.* **91**, 493 (1964).
956. A. Gemant, *Grace Hosp. Bull.* **44**, 47 (1966).
957. M. Demerec, *Nature* **159**, 604 (1947).
958. M. Demerec, *Genetics* **33**, 337 (1948).
959. R. W. Barrett and E. L. Tatum, *Cancer Res.* **11**, 234 (1951).
960. R. W. Barrett and E. L. Tatum, *Ann. N.Y. Acad. Sci.* **71**, 1072 (1958).
961. J. G. Carr, *Brit. J. Cancer* **1**, 152 (1947).
962. C. Auerbach, *Proc. Roy. Soc. Edinburgh* **60**, 559 (1940).
963. F. K. Zimmermann, *Z. Krebsforsch.* **72**, 65 (1969).
964. L. C. Strong, *Proc. Natl. Acad. Sci. U.S.* **31**, 290 (1945).
965. L. C. Strong, *Genetics* **32**, 108 (1946).
966. L. C. Strong, *Am. Naturalist* **81**, 50 (1947).
967. L. C. Strong, *A.M.A. Arch. Pathol.* **39**, 232 (1945).
968. C. A. Auerbach, *Proc. Roy. Soc. Edinburgh* **60**, 164 (1939).
969. S. Bhattacharya, *Nature* **162**, 573 (1948).
970. R. Latarjet, *Compt. Rend. Soc. Biol.* **142**, 453 (1948).
971. Union Chimique Belge., S. A., Belgian Patent 558,148 (1957); *Chem. Abstr.* **53**, 23102b (1959).
972. E. Windemuth, U.S. Patent 2,897,181 (1959); *Chem. Abstr.* **53**, 19462f (1959).
973. M. Uchida and M. Nagao, *Icagaku Zasshi* **60**, 496 (1957).
974. J. M. Mecco and J. W. Creely, U.S. Patent 2,774,648 (1956); *Chem. Abstr.* **51**, 6176c (1957).
975. C. Mascre and A. Lefebure, *Fonderie* **164**, 421 (1959); *Chem. Abstr.* **54**, 7492 (1960).
976. A. Hrubesch and W. France, German Patent 973,565 (1960); *Chem. Abstr.* **55**, 27876h (1961).
977. General Electric Co., Ltd., German Patent 972,571 (1959); *Chem. Abstr.* **55**, 11109c (1961).
978. A. A. Nitkin, C. C. Fite, Jr., and J. S. Gary, U.S. Patent 2,950,183 (1960); *Chem. Abstr.* **55**, 27876h (1961).
979. A. J. Clarke and A. M. Shepherd, *Nature* **208**, 502 (1965).
980. E. C. Cogbill and M. E. Hobbs, *Tobacco Sci.* **1**, 68 (1957).
981. R. C. Voss and H. Nicol, *Lancet* **II**, 435 (1960).
982. T. C. Tso, *Botan. Bull. Acad. Sinica* **7**, 28 (1966).
983. N. L. Kent and R. A.McChance, *Biochem. J.* **35**, 877 (1941).
984. W. F. von Oettingen, *Physiol. Rev.* **15**, 175 (1935).
985. L. T. Fairhall and P. A. Neal, *Natl. Inst. Health Bull.* **182**, 24 (1943).
986. G. C. Cotzias, *Physiol. Rev.* **38**, 503 (1958).
987. C. Cervinka, *Compt. Rend. Soc. Biol.* **102**, 262 (1929).
988. T. A. L. Davies and H. E. Harding, *Brit. J. Ind. Med.* **6**, 82 (1949).

989. H. Mella, *A.M.A. Arch. Neurol. Psychiat.* **11**, 405 (1924).
990. R. Penalver, *Ind. Med. Surg.* **24**, 1 (1955).
991. R. Penalver, *A.M.A. Arch. Ind. Health* **16**, 64 (1957).
992. D. M. Greenburg, H. D. Copp, and E. M. Cuthberson, *J. Biol. Chem.* **147**, 749 (1943).
993. J. A. Di Paolo, *Federation Proc.* **23**, 393 (1964).
944. P. A. Gallagher, *Nature* **216**, 391 (1967).
995. W. J. Rees and G. H. Sidrak, *Plant Soil* **14**, 101 (1961).
996. M. Demerec and J. Hanson, *Cold Spring Harbor Symp. Quant. Biol.* **16**, 215 (1951).
997. I. D. Steinman, V. N. Iyer, and W. Szybalski, *Arch. Biochem. Biophys.* **76**, 78 (1958).
998. N. N. Durham and O. Wyss, *J. Bacteriol.* **74**, 548 (1957).
999. M. Demerec, *Caryologia* **13**, Suppl., 201 (1954).
1000. R. W. Kaplan, *Naturwissenschaften* **49**, 457 (1962).
1001. R. A. Kehoe, J. Cholak, and R. V. Storey, *J. Nutr.* **19**, 581 (1940).
1002. I. B. Kogan and S. L. Makhover, *Gigiena i Sanit.* **2**, 52 (1954).
1003. J. J. Delfino and G. F. Lee, *Environ. Sci. Technol.* **3**, 761 (1969).
1004. J. J. Morgan, *J. Am. Water Works Assoc.* **57**, 107 (1965).
1005. R. R. Brooks, B. J. Presley, and I. R. Kaplan, *Talanta* **14**, 809 (1967).
1006. L. A. Muro and R. A. Goyer, *Arch. Pathol.* **87**, 660 (1969).
1007. H. Nishimura, "Chemistry and Prevention of Congenital Anomalies," p. 13. Thomas, Springfield, Illinois, 1964.
1008. E. L. Potter, "Pathology of the Fetus and Newborn," 2nd ed., p. 168. Year Book Publ., Chicago, Illinois, 1961.
1009. J. Deufel, *Chromosoma* **4**, 239 (1951).
1010. H. Jackson and A. R. Jones, *Nature* **220**, 591 (1968).
1011. M. Partington and A. J. Bateman, *Heredity* **19**, 191 (1964).
1012. S. Epstein, *Environ. Mutagen Soc. Newslet.* **2**, 33 (1969).
1013. Shell Chemicals, "Vapona in Slow Release Strips," Safe Use Manual No. 2/v, Sept., 1965.
1014. Shell International Chemical Co. Ltd., Safety Evaluation of Vapona Resin Strips, July, 1968.
1015. G. Löfroth, C. Kim, and S. Hussain, *Environ. Mutagen Soc. Newslet.* **2**, 21 (1969).
1016. Report of the Secretary's Commission on Pesticides and Their Relationship to Environmental Health, Part II, p. 612. U.S. Dept. of Health, Education and Welfare, U.S. Govt. Printing Office, 1969.

Author Index

Numbers in parentheses are reference numbers and indicate that an author's work is referred to although his name is not cited in the text. Numbers in italics show the page on which the complete reference is listed.

B

D

F

I

L

M

O

W

Y

Z

Subject Index

The numbers in parentheses refer to pages on which the formula, chemical or common name, Chemical Abstracts' registry number, and literature citations for point mutations and chromosome aberrations of the chemical mutagens are given.

E

F

G

N

Q

R

T

Printed in Dunstable, United Kingdom

91898141R00212